Applications of Manufacturing and Measurement Sensors

Applications of Manufacturing and Measurement Sensors

Guest Editors

Shichang Du
Yiping Shao
Delin Huang

MDPI

Basel • Beijing • Wuhan • Barcelona • Belgrade • Novi Sad • Cluj • Manchester

Guest Editors

Shichang Du
Department of Industrial Engineering and Management
Shanghai Jiao Tong University
Shanghai
China

Yiping Shao
College of Mechanical Engineering
Zhejiang University of Technology
Hangzhou
China

Delin Huang
School of Intelligent Manufacturing and Control Engineering
Shanghai Polytechnic University
Shanghai
China

Editorial Office
MDPI AG
Grosspeteranlage 5
4052 Basel, Switzerland

This is a reprint of the Special Issue, published open access by the journal *Sensors* (ISSN 1424-8220), freely accessible at: https://www.mdpi.com/journal/sensors/special_issues/ITR93T1V95.

For citation purposes, cite each article independently as indicated on the article page online and as indicated below:

Lastname, A.A.; Lastname, B.B. Article Title. *Journal Name* **Year**, *Volume Number*, Page Range.

ISBN 978-3-7258-4381-7 (Hbk)
ISBN 978-3-7258-4382-4 (PDF)
https://doi.org/10.3390/books978-3-7258-4382-4

Cover image courtesy of Yiping Shao

Contents

About the Editors

Shichang Du

Shichang Du received B.Sc. and M.Sc.E. degrees in Mechanical Engineering from the Hefei University of Technology, Hefei, China, in 2000 and 2003, respectively, and a Ph.D. degree in Industrial Engineering and Management from Shanghai Jiao Tong University, Shanghai, China, in 2008. He collaborated with the University of Michigan, Ann Arbor, MI, USA, from 2006 to 2007, as a visiting scholar. He is currently a professor with the Department of Industrial Engineering and Management, Shanghai Jiao Tong University. He has authored/co-authored 110 articles in journals, including *IEEE, ASME, IISE Transaction*, etc. His current research interests focus on quality control and smart manufacturing methods. His research has been supported by the National Natural Science Foundation of China, the Ministry of Science and Technology of China, and the Shanghai Association for Science and Technology. He works for several journals as an Area Editor, including *Computers and Industrial Engineering, Flexible Services and Manufacturing Journal, Precision Engineering, Measurement, Journal of Intelligent Manufacturing*, and *Computers in Industry*.

Yiping Shao

Yiping Shao received a B.E. degree in Industrial Engineering from the Beijing Institute of Technology, Beijing, China, in 2014, and a Ph.D. degree in Mechanical Engineering from Shanghai Jiao Tong University, Shanghai, China, in 2020. He is currently an assistant professor at the College of Mechanical Engineering, Zhejiang University of Technology. His main research areas include intelligent manufacturing and quality control, big data and machine vision, intelligent logistics equipment, etc. He has presided over 10 scientific projects at the National Natural Science Foundation of China and the Natural Science Foundation of Zhejiang Province and published over 30 academic papers at major conferences and in journals related to intelligent manufacturing. He serves as a member of the Early Career Advisory Board of *Measurement* and a reviewer for over 30 international journals.

Delin Huang

Delin Huang received a B.E. degree in Industrial Engineering from Northeastern University at Qinhuangdao, Qinhuangdao, China, in 2013, and a Ph.D. degree in Mechanical Engineering from Shanghai Jiao Tong University, Shanghai, China, in 2019. He is currently a lecturer at the School of Intelligent Manufacturing and Control Engineering, Shanghai Polytechnic University. His main research areas include intelligent manufacturing and quality control, industrial big data and machine vision, etc. He has presided over six scientific projects including for the Sailing Program of Shanghai Province and the Central University Basic Scientific Research Operation Cost Special Fund and published over 25 academic papers at major conferences and in journals related to intelligent manufacturing. He serves as a reviewer for over 20 international journals.

sensors

MDPI

Editorial

Advancements in Applications of Manufacturing and Measurement Sensors

Yiping Shao [1,2], Shichang Du [3,*] and Delin Huang [4]

1 College of Mechanical Engineering, Zhejiang University of Technology, Hangzhou 310023, China; syp123gh@zjut.edu.cn
2 Ningbo Yongxin Optics Co., Ltd., Ningbo 315040, China
3 School of Mechanical Engineering, Shanghai Jiao Tong University, No. 800 Dongchuan Road, Shanghai 200240, China
4 School of Intelligent Manufacturing and Control Engineering, Shanghai Polytechnic University, Shanghai 201209, China; dlhuang@sspu.edu.cn
* Correspondence: lovbin@sjtu.edu.cn

Manufacturing and measurement sensors are an integral part of advanced manufacturing technology, which requires sensors that can precisely capture and analyze various physical parameters during the manufacturing process [1]. These sensors have demonstrated remarkable achievements in the field of advanced manufacturing, especially in ensuring product quality and improving production efficiency [2,3]. However, the application of these sensors in the manufacturing domain has been accompanied by challenges. One of the main challenges is the complex and variable manufacturing environment, which requires sensors to maintain high reliability, stability, and accuracy under various adverse conditions [4]. Indeed, despite their widespread use, manufacturing and measurement sensors often face difficulties in adapting to extreme temperatures, high pressures, and strong electromagnetic interferences, which may affect the accuracy of the measurement data [3]. This situation raises concerns, especially in industries where high-precision manufacturing is crucial, as accurate sensor data are essential for ensuring product quality and process control [5].

In this regard, the development of advanced sensor technologies and intelligent algorithms has become a key focus in this field. For example, the emergence of smart sensors with self-calibration, self-diagnosis, and self-adaptation functions can effectively improve the performance of sensors in complex environments [6–8]. In particular, these advanced sensors and algorithms are designed to improve adaptability and accuracy in sensing applications, aiming to better meet the needs of modern manufacturing processes [9]. The primary objective is to enhance the reliability and effectiveness of manufacturing and measurement sensors in complex industrial scenarios [10]. This is of the utmost importance in modern manufacturing, where such improvements can have a significant impact, reducing production costs, improving product quality, and enhancing overall competitiveness [11]. In this context, several advanced sensor-based applications have been emerging in manufacturing, with applications in intelligent manufacturing systems, quality monitoring and control, and machine vision-based inspection [12–14].

Furthermore, the continuous expansion of the Internet of Things (IoT) in the industrial field has led to an increase in the availability of data from interconnected sensors. This rich data source provides a solid foundation for training and optimizing sensor-related models [15]. In addition, recent advances in sensor-manufacturing materials and processes have also improved the performance of sensors, which enable them to better meet the diverse requirements of modern manufacturing [16].

In this context, this topical collection includes eleven papers focused on the latest advancements in the field of manufacturing and measurement sensors. Each of the papers underwent a rigorous review process by multiple expert reviewers across several rounds of revision. The studies published in the current topical collection cover a wide range of topics, including advanced measurement sensors, measurement and sensing applications, manufacturing systems, applications of sensors in intelligent logistics equipment, quality monitoring and control, machine vision and applications, and methods of detection. The following provides a brief summary of these studies:

In Contribution 1, Luo, L. et al. introduce a high-precision capacitive pressure sensor for TPMS. By leveraging a CMUT structure and silicon–s ilicon bonding, the sensor achieves high accuracy, ultra-low power consumption, and cost-effectiveness. Rigorous tests show that it can precisely monitor tire pressure, ensuring automotive safety and reliability, which is crucial for the automotive manufacturing industry.

In Contribution 2, Wu, B. et al. explore the roundness of small cylindrical workpieces. Using the stitching linear scan method, they compare ruby ball and diamond styluses. Their experiments reveal that the ruby ball stylus is more suitable for small needle rollers, offering practical guidance for small-scale manufacturing quality control processes.

In Contribution 3, Zhu, Y. et al. develop a dual-band binocular stereo-imaging system. Integrating infrared and visible light capabilities, it can accurately measure distances and recognize objects. This system has proven effective in complex industrial environments, making it valuable for large-scale product inspection and robotic operations in intelligent manufacturing.

In Contribution 4, Du, B. et al. propose a 2D FSM angle measurement system based on diffuse reflection. The system can achieve large-angle, linear measurements with high precision. In optical equipment manufacturing, this allows for the accurate adjustment of optical components, which is essential for ensuring the quality of optical products.

In Contribution 5, Zhu, Y. et al. present a four-step infrared image algorithm. Through image segmentation, stretching coefficient calculation, layer separation, and image fusion, it significantly improves image quality and processing speed. This algorithm is effective in detecting defects in manufacturing processes, enhancing quality monitoring.

In Contribution 6, Roos-Hoefgeest, S. et al. establish a laser profilometer simulation model. By incorporating Perlin and Gaussian noise to mimic real-world speckle interference, the model accurately represents surface geometries. It serves as a powerful tool for sensor design and evaluation, which helps to improve measurement accuracy.

In Contribution 7, Yin, Y. et al. combine GNSS and IMU data using error state Kalman filtering and smoothing. This approach effectively reduces errors and enhances localization accuracy. In manufacturing logistics and automation, it ensures the precise positioning of equipment and products, which improves operational efficiency.

In Contribution 8, Jin, S. et al. analyze RV reducer rotation errors through deep Gaussian processes. Their analysis identifies key factors affecting reducer performance. In high-precision manufacturing, this knowledge helps manufacturers improve the reliability and accuracy of RV reducers.

In Contribution 9, Shao, Y. et al. put forward a point-cloud-driven pallet pose method. This method increases pose estimation accuracy by 35% and reduces feature extraction time by 30%. In manufacturing logistics, it strengthens the stability of the system, enhancing overall logistics efficiency.

In Contribution 10, Lyashenko, I.A. et al. design a high-precision tribometer. By overcoming motor backlash and signal drift, it can accurately measure adhesive contact forces. This is crucial for optimizing bonding processes in manufacturing, especially in electronics and composite material production.

In Contribution 11, Chen, X. et al. introduce a 3D metrology method using anamorphic lenses. Based on double optical centers, this method can efficiently reconstruct 3D data. This method simplifies manufacturing inspection processes and also has applications in car navigation systems, which expands its utility to different fields.

In summary, this topical collection addresses various crucial challenges and advancements in the field of manufacturing and measurement sensors. It introduces innovative sensor designs, advanced algorithms, and novel applications, which have great potential for enhancing manufacturing efficiency and product quality. We are extremely grateful to the Managing Team of *Sensors* journal for their unwavering support during the compilation of this collection. We also sincerely thank all the contributing authors for their excellent research, as well as the anonymous expert reviewers. Their dedicated efforts have been instrumental in ensuring the high quality of the selected submissions, making this collection a valuable resource for the industry.

Author Contributions: Conceptualization and supervision, S.D.; writing—original draft preparation, Y.S., S.D. and D.H.; writing—review and editing, Y.S., S.D. and D.H.; funding acquisition, Y.S. and S.D. All authors have read and agreed to the published version of the manuscript.

Funding: This work is supported by National Natural Science Foundation of China (Grant No. 52405565, 52275499, 92467101) and National Key Research and Development Program of China (Grant No. 2022YFF0605700).

Conflicts of Interest: The authors declare no conflicts of interest.

List of Contributions:

1. Luo, L.; Wang, Z.; Chen, J.; Hui, A.G.; Rogikin, A.M.; Liu, R.; Zhou, Y.; Jiang, Z.; He, C. An Investigation into High-Accuracy and Energy-Efficient Novel Capacitive MEMS for Tire Pressure Sensor Application. *Sensors* **2024**, *24*, 8037.
2. Wu, B.; Zeng, C.; Li, Q. Comparison Analysis of Roundness Measurement of Small Cylindrical Workpieces with Different Styluses. *Sensors* **2024**, *24*, 3819.
3. Zhu, Y.; Zhang, D.; Zhou, Y.; Jin, W.; Zhou, L.; Wu, G.; Li, Y. A Binocular Stereo-Imaging-Perception System with a Wide Field-of-View and Infrared- and Visible Light-Dual-Band Fusion. *Sensors* **2024**, *24*, 676.
4. Du, B.; Lv, Y.; Liu, L.; Liu, Y. High-Range and High-Linearity 2D Angle Measurement System for a Fast Steering Mirror. *Sensors* **2023**, *23*, 9192.
5. Zhu, Y.; Zhou, Y.; Jin, W.; Zhang, L.; Wu, G.; Shao, Y. A Low-Delay Dynamic Range Compression and Contrast Enhancement Algorithm Based on an Uncooled Infrared Sensor with Local Optimal Contrast. *Sensors* **2023**, *23*, 8860.
6. Roos-Hoefgeest, S.; Roos-Hoefgeest, M.; Álvarez, I.; González, R.C. Simulation of Laser Profilometer Measurements in the Presence of Speckle Using Perlin Noise. *Sensors* **2023**, *23*, 7624.
7. Yin, Y.; Zhang, J.; Guo, M.; Ning, X.; Wang, Y.; Lu, J. Sensor Fusion of GNSS and IMU Data for Robust Localization via Smoothed Error State Kalman Filter. *Sensors* **2023**, *23*, 3676.
8. Jin, S.; Shang, S.; Jiang, S.; Cao, M.; Wang, Y. Sensitivity Analysis of RV Reducer Rotation Error Based on Deep Gaussian Processes. *Sensors* **2023**, *23*, 3579.
9. Shao, Y.; Fan, Z.; Zhu, B.; Lu, J.; Lang, Y. A Point Cloud Data-Driven Pallet Pose Estimation Method Using an Active Binocular Vision Sensor. *Sensors* **2023**, *23*, 1217.
10. Lyashenko, I.A.; Popov, V.L.; Pohrt, R.; Borysiuk, V. High-Precision Tribometer for Studies of Adhesive Contacts. *Sensors* **2023**, *23*, 456.
11. Chen, X.; Zhang, J.; Xi, J. 3D Metrology Using One Camera with Rotating Anamorphic Lenses. *Sensors* **2022**, *22*, 8407.

References

1. Babiceanu, R.F.; Seker, R. Big Data and virtualization for manufacturing cyber-physical systems: A survey of the current status and future outlook. *Comput. Ind.* **2016**, *81*, 128–137. [CrossRef]

2. Huang, Y.; Fan, X.; Chen, S.-C.; Zhao, N. Emerging Technologies of Flexible Pressure Sensors: Materials, Modeling, Devices, and Manufacturing. *Adv. Funct. Mater.* **2019**, *29*, 1808509. [CrossRef]
3. Perera, Y.S.; Ratnaweera, D.A.A.C.; Dasanayaka, C.H.; Abeykoon, C. The role of artificial intelligence-driven soft sensors in advanced sustainable process industries: A critical review. *Eng. Appl. Artif. Intell.* **2023**, *121*, 105988. [CrossRef]
4. Li, Q.T.; Zeng, W.; Li, Y.Q. Metal oxide gas sensors for detecting NO2 in industrial exhaust gas: Recent developments. *Sens. Actuators B-Chem.* **2022**, *359*, 131579. [CrossRef]
5. Yi, J.H.; Xianyu, Y.L. Gold Nanomaterials-Implemented Wearable Sensors for Healthcare Applications. *Adv. Funct. Mater.* **2022**, *32*, 2113012. [CrossRef]
6. Abdallah, M.; Joung, B.-G.; Lee, W.J.; Mousoulis, C.; Raghunathan, N.; Shakouri, A.; Sutherland, J.W.W.; Bagchi, S. Anomaly Detection and Inter-Sensor Transfer Learning on Smart Manufacturing Datasets. *Sensors* **2023**, *23*, 486. [CrossRef]
7. Hassan, M.S.; Zaman, S.; Dantzler, J.Z.R.; Leyva, D.H.; Mahmud, M.S.; Ramirez, J.M.; Gomez, S.G.; Lin, Y. 3D Printed Integrated Sensors: From Fabrication to Applications-A Review. *Nanomaterials* **2023**, *13*, 3148. [CrossRef] [PubMed]
8. Hossain, M.J.; Tabatabaei, B.T.; Kiki, M.; Choi, J.-W. Additive Manufacturing of Sensors: A Comprehensive Review. *Int. J. Precis. Eng. Manuf.-Green Technol.* **2025**, *12*, 277–300. [CrossRef]
9. McCorry, M.C.; Reardon, K.F.; Black, M.; Williams, C.; Babakhanova, G.; Halpern, J.M.; Sarkar, S.; Swami, N.S.; Mirica, K.A.; Boermeester, S.; et al. Sensor technologies for quality control in engineered tissue manufacturing. *Biofabrication* **2023**, *15*, 012001. [CrossRef] [PubMed]
10. Sen, S.; Husom, E.J.; Goknil, A.; Politaki, D.; Tverdal, S.; Nguyen, P.; Jourdan, N. Virtual sensors for erroneous data repair in manufacturing a machine learning pipeline. *Comput. Ind.* **2023**, *149*, 103917. [CrossRef]
11. Xu, D.T.; Zhang, Z.S.; Shi, J.F. A New Multi-Sensor Stream Data Augmentation Method for Imbalanced Learning in Complex Manufacturing Process. *Sensors* **2022**, *22*, 4042. [CrossRef] [PubMed]
12. Shao, Y.P.; Chen, J.; Gu, X.L.; Lu, J.S.; Du, S.C. A novel curved surface profile monitoring approach based on geometrical-spatial joint feature. *J. Intell. Manuf.* **2024**. [CrossRef]
13. Shao, Y.P.; Xu, F.C.; Chen, J.; Lu, J.S.; Du, S.C. Engineering surface topography analysis using an extended discrete modal decomposition. *J. Manuf. Process.* **2023**, *90*, 367–390. [CrossRef]
14. Zhao, C.; Lui, F.C.; Du, S.C.; Wang, D.; Shao, Y.P. An Earth Mover's Distance based Multivariate Generalized Likelihood Ratio Control Chart for Effective Monitoring of 3D Point Cloud Surface. *Comput. Ind. Eng.* **2023**, *175*, 108911. [CrossRef]
15. Chai, B.X.; Gunaratne, M.; Ravandi, M.; Wang, J.Z.; Dharmawickrema, T.; Di Pietro, A.; Jin, J.; Georgakopoulos, D. Smart Industrial Internet of Things Framework for Composites Manufacturing. *Sensors* **2024**, *24*, 4852. [CrossRef] [PubMed]
16. Jiang, Y.J.; Islam, M.N.; He, R.; Huang, X.Z.; Cao, P.F.; Advincula, R.C.; Dahotre, N.; Dong, P.; Wu, H.F.; Choi, W. Recent Advances in 3D Printed Sensors: Materials, Design, and Manufacturing. *Adv. Mater. Technol.* **2023**, *8*, 2200492. [CrossRef]

sensors

MDPI

Article

An Investigation into High-Accuracy and Energy-Efficient Novel Capacitive MEMS for Tire Pressure Sensor Application

Liang Luo [1], Ziyuan Wang [2], Jianwei Chen [3], Aisn Gioronara Hui [4], Allwins Moore Rogikin [5], Rongzhen Liu [6], Yao Zhou [7], Zhujin Jiang [8] and Changde He [6,*]

1 Department of Engineering Mechanics, Tsinghua University, Beijing 100084, China; medinathu@163.com
2 School of Integrated Circuits, Southeast University, Nanjing 214135, China; 230228414@seu.edu.cn
3 Shanghai Institute of Ceramics, Chinese Academy of Sciences, Shanghai 201800, China; chenjw@mail.sic.ac.cn
4 Einsteck Inc., Palo Alto, CA 94305, USA; einsteck@126.com
5 Golden Rogikin Limited, Kowloon 999077, Hong Kong; moore24@126.com
6 School of Instrument and Electronics, North University of China, Taiyuan 030051, China; 13027020313@163.com
7 School of Modern Post, Xi'an University of Posts and Telecommunications, Xi'an 710061, China; vejango123@163.com
8 School of Mechanical and Electric Engineering, Soochow University, Suzhou 215021, China; zjjiang2024@suda.edu.cn
* Correspondence: hechangde@nuc.edu.cn

Abstract: Tire pressure monitoring systems (TPMSs) are essential for maintaining driving safety by continuously monitoring critical tire parameters, such as pressure and temperature, in real time during vehicle operation. Among these parameters, tire pressure is the most significant, necessitating the use of highly precise, cost-effective, and energy-efficient sensing technologies. With the rapid advancements in micro-electro-mechanical system (MEMS) technology, modern automotive sensing and monitoring systems increasingly rely on MEMS sensors due to their compact size, low cost, and low power consumption. This study presents a novel high-precision capacitive pressure sensor based on a capacitive micromachined ultrasonic transducer (CMUT) structure and a silicon–silicon direct bonding process. The proposed design offers exceptional performance with high accuracy, ultra-low power consumption, and reduced production costs, making it an optimal solution for enhancing the precision and efficiency of TPMS. Leveraging its low power requirements, capacitive sensing technology emerges as a superior choice for energy-efficient systems in the automotive industry.

Keywords: MEMS; tire pressure monitoring system; capacitive; silicon–silicon bonding; ultra-low power consumption

1. Introduction

TPMSs are essential for real-time monitoring of tire temperature and pressure, offering timely warnings about air leaks, low pressure, or high temperatures to pre-emptively signal deteriorating tire conditions and ensure driving safety. Additionally, maintaining optimal tire pressure is crucial for prolonging tire life, minimizing wear on the suspension system, improving fuel economy and CO_2 reduction [1–3]. This study introduces a micro tire pressure monitoring system characterized by ultra-low power consumption and high precision. The system aims to rectify the inadequate pressure measurement accuracy of existing sensor chips [4,5], enhancing measurement accuracy up to ±0.01 Bar (1 kPa) with a working current at the nanoampere level. This breakthrough is a direct result of advancements in integrated circuit (IC) technology and micro-electro-mechanical system (MEMS) technology, which allow for the production of highly precise micromechanical structures on a large scale [6,7]. The CMUT technology, driven by MEMS advancements, leverages semiconductor manufacturing processes to facilitate the mass production of transducers. This development guarantees the high reliability, uniformity, and cost-efficiency of the products, offering a solution for creating large-scale two-dimensional array devices for ultrasonic transducers [8,9].

This article introduces a cutting-edge, high-precision capacitive pressure sensing technology based on a CMUT structure. Utilizing the latest advancements in MEMS and wafer bonding techniques, the proposed scheme employs a capacitance detection-based pressure sensor [10]. The design incorporates an array cavity structure that provides precise and reliable tire pressure monitoring. The sensor stands out for its high measurement accuracy, robust stability, quick response time, and low power consumption, all packaged in a compact size with outstanding environmental adaptability. The theoretical accuracy of this system is pinpointed at ± 0.01 Bar (1 kPa), while achieving ultra-low power consumption. Capacitive sensors, such as the one described here, are especially suited to low-power systems due to their power-efficient measurement principles, offering marked improvements in precision, energy efficiency, and noise reduction over resistive pressure sensors, which do not require a DC current bias. Notably, this capacitive tire pressure sensor is designed for up to 10 years of continuous operation, maintaining ultra-low power use throughout its extended service life [11,12].

2. Materials and Methods

The design of CMUT devices is heavily influenced by the choice of membrane shape, with common geometries including square, circular, and regular hexagonal. The anisotropic nature of silicon wafers in wet process technology usually results in square membrane structures, as this method typically produces square cavities. Despite the higher plot ratio appeal of regular hexagonal structures, both square and hexagonal films can lead to reduced manufacturing yields. This reduction is due to uneven stress distributions that occur during processing, which can affect device sensitivity. Square membranes, in particular, are susceptible to stress concentration at their sharp corners, potentially causing membrane rupture and thus yield loss. To mitigate these challenges, a circular thin-film structure has been proposed as a viable alternative [13–16]. The optimal CMUT unit, as structured in a top-down arrangement in Figure 1, includes an upper electrode, an insulation layer to prevent Ohmic contact, the core vibrating film, a cavity, peripheral support for structure and fixation, another insulation layer, a supportive substrate plane, and a lower electrode. R_i is the radius of insulation layer, and re is the radius of top electrode. This design approach aims to minimize stress concentration, thereby enhancing device performance and production yield [7,16–18].

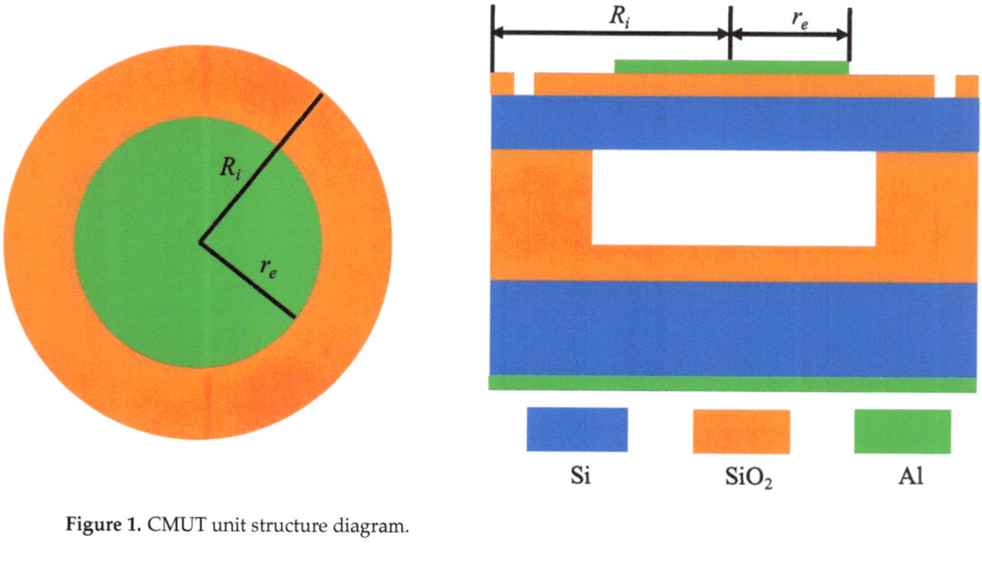

Figure 1. CMUT unit structure diagram.

2.1. Parallel Plate Capacitance Model

CMUT can be seen as a parallel plate capacitor with a moving upper plate and a fixed lower plate. According to the unit structure and working principle of CMUT, the membrane can be seen as the upper moving plate, and CMUT can be equivalent to a parallel plate capacitance model composed of a mass block, spring, and parallel plate capacitance, as shown in Figure 2. m is the mass of the mass block; k is the spring elasticity coefficient; V is the DC bias voltage applied to the film and substrate; g_0 is the original cavity height, and x is the initial displacement of the film under DC bias voltage.

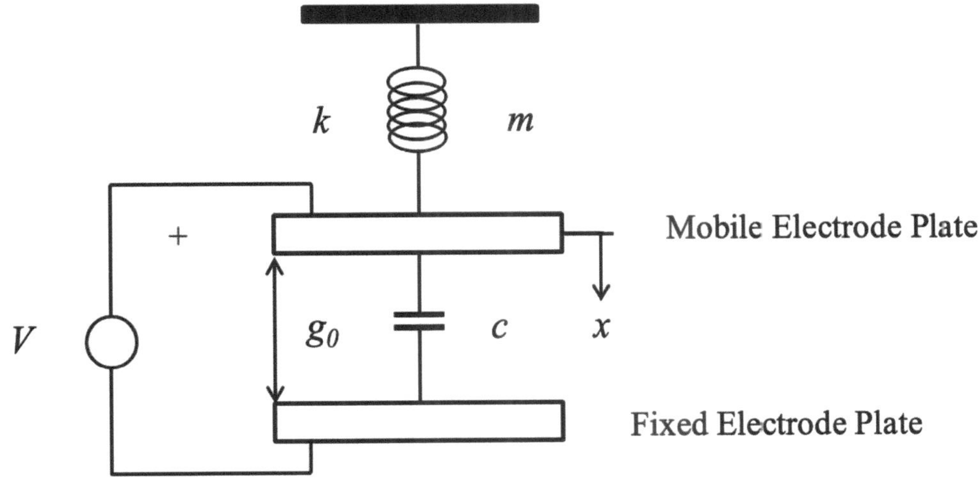

Figure 2. CMUT flat plate capacitor model.

The force analysis of the upper plate can be obtained from Newton's second law:

$$F_{mass} + F_e + F_s = 0 \tag{1}$$

Among them, F_{mass} represents the total external force applied to the thin film; F_e represents the electrostatic force between two parallel plates caused by applied bias, and F_s is the spring restoring force. The expression for electrostatic force F_e is [17,19,20]

$$F_e = -\frac{d}{d_x}\left(\frac{CV^2}{2}\right) = -\frac{1}{2}V^2\left[\frac{d}{dx}\left(\frac{\varepsilon_0 A}{g_0 - x}\right)\right] = \frac{\varepsilon_0 A V^2}{2(g_0 - x)^2} \tag{2}$$

where A is the equivalent area of the capacitor; ε_0 is the vacuum dielectric constant with a value of $8.854187817 \times 10^{-12}\ F/m$. The spring restoring force F_s can be expressed as

$$F_s = -kx \tag{3}$$

2.2. Working Principle

The schematic diagram of the CMUT unit structure used for pressure measurement is as follows in Figure 3 [16,21]. The effective radius of the pressure silicon film is r_m; the thickness is h, and it is used as a pressure sensing element. The height of the vacuum chamber is d_0, and the silicon substrate is used as the bottom electrode of the CMUT. When external pressure is applied to the CMUT capacitive structure, the d_0 decreases, leading to an increase in capacitance. As a result, any applied pressure causes a corresponding change in capacitance.

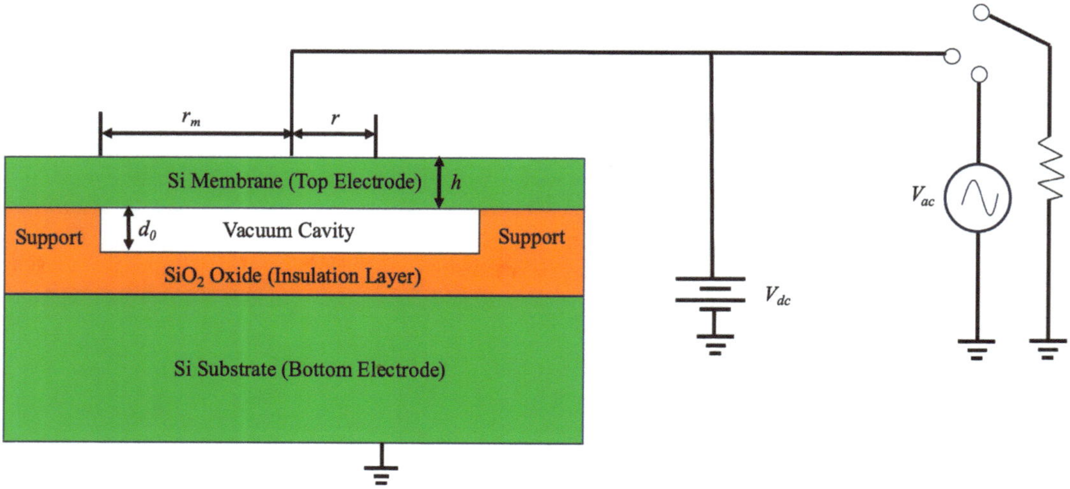

Figure 3. CMUT pressure unit structure.

The resonant characteristics of the pressure film in a capacitive pressure sensor based on the CMUT structure depend not only on its mechanical properties and geometric shape but also on the DC bias and measured pressure. When a DC bias voltage V_{bias} is applied, the membrane deflects towards the bottom plate, causing a shift in the resonant frequency of the membrane. Research has shown that within a lower bias voltage range, the resonant frequency varies linearly with pressure. If a bias voltage is applied to the electrode, the relationship between the resonant frequency and the pressure acting on the membrane under bias voltage is as follows:

$$f_p = f_0(a - bP) \tag{4}$$

$$f_0 = \frac{0.83}{r_m}\sqrt{\frac{Eh^3}{m(1-v^2)}} \tag{5}$$

In Equation (4), f_p is the resonant frequency under applied pressure P; f_0 is the natural frequency of the membrane, shown in Equation (5) [21,22], and a and b are two coefficients determined by the mechanical properties and geometric shape of the membrane, as well as the bias voltage V_{bias}. a reflects the influence of DC bias voltage on resonant frequency, while b determines the resonant frequency shift caused by pressure P under bias voltage. The corresponding pressure sensitivity is defined as follows:

$$S_P = \frac{1}{af_0}\frac{\partial f_P}{\partial P} = -\frac{b}{a}\left(\mathrm{ppmPa}^{-1}\right) \tag{6}$$

Under the same external pressure, the membrane achieves a greater resonance frequency shift under DC bias. Compared with membranes without DC bias voltage, known as under bias voltage, pressure will cause greater changes in membrane offset and stress, which can be analyzed by the following two formulas [21,22]:

$$\Delta w_0(V_{bias}, P) = \frac{\Delta P}{64[D - Q_1 r_m^4/(42d_0)]}\left(r_m^2 - r^2\right)^2 \tag{7}$$

$$\Delta w_0(P) = \frac{\Delta P}{64D}\left(r_{\mathrm{m}}^2 - r^2\right)^2 \tag{8}$$

$$Q_1 = \frac{\varepsilon_0 V_{bias}^2}{2d_0^2} \tag{9}$$

$$D = \frac{Eh^3}{12(1-v^2)} \tag{10}$$

Δw_0 is the deflection change in the clamping film; ΔP is the pressure change, and Q_1 represents the effect of the applied bias voltage V_{bias} in Equation (9) [21]. D represents the flexural stiffness of the CMUT membrane, which is often modeled as a thin plate. The stiffness D is given in Equation (10), where E is the Young's modulus of the membrane material; h is the thickness of the membrane, and v is the Poisson's ratio [21].

Equation (7) describes the change in membrane deflection caused by pressure when a bias voltage is applied, while Equation (8) represents the membrane deflection due only to measured pressure. As a result, DC-biased CMUTs achieve an amplified frequency shift, enhancing pressure sensitivity. Under identical pressure conditions, the biased membrane exhibits greater deflection than the unbiased membrane, resulting in a more pronounced frequency shift. Thus, DC-biased CMUTs can achieve amplified frequency shifts, improving pressure sensitivity [21–28].

2.3. Manufacturing Process

The fabrication of CMUTs hinges on two pivotal processing techniques. The first technique, known as the sacrificial layer release process, is the cornerstone of traditional CMUT manufacturing process. The second key process in CMUT production is wafer bonding technology, an approach refined alongside advancements in MEMS technology.

The two primary CMUT manufacturing processes each present distinct advantages. The sacrificial layer release process is well suited for initial stages of CMUT fabrication, while wafer bonding technology offers a more advanced, cleaner, and controlled approach for creating superior CMUT devices [7,18].

Traditional CMUT manufacturing relies on the sacrificial layer release method. Despite its widespread adoption, it has several limitations including a complex process flow, inconsistency in cavity size and distribution, and a high potential for contamination during production. However, as manufacturing technologies have evolved, wafer bonding has emerged as a preferred technique in recent years, addressing many of the issues associated with the sacrificial layer release method. Wafer bonding not only streamlines the fabrication process but also produces more uniformly sized cavities with reduced contamination, making it ideal for the production of large-area CMUT two-dimensional arrays. Consequently, wafer bonding has become the method of choice for CMUT two-dimensional array preparation [7,18].

Currently, wafer bonding techniques can be classified into direct bonding, anodic bonding, and intermediate layer bonding as shown in Figure 4. Among these, low-temperature direct bonding is deemed most appropriate for CMUT two-dimensional array fabrication, as no intermediate layer is permissible when forming CMUT cavities. Options for bonding without an intermediate layer often necessitate chemical or plasma activation, which can complicate and hinder the bonding process. Thus, the simplicity and reliability of the silicon–silicon direct bonding process have made it the preferred method for assembling CMUT two-dimensional arrays [7,18,29–31].

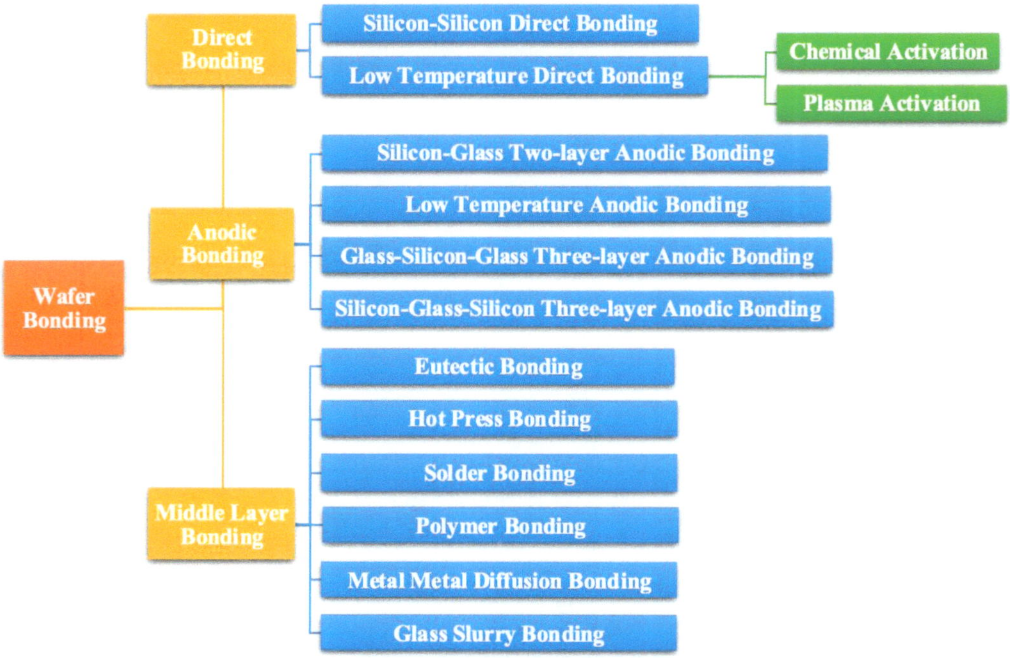

Figure 4. Classification of wafer bonding.

In our design, we have implemented a low-temperature direct bonding process augmented by microcapacitor micro–nano technology. This approach has substantially increased the yield and consistency of microcapacitor structures and concurrently reduced the costs associated with manufacturing. For the CMUT pressure sensor assembly, silicon (Si) has been selected as the material for the pressure sensing membrane, silicon dioxide (SiO_2) as the isolation and insulation layer, and aluminum (Al) as the electrode material [32–34].

The fabrication steps of the CMUT microcapacitor pressure sensing structure is delineated in Figure 5. First, a suitable SOI (Silicon-On-Insulator) wafer is selected, and a SiO_2 (Silicon Dioxide) cavity is etched into the silicon, as shown in Figure 5a. Using a low-temperature direct bonding process, an SOI-bonded silicon wafer is formed, as depicted in Figure 5b. Next, the silicon layer and oxide layer of the SOI substrate are removed, resulting in a thin silicon film, as seen in Figure 5c. The insulation layer is oxidized; the top metal electrode is prepared; isolation trenches are etched, and the oxide layer on the back of the silicon is removed. Finally, the bottom electrode is fabricated to form an ohmic contact, completing the device structure, as illustrated in Figure 5d. This process leverages the uniformity of the SOI device layer, ensuring consistency in the sensor's vibrating membrane structure.

The following are key requirements for low-temperature direct bonding: Surface Cleanliness: wafers must be free of contaminants for effective bonding; Surface Roughness: roughness should be smaller than 0.25 μm for hydrophobic wafers and no more than 0.5 μm for hydrophilic wafers; Total Thickness Variation (TTV): ideally around 2 μm; Surface Activation: required to improve bonding success. The bonding process has two stages: pre-bonding and high-temperature annealing. During pre-bonding, wafers are aligned and weakly bonded. High-temperature annealing then forms covalent bonds, ensuring a strong, permanent connection. Low-temperature direct bonding is used to bond the silicon wafer to the SOI, following these steps: Cleaning: both silicon and SOI wafers are cleaned with the RCA standard method to remove dust and residues that could hinder bonding; Surface Activation: the cleaned surfaces are treated with O_2 plasma to enhance bonding success; Pre-Bonding: at room temperature, the silicon and SOI wafers are aligned and pre-bonded in a wafer bonder; High-Temperature Annealing: the pre-bonded wafers are then annealed at 1050 °C in an inert gas for 20 h, creating strong covalent bonds between the silicon and SOI. This process is essential for creating a stable, reliable bond between the silicon wafer and SOI, crucial for the structural integrity of the final device [32–34].

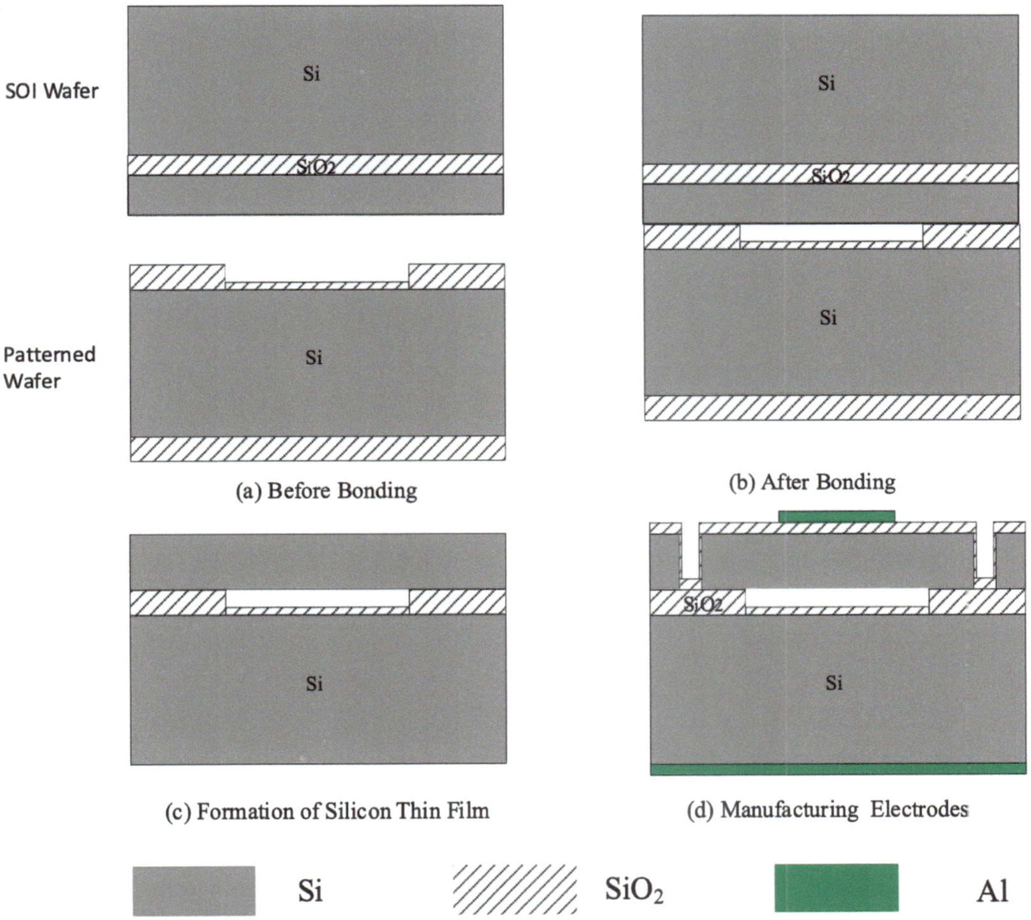

Figure 5. Schematic diagram of low-temperature direct wafer bonding manufacturing process.

3. Results

3.1. COMSOL Multiphysics Simulation (COMSOL. Inc., Burlington, MA, USA)

The CMUT capacitive pressure sensor device design and parameters, which are shown in Figure 6 and Table 1, respectively, akin to other MEMS devices, encompass the integration of multiple physical field factors. To assess the sensor's performance accurately, it is imperative to employ robust tools capable of coupling analysis between these physical fields. Such tools must meticulously delineate the interactions among the various physical fields to ensure a comprehensive understanding of the sensor's operational dynamics. This detailed analysis is crucial for optimizing the sensor's functionality and reliability in applications.

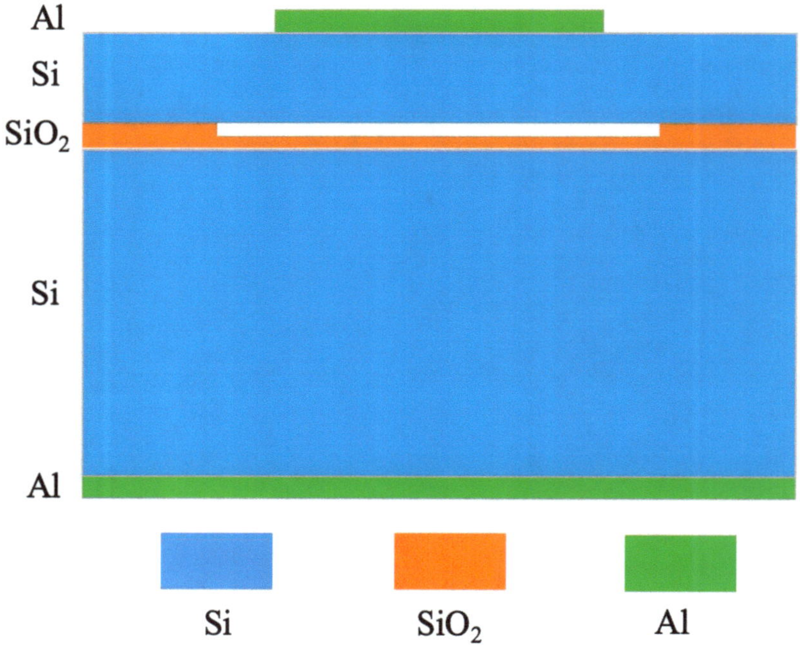

Figure 6. Design of CMUT pressure sensing unit structure.

Table 1. Design parameters of CMUT pressure structure sample.

Structural Parameters	Size
Membrane Radius (μm)	46
Membrane Thickness (μm)	2
Electrode Radius (μm)	23
Electrode Thickness (μm)	0.5
Cavity Height (μm)	0.2

Eigen frequency analysis is an essential step in understanding the displacement vibration characteristics of CMUT thin films and identifying their resonant frequencies. The resonant frequency serves as a critical parameter in evaluating the performance of a CMUT device. Through the construction of a three-dimensional model of a CMUT and examination of its characteristic frequencies, it is possible to discern the modal vibration modes and the resonant frequency. This information is pivotal as it directly influences the transducer's efficiency and efficacy. A visualization of these modal patterns, typically the first six modes of vibration, where the modes and frequencies shown in Table 2, can be graphically represented and is often illustrated in technical figures to provide a clear understanding of the CMUT's operational dynamics, as shown in Figure 7 [35–39].

Table 2. The first six modes and frequencies of CMUT unit structure.

Mode	Eigen Frequency/MHz	Mode	Eigen Frequency/MHz
First Mode	3.8412	Fourth Mode	13.648
Second Mode	8.2585	Fifth Mode	13.827
Third Mode	8.285	Sixth Mode	15.877

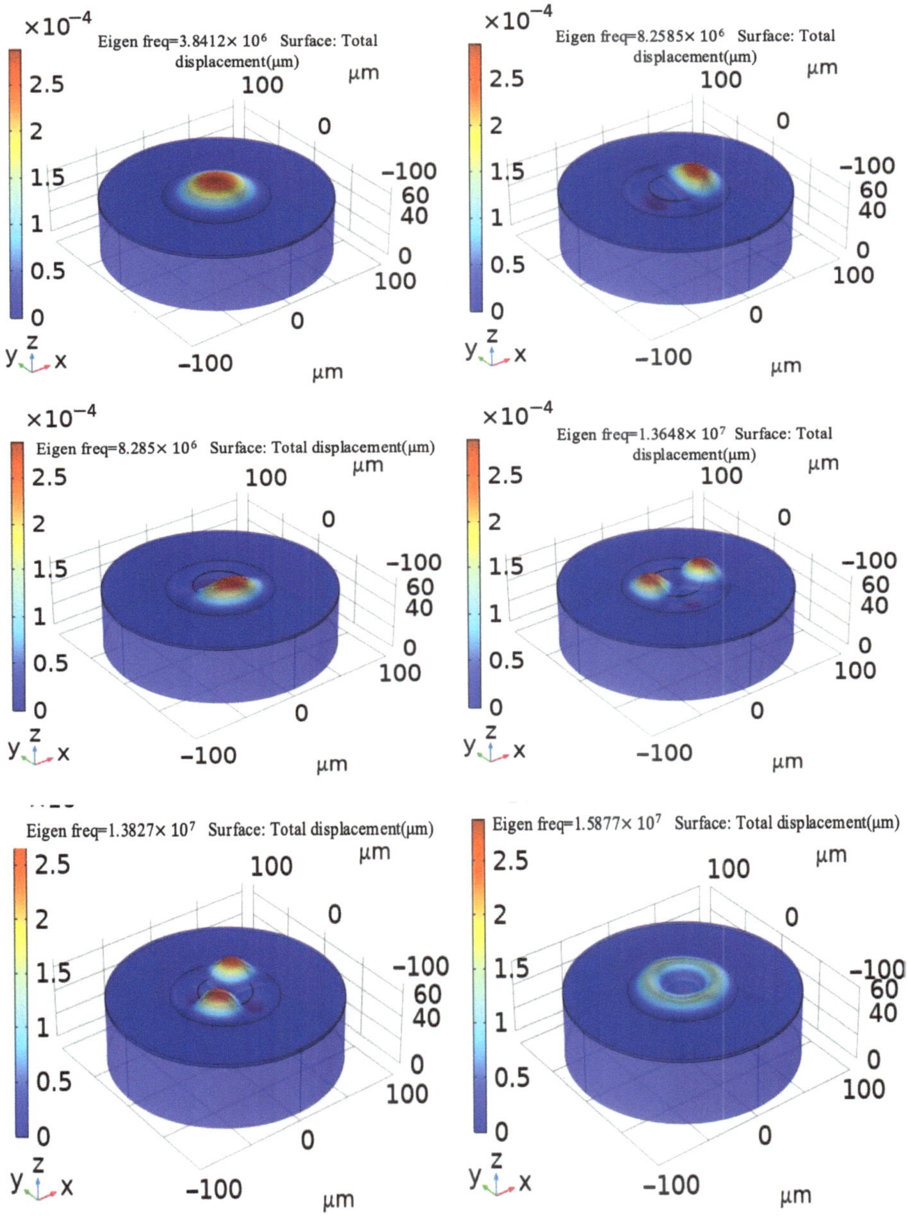

Figure 7. Analysis of the first six modes of CMUT pressure sensing unit structure.

By maintaining a constant membrane thickness and adjusting the membrane radius, we can analyze the resonance frequencies of a CMUT at various radii. This approach enables us to observe how changes in membrane size affect resonance frequency, providing valuable guidance for CMUT pressure sensor design optimization. By strategically adjusting membrane radius, we can strike an optimal balance between sensitivity and operational frequency, tailoring CMUT performance to meet specific needs.

3.2. Testing

To evaluate the performance of the capacitive pressure sensor, the testing device shown in Figure 8c was installed within a bicycle tire in Figure 8a, which generally operates at pressures higher than those of automotive tires, thereby satisfying the pressure range requirements for automotive applications. Standard mountain bike tires are rated for pressures up to 2.8–4.5 Bar, while road bicycle tires can withstand pressures exceeding 6.8 Bar. The testing procedure began by creating a small opening in an intact, leak-free inner tube of a bicycle tire. The pressure sensor was then inserted into a small hole, and multiple layers of tire patch adhesive were then applied to ensure a secure and airtight seal around the sensor. The sensor leads were carefully extended through the opening for external connection. After allowing the patch adhesive to dry for 5–10 min, the inner tube was inflated using a Michelin tire pump (Michelin, Shanghai, China) equipped with both pressure control and display features in Figure 8a, facilitating accurate pressure monitoring during testing. This testing setup provides a robust and controlled method for assessing the pressure sensor's accuracy and reliability under conditions analogous to those in automotive tire environments.

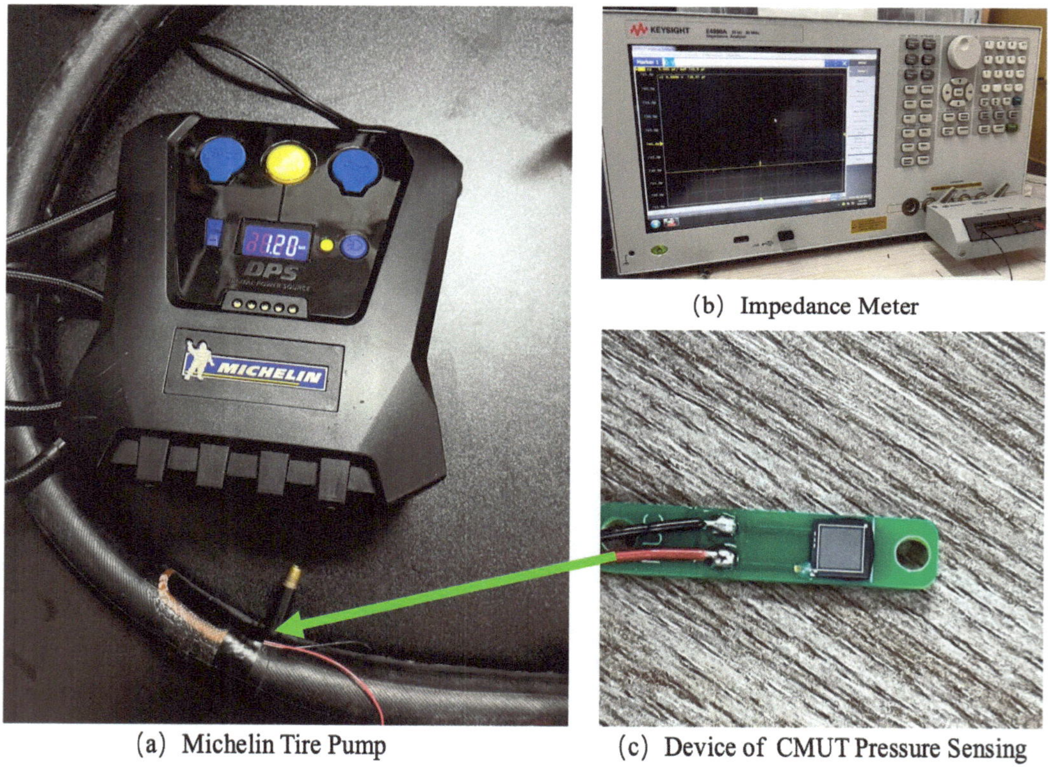

Figure 8. Photo of testing setup and CMUT pressure sensing device.

When measuring capacitance with a Keysight E4990A Impedance Analyzer (Keysight Technologies, Santa Rosa, CA, USA) as shown in Figure 8b, an AC voltage ranging from −20 V to +20 V is applied at a frequency of 1 kHz, which is far from the CMUT pressure sensing design resonant frequency. The testing data are listed in Table 3.

Table 3. Test data of CMUT pressure sensing structure for tire pressure measurement.

Pressure (Pa)	Capacitance (F)
0	7.8578×10^{-10}
55,000	7.9174×10^{-10}
100,000	7.973×10^{-10}
120,000	8.008×10^{-10}
150,000	8.0452×10^{-10}
180,000	8.0917×10^{-10}
200,000	8.13093×10^{-10}
230,000	8.1833×10^{-10}

4. Discussion

In this CMUT pressure sensing array design, each unit features a pressure film with a thickness of 2 μm and a cavity height of 0.2 μm within a 30 × 30 array structure. The capacitive sensor's nonlinear properties yield an average sensitivity of 154.6 aF/kPa, indicating that sensitivity fluctuates with pressure changes. At 230 kPa and 35.56 fF, maximum capacitance change is expected. At 1 kPa resolution, the calculation for the minimum charge variation for a single CMUT unit capacitor at 1.5V is detailed in the study.

$$\Delta Q = \Delta C \cdot U \tag{11}$$

The calculations reveal that the minimal change in capacitance due to pressure application on a CMUT device corresponds to a displacement of 1449 electrons across the capacitor plates, given the electron charge of 1.6×10^{-19} coulombs. To boost the system's sensitivity, a matrix of sensor units can be created by connecting multiple capacitor units in parallel [40].

Moreover, CMUT technology offers benefits such as ease of mass production, circuit integration, structural robustness, and reliable operation in adverse conditions, making it exemplary for various applications.

Particularly for tire pressure monitoring sensors, the emphasis on power consumption is greater than that on accuracy. CMUT pressure sensitivity and precision meet measurement standards even without a bias voltage, significantly reducing power consumption. This low-power mode of operation enables the sensor to last up to 10 years with a low energy consumption, demonstrating CMUT potential for long-term, energy-efficient applications [40].

The pressure measurement sensitivity of the simulated data for the CMUT-based pressure sensor designed in this study is 119 fF/kPa, with the result of the simulation shown in Figure 9, while the actual measurement sensitivity for tire pressure data is 139 fF/kPa, testing results are shown in Figure 10. The close match between the simulation and test data validates the design approach for the CMUT pressure sensor structure.

We utilized a CA-330PRO Dynamic Current Analyzer (Beijing Chipment Technology Co., Ltd., Beijing, China) to measure the power consumption of the CMUT pressure sensing device. The equipment specifications are as follows:

DC Voltage: from 0.0002 mV to 0.006 V;

DC Current: from 0.02 μA to 0.002 A;

Resistance: from 0.0002 Ω to 0.015 MΩ;

Figure 11 highlights the exceptionally low power consumption of the CMUT pressure sensing device. The measured power is 1.47 μW, achieved with a supply voltage of 9 V and a resulting current of 160 nA. This ultra-low power operation underscores the device's suitability for applications where energy efficiency is critical.

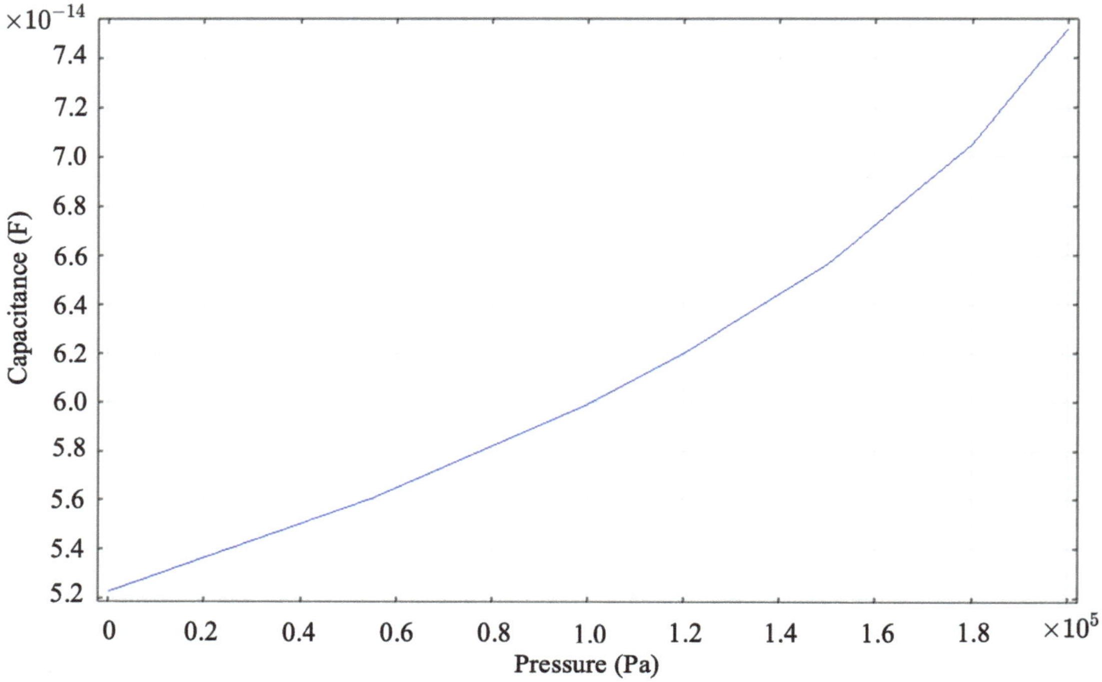

Figure 9. Simulation of capacitance and pressure relationship of CMUT pressure sensing structure.

The slope data presented in Table 4 show a clear dependence on the bias voltage. As the bias voltage increases, the slope values also grow larger. This relationship demonstrates that the slope is directly proportional to the pressure measurement sensitivity of the device. Higher bias voltages result in improved sensitivity, making the device more effective in detecting minute pressure changes.

Table 4. DC bias effect of the CMUT pressure sensing device.

DC Bias (V)	Slope
0	1.16367×10^{-16}
4	1.17213×10^{-16}
8	1.194×10^{-16}
12	1.30619×10^{-16}
16	1.47752×10^{-16}
20	1.6166×10^{-16}

Figure 12 further supports this observation by illustrating the capacitance variation in the CMUT pressure sensing device under different bias voltages. The results reveal that as the pressure changes, the capacitance becomes increasingly sensitive, particularly at higher bias voltages. This enhanced sensitivity is a direct consequence of the DC bias applied to the CMUT, which optimizes its operational characteristics for precise pressure measurement. The combination of the slope data and the capacitance behavior under varying bias voltages underscores the effectiveness of DC-biased CMUTs in enhancing pressure sensitivity. This makes them well suited for applications requiring high-precision and low-power pressure sensing, such as medical diagnostics, industrial monitoring, and other scenarios demanding real-time, ultra-sensitive pressure detection.

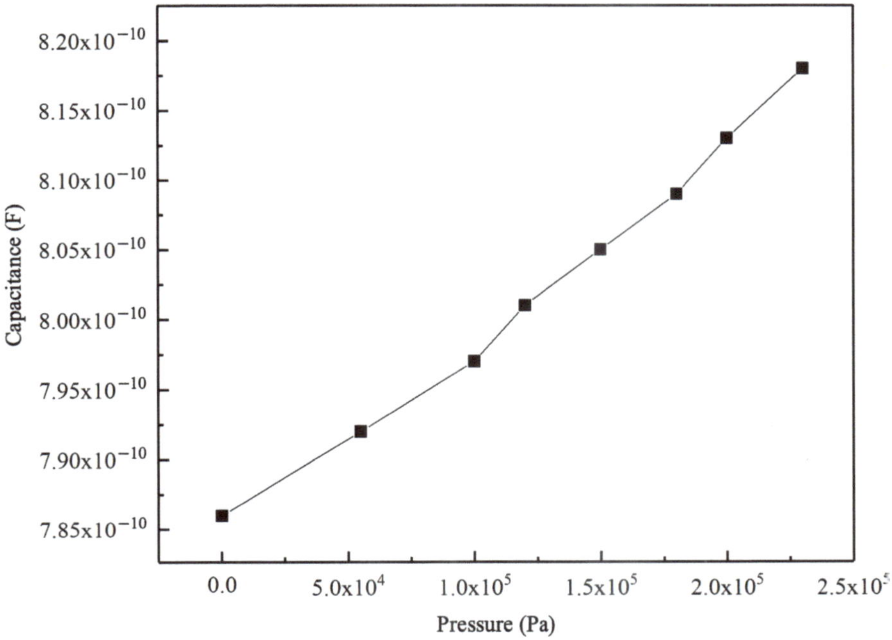

Figure 10. CMUT pressure device testing results showing relationship between capacitance and pressure.

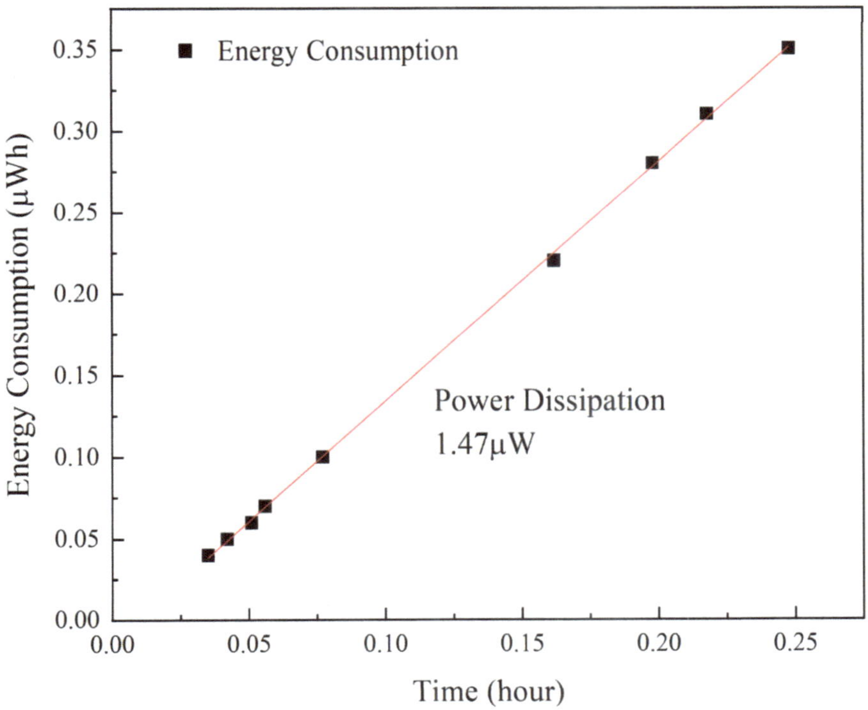

Figure 11. Power dissipation testing data of CMUT pressure device.

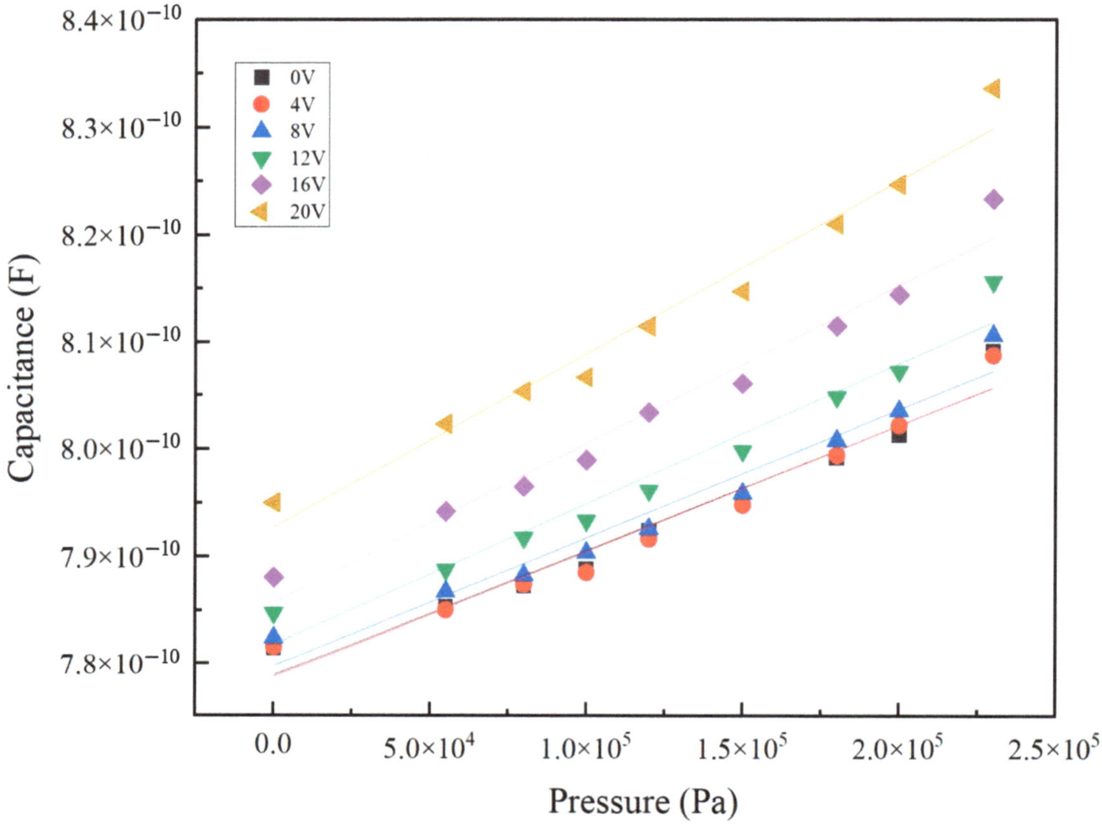

Figure 12. Capacitance changes with pressure of CMUT pressure sensing device at different DC bias.

5. Conclusions

Cost-Effective manufacturing through low-temperature direct bonding: The CMUT pressure sensor is fabricated using silicon–silicon bonding technology, a method that combines high sensitivity with low manufacturing costs. This technique not only ensures robust structural integrity and excellent hermetic sealing but also simplifies the manufacturing process, making it suitable for large-scale production. The cost advantages of silicon–silicon bonding make CMUT sensors an economical choice for applications that require large sensor arrays, such as industrial pressure monitoring and consumer electronics. Versatile applications with ultra-low power consumption: The CMUT structure is highly versatile, functioning both as an ultrasonic transducer and a pressure sensor. This multifunctionality broadens its applicability across various fields, including medical ultrasound imaging and pressure monitoring. Furthermore, its ultra-low power consumption makes it ideal for devices with limited energy resources, such as tire pressure monitoring systems (TPMSs), where battery life is critical. The CMUT's low power requirements make it a valuable component in the next generation of smart, autonomous systems that prioritize energy efficiency. Integration potential in IoT and automotive systems: With its high sensitivity, low power consumption, and cost-effective production, the CMUT pressure sensor is well suited for integration into IoT-enabled and automotive systems. In automotive TPMS, for example, CMUT sensors provide real-time pressure data with minimal power use, enhancing safety without imposing additional strain on the vehicle's electrical system. In IoT applications, CMUT-based pressure sensors deliver precise, reliable data for smart home systems, environmental monitoring, and industrial automation, contributing to more intelligent and responsive systems. With its advanced silicon fabrication, cost efficiency, and versatile design, CMUT technology offers a powerful solution for high-performance, affordable, and energy-efficient sensing applications.

Author Contributions: Conceptualization, L.L.; methodology, L.L.; software, L.L., Z.W., R.L. and Y.Z.; validation, L.L., C.H. and J.C.; formal analysis, L.L., Z.W., J.C. and A.M.R.; investigation, L.L.; resources, J.C., A.M.R. and Z.J.; data curation, L.L., J.C. and C.H.; writing—original draft preparation, L.L.; writing—review and editing, A.G.H. and C.H.; visualization, L.L. and A.G.H.; supervision, C.H.; project administration, L.L. and C.H.; funding acquisition, J.C. All authors have read and agreed to the published version of the manuscript.

Funding: This research was funded by the National Key Research and the Development Program of China under Grant 2021YFB3203002.

Institutional Review Board Statement: Not applicable.

Informed Consent Statement: Not applicable.

Data Availability Statement: Data are contained within the article.

Acknowledgments: The authors would like to thank the editor and referees for their valuable comments.

Conflicts of Interest: Author Aisn Gioronara Hui was employed by the company Einsteck Inc., and Author Allwins Moore Rogikin was employed by the company Golden Rogikin Limited. The remaining authors declare that the research was conducted in the absence of any commercial or financial relationships that could be construed as a potential conflict of interest.

References

1. Kubba, A.E.; Jiang, K. A Comprehensive Study on Technologies of Tyre Monitoring Systems and Possible Energy Solutions. *Sensors* **2014**, *14*, 10306–10345. [CrossRef] [PubMed]
2. NXP. A Global Solution for Tire Pressure Monitoring Systems. Available online: https://www.nxp.com/docs/zh/white-paper/TPMSWP.pdf (accessed on 15 December 2024). (In Chinese)
3. van Zyl, P.S.; van Goethem, S.; Kanarachos, S.; Rexeis, M.; Hausberger, S.; Smokers, R. Study on Tyre Pressure Monitoring Systems (TPMS) as a Means to Reduce Light-Commercial and Heavy-Duty Vehicles Fuel Consumption and CO_2 Emissions, TNO Report, TNO 2013 R10986, 29 July 2013. Available online: https://climate.ec.europa.eu/document/download/a72c3926-a097-4272-8bb6-8bb4f200dcb8_en?filename=tno_2013_final_report_en.pdf (accessed on 30 June 2024).
4. Jena, S.; Gupta, A. Review on Pressure Sensors: A Perspective from Mechanical to Micro-electro-mechanical systems. *Sens. Rev.* **2021**, *41*, 320–329. [CrossRef]
5. Mitrakos, V.; Macintyre, L.; Denison, F.C.; Hands, P.J.; Desmulliez, M.P. Design, Manufacture and Testing of Capacitive Pressure Sensors for Low-Pressure Measurement Ranges. *Micromachines* **2017**, *8*, 41. [CrossRef]
6. Xin, Q.; Jingfeng, G.; Junjie, G.; Ri, B.; Mingxing, Y.; Pian, Z. Automobile tire pressure monitoring technology and development trend. *J. Physics Conf. Ser.* **2019**, *1314*, 012100. [CrossRef]
7. Ergun, A.S.; Yaralioglu, G.G.; Oralkan, O.; Khuri-Yakub, B.T. MEMS/NEMS Techniques and Applications: Techniques and Applications of Capacitive Micromachined Ultrasonic Transducers. In *MEMS/NEMS Handbook*; Springer: Boston, MA, USA, 2006; Chapter 7. [CrossRef]
8. Ergun, A.S.; Yaralioglu, G.G.; Khuri-Yakub, B.T. Capacitive Micromachined Ultrasonic Transducers: Theory and Technology. *J. Aerosp. Eng.* **2003**, *16*, 76–84. Available online: https://ascelibrary.org/doi/abs/10.1061/(ASCE)0893-1321(2003)16:2(76) (accessed on 23 December 2020). [CrossRef]
9. Brenner, K.; Ergun, A.S.; Firouzi, K.; Rasmussen, M.F.; Stedman, Q.; Khuri-Yakub, B.P. Advances in Capacitive Micromachined Ultrasonic Transducers. *Micromachines* **2019**, *10*, 152. [CrossRef]
10. Sharma, A.; Singh, J. Design and Analysis of High Performance MEMS Capacitive Pressure Sensor for TPMS. In Proceedings of the 2013 International Conference on Control, Automation, Robotics and Embedded Systems (CARE), Jabalpur, India, 16–18 December 2013; pp. 1–5.
11. Song, P.; Ma, Z.; Ma, J.; Yang, L.; Wei, J.; Zhao, Y.; Zhang, M.; Yang, F.; Wang, X. Recent Progress of Miniature MEMS Pressure Sensors. *Micromachines* **2020**, *11*, 56. [CrossRef] [PubMed]
12. Han, X.; Huang, M.; Wu, Z.; Gao, Y.; Xia, Y.; Yang, P.; Fan, S.; Lu, X.; Yang, X.; Liang, L.; et al. Advances in high-performance MEMS pressure sensors: Design, fabrication, and packaging. *Microsyst. Nanoeng.* **2023**, *9*, 156. [CrossRef]
13. Tian, J.; Wang, C.; Xu, L. A new type of silicon micro-microphone. In Proceedings of the 16th International Congress on Acoustics, Kyoto, Japan, 4–9 April 2004.
14. Tian, J.; Wang, C.H.; Xu, L.; Qiao, D.H.; Ma, J.; Hao, Z.H.; Wei, J.H. Silicon Condenser Microphone with Round Membranes. *Chin. J. Sens. Actuators* **2006**, *12*, 2297–2299.

15. Gupta, S.; Kothari, P.; Sharma, R.; Mukhiya, R. Design, modelling and FEM-based validation of composite membrane circular cMUT cell. *Microsyst. Technol.* **2023**, *29*, 1589–1600. [CrossRef]
16. Ding, Q.; Wang, H.; Zhang, H.; Huang, X.; Sun, X.; Qin, Z.; Ren, R.; Zhu, J.; He, C.; Zhang, W. Research on Novel CMUTs for Detecting Micro-Pressure with Ultra-High Sensitivity and Linearity. *Micromachines* **2021**, *12*, 1340. [CrossRef]
17. Wang, H. *Working Mechanism and Application Basis of Capacitive Micromachined Ultrasonic Transducers*; Science Press: Beijing, China, 2018.
18. Joseph, J.; Ma, B.; Khuri-Yakub, B.T. Applications of Capacitive Micromachined Ultrasonic Transducers: A Comprehensive Review. *IEEE Trans. Ultrason. Ferroelectr. Freq. Control* **2021**, *69*, 456–467. [CrossRef]
19. Ladabaum, I.; Jin, X.; Soh, H.; Atalar, A.; Khuri-Yakub, B. Surface Micromachined Capacitive Ultrasonic Transducers. *IEEE Trans. Ultrason. Ferroelectr. Freq. Control* **1998**, *45*, 678–690. [CrossRef] [PubMed]
20. Satir, S.; Zahorian, J.; Degertekin, F.L. A Large-Signal Model for CMUT Arrays with Arbitrary Membrane Geometry Operating in Non-Collapsed Mode. *IEEE Trans. Ultrason. Ferroelectr. Freq. Control* **2013**, *60*, 2426–2439. [CrossRef]
21. Li, Z.; Zhao, L.; Jiang, Z.; Akhbari, S.; Ding, J.; Zhao, Y.; Zhao, Y.; Lin, L. Capacitive micromachined ultrasonic transducer for ultra-low pressure measurement: Theoretical study. *AIP Adv.* **2015**, *5*, 127231. [CrossRef]
22. Li, Z.; Zhao, L.; Jiang, Z.; Yihe, Z.; Zhao, Y. Capacitive Micromachined Ultrasonic Transducer for Ultra-low Pressure Detection. In Proceedings of the 9th IEEE International Conference on Nano/Micro Engineered and Molecular Systems (NEMS), Waikiki Beach, HI, USA, 13–16 April 2014; pp. 600–603.
23. Ho, M.-C.; Park, K.K.; Eckhoff, K.; Kupnik, M.; Khuri-Yakub, B.T. Air-coupled CMUTs operating at ambient pressures ranging from 1 to 20 atm. In Proceedings of the 2013 IEEE International Ultrasonics Symposium (IUS), Prague, Czech Republic, 21–25 July 2013; pp. 1412–1415.
24. Bitsie, F.; Eaton, W.P.; Plummer, D.W.; Smith, J.H. A new analytical solution for diaphragm deflection and its application to a surface micromachined Pressure sensor. In Technical Proceedings of the 1999 International Conference on Modeling and Simulation of Microsystems, San Juan, PR, USA, 19 April 1999; pp. 640–643.
25. Wang, J.; Yu, Y.; Pun, S.H.; Gao, Y.; Che, U.K.; Mak, P.U. An Accurate Design to Calculate and Measure the Capacitance of CMUT for Pressure Sensors. In Proceedings of the 2016 IEEE International Conference on Consumer Electronics-China (ICCE-China), Guangzhou, China, 19–21 December 2016; pp. 1–4.
26. Zhang, X.; Yu, L.; Guo, Q.; Li, D.; Zhang, H.; Yu, H. Resonance frequency analysis of a dual-frequency capacitive micromechanical ultrasonic transducer for detecting high and low pressures simultaneously with high sensitivity and linearity. *J. Phys. D Appl. Phys.* **2020**, *53*, 035401. [CrossRef]
27. Zhang, X.; Yu, L.; Yu, H.; Li, D. Design of Resonant Vacuum Pressure Sensor with CMUT for High Sensitivity and Linearity. In Proceedings of the 23rd International Congress on Acoustics: Integrating 4th EAA Euroregio 2019, Aachen, Germany, 9–13 September 2019; Available online: https://pub.dega-akustik.de/ICA2019/data/articles/000874.pdf (accessed on 15 June 2024).
28. Li, Z.; Zhao, L.; Ye, Z.; Wang, H.; Zhao, Y.; Jiang, Z. Resonant frequency analysis on an electrostatically actuated microplate under uniform hydrostatic pressure. *J. Phys. D Appl. Phys.* **2013**, *46*, 195108. [CrossRef]
29. Zheng, Z.; Yao, Y.; Liu, J.A.; Sun, Y.; Yeow, J.T. Highly sensitive CMUT-based humidity sensors built with nitride-to-oxide wafer bonding technology. *Sens. Actuators B Chem.* **2019**, *294*, 123–131. [CrossRef]
30. Park, K.K.; Lee, H.; Kupnik, M.; Khuri-Yakub, B.T. Fabrication of Capacitive Micromachined Ultrasonic Transducers via Local Oxidation and Direct Wafer Bonding. *J. Microelectromech. Syst.* **2011**, *20*, 95–103. [CrossRef]
31. Ergun, A.S.; Huang, Y.; Zhuang, X.; Oralkan, Ö.; Yaralıoglu, G.G.; Khuri-Yakub, B.T. Capacitive Micromachined Ultrasonic Transducers: Fabrication Technology. *IEEE Trans. Ultrason. Ferroelectr. Freq. Control* **2005**, *52*, 2242–2258. [CrossRef] [PubMed]
32. Wang, Z.; He, C.; Zhang, W.; Li, Y.; Gao, P.; Meng, Y.; Zhang, G.; Yang, Y.; Wang, R.; Cui, J.; et al. Fabrication of 2-D Capaci-tive Micromachined Ultrasonic Transducer (CMUT) Array through Silicon Wafer Bonding. *Micromachines* **2022**, *13*, 99. [CrossRef]
33. He, C.; Zhang, B.; Xue, C.; Zhang, W.; Zhang, S. Wafer-Bonding Fabricated CMUT Device with Parylene Coating. *Micromachines* **2021**, *12*, 516. [CrossRef] [PubMed]
34. Qi, D. Research of Micro Pressure Sensor Based on Capacitive Micromachined Ultrasonic Transducer. Master's Thesis, North University of China, Taiyuan, China, 2022.
35. Wang, H.; Lv, Y.; Wang, C.; Wang, X.; He, C.; Xue, C.; He, S. Simulation Analysis and Performance Testing Investigation of Capacitive Micromachined Ultrasonic Transducer. *Int. J. Pattern Recognit. Artif. Intell.* **2018**, *32*, 1858004. [CrossRef]
36. Wang, H.; Wang, X.; He, C.; Xue, C. Reception characteristics investigation and measurement of capacitive micromachined ultrasonic transducer. *Sens. Rev.* **2020**, *40*, 237–246. [CrossRef]
37. Teng, M.F.; Hariz, A.J. Characterisation and Modelling of MEMS Ultrasonic Transducers. *J. Phys. Conf. Ser.* **2006**, *34*, 949–954. [CrossRef]
38. Anbalagan, S.A.; Uma, G.; Umapathy, M. Modeling and simulation of Capacitive Micromachined Ultrasonic Transducer (CMUT). *J. Phys. Conf. Ser.* **2006**, *34*, 595–600. [CrossRef]

39. Ravi, A.; Krishna, R.M.A.; Christen, J.B. Modeling and Simulation of Dual Application Capacitive MEMS Sensor. In Proceedings of the 2014 COMSOL Conference in Boston, Boston, MA, USA, 8–10 October 2014.
40. Kolle, C.; Scherr, W.; Hammerschmidt, D.; Pichler, G.; Motz, M.; Schaffer, B.; Forster, B.; Ausserlechner, U. Ultra Low-Power Monolithically Integrated Capacitive Pressure Sensor for Tire Pressure Monitoring. In Proceedings of the SENSORS, IEEE 2004, Vienna, Austria, 24–27 October 2004; pp. 244–247.

sensors

MDPI

Communication

Comparison Analysis of Roundness Measurement of Small Cylindrical Workpieces with Different Styluses

Borong Wu [1], Chuang Zeng [2] and Qiaolin Li [2,*]

[1] Chongqing University-University of Cincinnati Joint Co-op Institute, Chongqing University, Chongqing 400044, China; 20206107@cqu.edu.cn
[2] Shenzhen International Graduate School, Tsinghua University, Shenzhen 518055, China; zengc21@mails.tsinghua.edu.cn
* Correspondence: li.qiaolin@sz.tsinghua.edu.cn

Abstract: To investigate the high-accuracy roundness metrology of a needle roller 1.5 mm in diameter and 5.8 mm in length using the stitching linear scan method, a ruby ball stylus with a tip radius of 150 μm and a diamond stylus with a tip radius of 2 μm were employed to perform experiments under the same conditions. The precision coordinate data, derived from the needle roller's cross-sectional circumference, were segmented into uniform eighths, each scanned with the stylus of a roughness measuring machine. The roundness profile of the needle roller was obtained by stitching the arc profiles, which were characterized according to the precision coordinate data of the arcs. The cross-correlation function, Euclidean distance, residual sum of squares, position error, and curvature of the measured arcs were used to evaluate the results, which can reflect the performance of the stylus. A comparison of the results obtained using the ruby ball stylus versus the diamond stylus demonstrates the ruby ball stylus' greater suitability for use in the roundness metrology of the needle roller bearing examined in this paper.

Keywords: roundness measurement; small cylindrical workpieces; ruby ball stylus; diamond stylus; stitching linear scan method

1. Introduction

As fundamental components in manufacturing processes, cylindrical workpieces are pivotal in various industrial fields, including agricultural machinery, intelligent automobiles, additive manufacturing, metallurgical industries, marine vessels, and microelectronic processing [1–6]. Their multifunctionality makes them indispensable components within mechanical systems, such as gears, rotating cylinders, crankshafts, and other precision machines [7–9]. Among these, roundness serves as a critical quality control parameter that significantly influences the functionality, performance, and lifetime of these components [10–12]. An accurate roundness measurement not only validates compliance with design specifications but also has a direct impact on product performance [13]. However, current measurement techniques have limitations, particularly when it comes to accurately evaluating small workpieces. Common roundness measurement methods include the two-point method and the use of roundness gauges. These methods rely on the measurement personnel's skill and are susceptible to errors caused by workpiece eccentricity and tilt, especially as the dimensions of the cylindrical workpiece being measured decrease, complicating the accuracy of the roundness measurement. Workpieces less than 3 mm in diameter present particular difficulties in this regard [14–16].

Over the years, many roundness measurement methods have been proposed. For example, Ma Yu-zhen [17] introduced a spindle roundness measurement method based on a capacitive scanning system that uses three capacitive displacement probes to eliminate spindle errors through averaging effects, thereby achieving roundness error separation and confirming the effectiveness of the measurement system. Bin Duan [18] proposed a novel roundness measurement method based on single-probe scanning that simplifies mathematical processing without requiring complex forward and backward Fourier transforms. Through simulation analysis, the method's feasibility and repeatability were verified, and the roundness measurement's accuracy was significantly improved. Jie Liang [19] introduced a method for measuring roundness using a laser profile scanner and merging the line point cloud data, using two-dimensional laser displacement sensors to collect line point cloud data and merging them radially and circumferentially to achieve micrometer-level high-precision measurement. Jiao Bai et al. [20] proposed a three-probe measurement method based on error separation technology. They used the Monte Carlo method to analyze the effects of sensor angle, off-axis distance, and harmonic suppression on error separation. Using a spectral confocal displacement sensor for roundness measurement, high-precision measurements in the sub-micron range were achieved. Erik Oertel [21] proposed a roundness measurement method based on an atomic force microscope, which enables the high-precision measurement of small, complex components through surface scanning and stitching algorithms. Eiki Okuyama [22] proposed a combined method, using the three-point method for roundness contour measurement and the integral method for straightness contour measurement. Through the three-point method, the roundness contour of the workpiece and the radial movement of the rotary table were separated, and reverse filtering was used to restore the original roundness contour, thus ensuring measurement stability. A Küng [23] proposed a roundness measurement method based on a three-dimensional coordinate measuring machine, reducing the motion mass and ensuring low isotropic stiffness through the innovative design of the flexible hinge–parallel motion structure, further improving the roundness measurement accuracy. These methods include the use of rotary scanning roundness testers, two-point or three-point methods, and coordinate measuring machines. Although these methods have proven effective in many cases, the influence of workpiece eccentricity on roundness measurement becomes more pronounced when measuring small cylindrical workpieces, especially when the diameter is less than 3 mm, making accurate measurement unattainable [24]. The non-contact measurement methods proposed by relevant scientists, particularly that of roundness measurement using spectral confocal technology, can eliminate the errors introduced by contact-based measurements, improving measurement accuracy [25–28]. However, accurate measurement remains a challenge, especially when dealing with small, weakly reflective objects. Therefore, research into contact-based methods remains crucial. To overcome the difficulties of measuring small-sized workpieces, relevant teams have proposed a stitching linear scan method using hardware filters [29–34]. The efficient filtering of high-frequency data and its impact mitigation have been achieved by replacing a diamond stylus with a tip radius of 2 μm with a ruby stylus with a tip radius of 150 μm. This substitution has allowed for the precise measurement of roundness in cylindrical rollers, which have dimensions of 10 mm in length and 1.5 mm in diameter. A roundness profile was reconstructed by splicing a series of circular arcs to derive roundness values, thereby facilitating the high-precision measurement of small cylindrical workpieces. This method has been shown to have exceptional measurement accuracy and repeatability for several critical measurement outcomes during the assessment of small-sized workpieces, in contrast to the conventional stitching linear scan method. Therefore, it is essential to investigate how using different-sized styluses impacts the results of workpiece roundness measurement through comparative analysis, especially for small-sized workpieces.

In this paper, experiments in roundness metrology were conducted on a small cylindrical workpiece, 1.5 mm in diameter and 5.8 mm in length, to investigate the efficacy of the stitching linear scan method for measuring roundness, with both diamond and ruby ball styluses employed, respectively. Quantitative analyses were conducted using the cross-correlation function, Euclidean distance, residual sum of squares, position error, and curvature. The measuring performance was evaluated by comparing the results obtained with the ruby ball stylus to the results obtained with the diamond stylus.

2. Principle and Experiments

The surface topography of the measured needle roller on a nano-scale can be conceptualized as a form which can be reconstructed with coordinates. Thus, a profilometer is employed to obtain these coordinates. As shown in Figure 1b, a Taylor Hobson Form Talysurf PGI 420 (VirtualExpo, Marseille, France) roughness measuring machine, with specifications shown in Table 1, is used to perform the experiments. As shown in Figure 1a, the needle roller is attached with a rotating jig and positioned on a V-block mounted on a multi-degree-of-freedom stage. Prior to conducting experiments, alignment through the adjustment of the stage's position around the X and Z axes must be achieved. As shown in Figure 1c, the stylus of the roughness measuring machine scans across the needle roller surface and returns to the initial position. Arc coordinate data can be obtained. Subsequently, the rotating jig is manipulated to rotate 45°, causing the needle roller to also rotate by 45°, after which the stylus performs a second scan to obtain the subsequent arc coordinate data. This process is repeated until the entire cross-section circle has been scanned. As shown in Figure 2, eight arc coordinate data are obtainable. Of note, there is an overlapping portion for adjacent arc stitching.

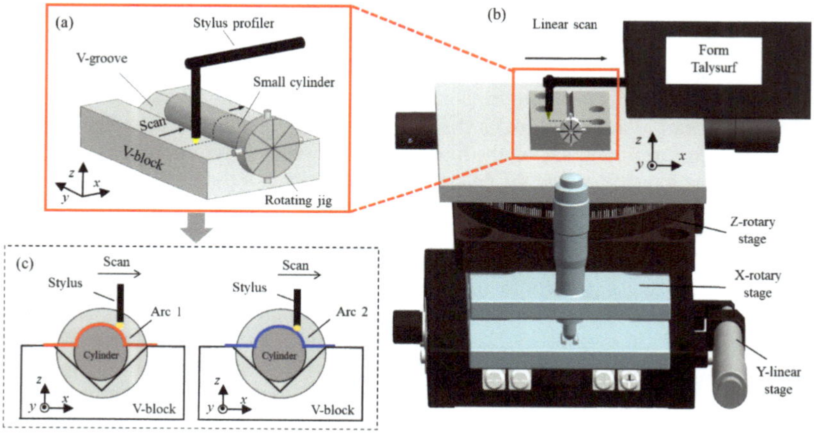

Figure 1. The linear scan on the surface of the needle roller by a profilometer: (**a**) The stylus scans across the surface of the needle roller; (**b**) the schematic of linear scan by a profilometer; (**c**) the arc coordinate data obtained by the stylus scanning.

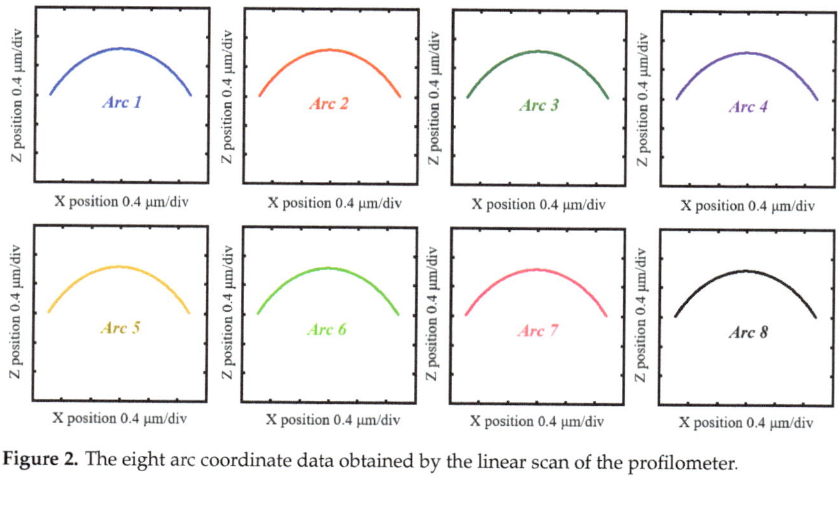

Figure 2. The eight arc coordinate data obtained by the linear scan of the profilometer.

Table 1. The specification of Form Talysurf PGI 420.

Item	Resolution (Z)	Measuring Range (Z)	Measuring Range (X)	Sampling Step (X)	Scanning Speed
Value	3.2	4	120	0.125	0.1
Unit	nm	mm	mm	µm	mm/s

As shown in Figure 3, two sets of experiments were conducted under identical conditions. One employed a diamond stylus with a tip radius of 2 µm, while the other used a ruby ball stylus with a tip radius of 150 µm. This enabled the characterization of arc profiles through data processing. Figures 4 and 5 show the eight arc profiles characterized according to the arc coordinate data obtained using a diamond stylus and a ruby ball stylus, respectively.

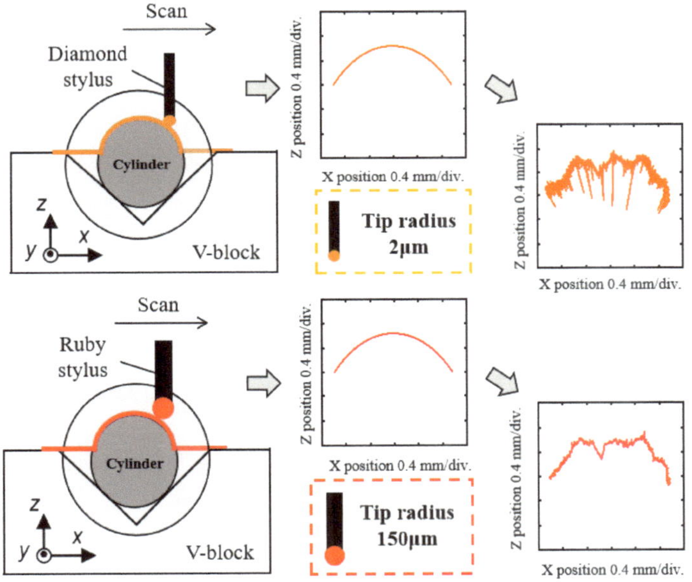

Figure 3. The scheme of linear scanning with different ball sizes.

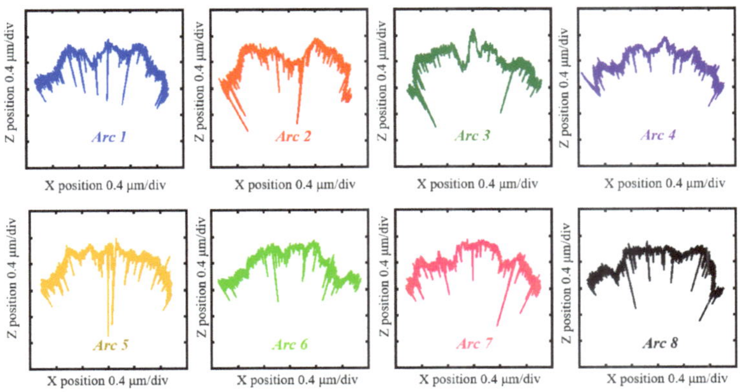

Figure 4. The eight arc profiles characterized according to the arc coordinate data obtained using the diamond stylus.

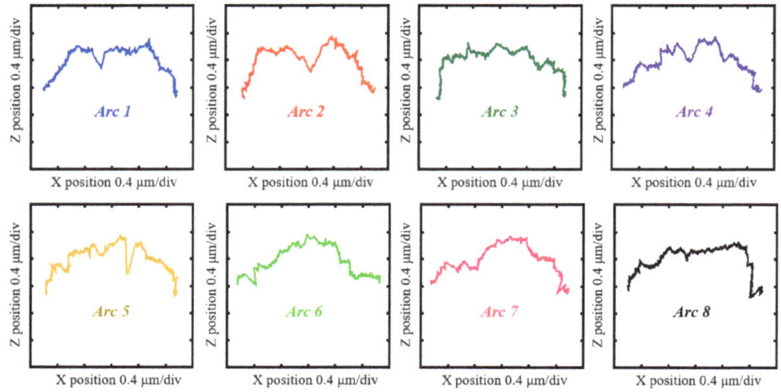

Figure 5. The eight arc profiles characterized according to the arc coordinate data obtained using the ruby ball stylus.

The roundness profile of the needle roller is effectively reconstructed through the assembly of the eight arc profiles. Figure 6 shows the arc profile transformation, from the X-Z to the $\theta - \Delta R$ coordinate system, which enhances the visualized matching of adjacent arc profiles. Figure 6a,b show the matching of adjacent arc profiles from the diamond and ruby styluses, respectively. The cross-correlation function serves to evaluate the circumferential matching of the overlapping portions between adjacent arc profiles, as defined in Equation (1). In this context, C represents the coefficient, $g(\theta)$ denotes the plotted arc profile, and θ denotes the angular position of the arc profiles. The Euclidean distance, as defined by Equation (2), evaluates the radial matching for the overlapping portion between adjacent arc profiles, where Δr is the difference between the radius and the reference radius in accordance with each coordinate on each arc coordinate data.

$$C_{i\sim i+1} = \int_{-\infty}^{\infty} g_i(\theta) g_{i+1}(\theta + \Delta\theta_{i\sim i+1}) d_\theta \quad (1)$$

$$D_{i\sim i+1} = \frac{1}{n_c} \sum_{i,j=1}^{n_c} \sqrt{\left(\theta_{i,j} - \theta_{i+1,j}\right)^2 + \left(\Delta r_{i,j} - \Delta r_{i+1,j}\right)^2} \quad (2)$$

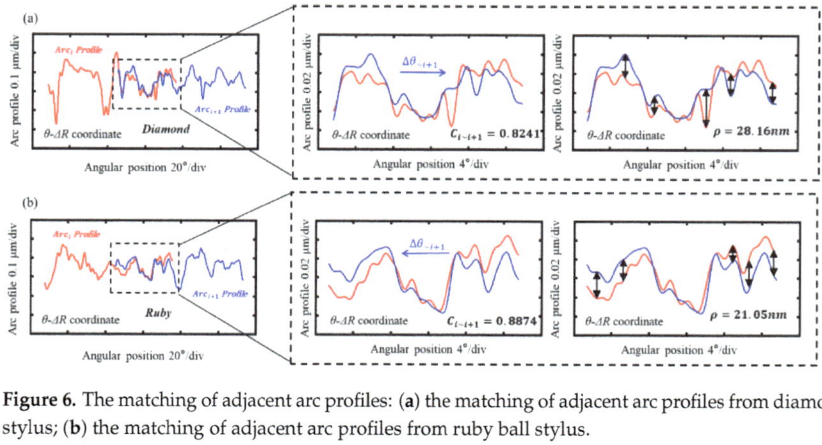

Figure 6. The matching of adjacent arc profiles: (**a**) the matching of adjacent arc profiles from diamond stylus; (**b**) the matching of adjacent arc profiles from ruby ball stylus.

In general, a higher coefficient value is indicative of a superior circumferential alignment, while a smaller Euclidean distance is indicative of an improved radial alignment.

Only if both the circumferential matching and radial matching are good may the matching be considered good. The roundness profile and its corresponding value are affected by two types of roundness matching. A position error, θ_E, occurs during the scanning with the diamond and ruby ball styluses, as shown in Figure 7(a1,b1). Therefore, the trajectory of the stylus on the surface of the needle roller is not always a circle but an ellipse. The position error can be calculated by Equation (3), where d_{mean} is the small cylinder's diameter and $a_{ellipse}$ is the ellipse's long axis.

$$\theta_E = d_{mean} / a_{ellipse} \tag{3}$$

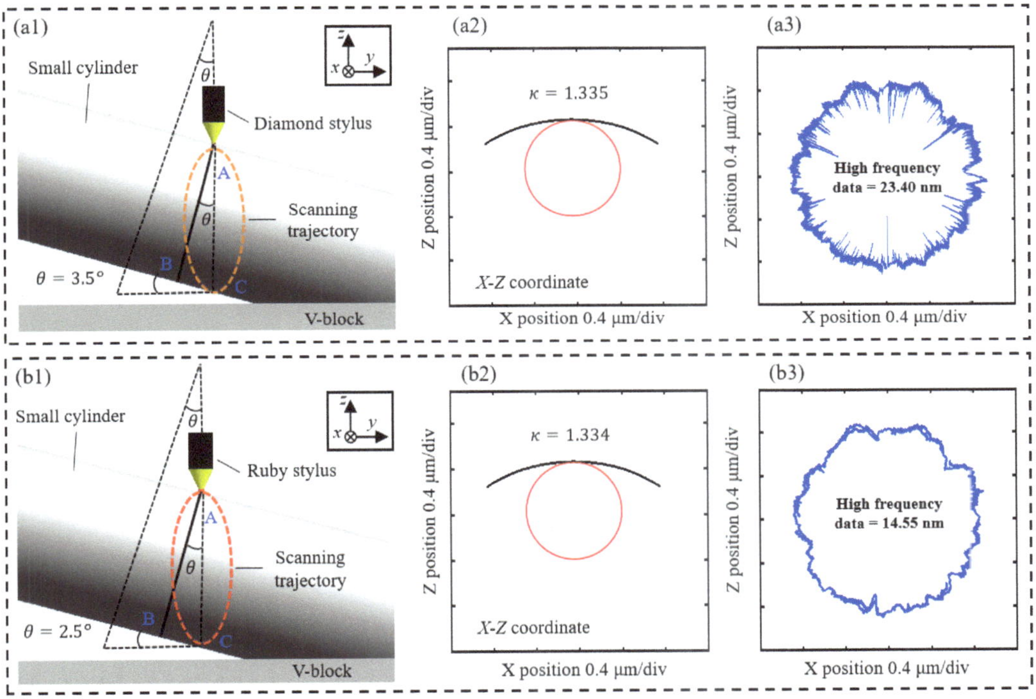

Figure 7. The evaluation of arc coordinate data quality: (**a1**,**b1**) the position error when scanning; (**a2**,**b2**) the curvature of the arc obtained; (**a3**,**b3**) the high-frequency data of the arc obtained.

The quality of the eight arc coordinate data can be evaluated by calculating the mean curvature K, as shown in Figure 7(a2,b2). The high-frequency data that can affect the stitching of the adjacent arc profile, as well as the final roundness profile, can be calculated by the residual sum of squares between the reference radius and radius in accordance with each coordinate on each arc coordinate data, as shown in Figure 7(a3,b3).

3. Comparison Analysis

Two sets of roundness measurements were performed using a ruby ball and a diamond stylus, respectively, and the measurement performance of the two styluses was quantitatively analyzed using the evaluation methods described above. Figure 8a shows the coefficient for adjacent arc profiles, characterized by the arc coordinate data from the ruby ball stylus, which has a mean matching value of 0.845. This value is lower than the 0.798 coefficient obtained from the diamond stylus, indicating a superior circumferential alignment. Figure 8b shows that the mean Euclidean distance for the overlapping portions of adjacent arc profiles, characterized by the arc coordinate data scanned using the diamond stylus, is 26.44 nm, which is greater than the 24.59 nm with respect to the ruby ball stylus. This indicates that the stitched roundness profile obtained with the diamond stylus exhibits a greater degree of deviation compared to that obtained when using the ruby ball stylus. Figure 8c shows that the high-frequency data of the eight arc profiles with respect to the ruby ball stylus are 14.55 nm, which is less than the result of 23.40 nm with respect to the diamond stylus, indicating a smoother measured profile. As the fitting arc coordinates obtain the arc profiles, the curvature and position errors can reflect the data from the eight arc coordinates.

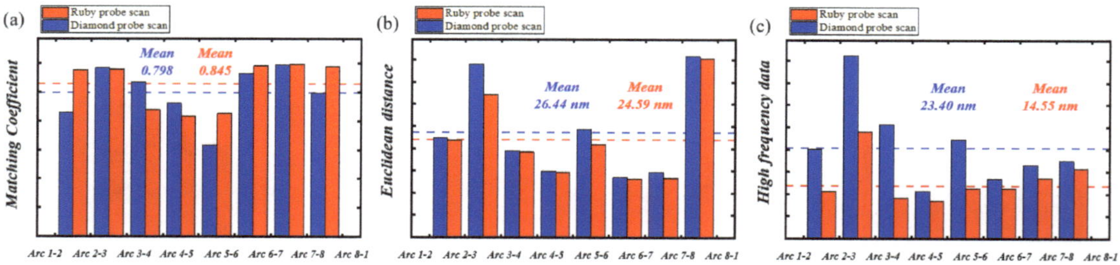

Figure 8. The evaluation results of the diamond stylus and ruby ball stylus: (**a**) the histogram of matching coefficient; (**b**) the Euclidean distance; (**c**) the high frequency data.

The accuracy of the stitching of a roundness profile can be affected by a number of factors, including the matching coefficient of adjacent arcs, the Euclidean distance of overlapping portions of arc profiles, and the high-frequency data of the profiles, with these quantitative results presented in Table 2 for comparison. Figure 9 shows the final roundness profiles for the needle roller, with dimensions 1.5 mm in diameter and 5.8 mm in length. Figure 9(a1,b1) show the stitched roundness profile, as measured by the diamond stylus. Figure 9(a2,b2) show the roundness profile following the integration of the overlapping portion of adjacent arc profiles. Figure 9(a3,b3) show the roundness profile at the 50UPR filter. Similarly, Figure 9(c1–c3,d1–d3) show the roundness profiles with respect to the ruby ball stylus. Table 3 presents the quantitative data, which show that the ruby ball stylus's measurement data outperforms that of the diamond ball stylus.

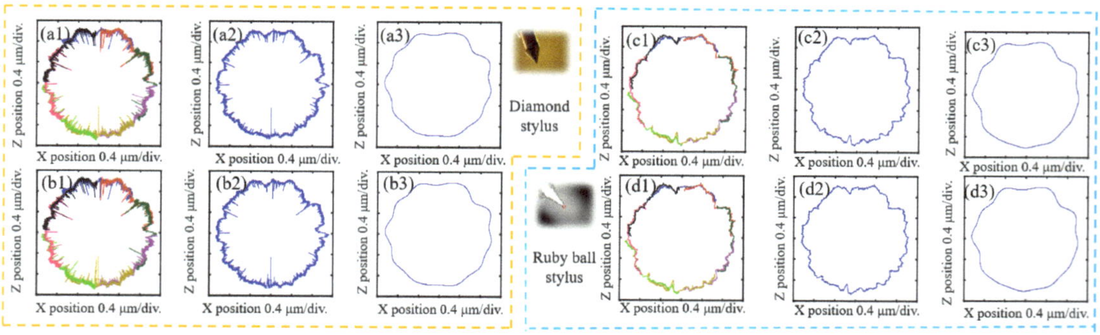

Figure 9. Results comparison: (**a1–a3,b1–b3**) results from two experiments using the diamond stylus; (**c1–c3,d1–d3**) results from two experiments using the ruby ball stylus.

Table 2. The evaluation results.

	Matching Coefficient	Euclidean Distance	High-Frequency Data	Position Error	Mean Curvature
Diamond stylus	0.78	26.44 nm	23.40 nm	3.5°	1.335
Ruby ball stylus	0.85	24.48 nm	14.55 nm	2.3°	1.334

Table 3. The final diameter and roundness results.

	Tip Radius	Diameter of Needle Roller	Roundness of Needle Roller
Diamond stylus	2 μm	1.497 mm	0.16 μm
Ruby ball stylus	150 μm	1.499 mm	0.14 μm

4. Conclusions

A roughness measuring machine was employed in experiments on the roundness metrology of a needle roller measuring 1.5 mm in diameter and 5.8 mm in length, using the stitching linear scan method. The needle roller's roundness profile was reconstructed by stitching a series of arc profiles, according to the arc coordinate data obtained using a diamond stylus with a 2 μm tip radius and a ruby ball stylus with a 150 μm tip radius. The position error of the measurement and the curvature of the obtained arc were used to evaluate the arc coordinate data quality. The mean matching coefficient of adjacent arc profiles, the mean Euclidean distance of overlapping portions of adjacent arc profiles, and the high-frequency data were calculated to evaluate the roundness profile stitching quality. The results of the evaluation show that the arc coordinate data and roundness profile, with respect to the ruby ball stylus, demonstrate greater accuracy compared to the corresponding data from the diamond stylus. This indicates that the ruby ball stylus is more effective for measuring the roundness of small-sized needle rollers. The final roundness measurements of the needle roller, as determined by the diamond stylus and the ruby ball stylus, were found to be 0.16 μm and 0.14 μm, respectively. However, as the sole focus of this experiment was on the roundness measurements of a small needle roller with a diameter less than 3 mm and a length less than 50 mm, using styluses with different tip radii, other parameters, such as the stylus material, scanning speed, workpiece radius, and scanning rate, were not examined. Consequently, the inferences derived from this study are subject to certain constraints. In the future, experiments will be conducted to explore the performance of different styluses in measuring the roundness of rollers of various sizes. Additionally, experiments will be conducted to integrate other parameters, including conducting experiments on small-sized workpieces using large-sized styluses at low sampling rates.

Author Contributions: Conceptualization, Q.L.; Software, C.Z.; Validation, B.W.; Formal analysis, B.W.; Resources, B.W.; Data curation, C.Z.; Writing—original draft, B.W.; Writing—review & editing, Q.L.; Visualization, B.W.; Supervision, Q.L.; Project administration, Q.L. All authors have read and agreed to the published version of the manuscript.

Funding: This research received no external funding.

Institutional Review Board Statement: Not applicable.

Informed Consent Statement: Informed consent was obtained from all subjects involved in the study.

Data Availability Statement: Data are contained within the article.

Conflicts of Interest: The authors declare no conflict of interest.

References

1. Polák, P.; Mikuš, R.; Kročko, V. Monitoring of conditions of agricultural machines' parts in operation. *Res. Agric. Eng.* **2014**, *60*, S57–S65. [CrossRef]
2. Aliakbari, K. Failure Analysis of Ductile Iron Crankshaft in Four-Cylinder Diesel Engine. *Int. J. Met.* **2021**, *15*, 1223–1237. [CrossRef]
3. Sanaei, N.; Fatemi, A.; Phan, N. Defect characteristics and analysis of their variability in metal L-PBF additive manufacturing. *Mater. Des.* **2019**, *182*, 108091. [CrossRef]
4. Adamczak, S.; Zmarzły, P.; Janecki, D. Theoretical and Practical Investigations Of V-Block Waviness Measurement of Cylindrical Parts. *Metrol. Meas. Syst.* **2015**, *22*, 181–192. [CrossRef]
5. Khaeroman, K.; Haryadi, G.D.; Ismail, R.; Kim, S.J. Failure analysis and evaluation of a six cylinders crankshaft for marine diesel generator. *AIP Conf. Proc.* **2017**, *1788*, 030064.
6. Lee, D. Experimental Investigation of Laser Ablation Characteristics on Nickel-Coated Beryllium Copper. *Metals* **2018**, *8*, 211. [CrossRef]
7. Bleicher, F.; Biermann, D.; Drossel, W.G.; Moehring, H.C.; Altintas, Y. Sensor and actuator integrated tooling systems. *CIRP Ann.* **2023**, *72*, 673–696. [CrossRef]
8. Swiderski, J.; Makiela, W.; Dobrowolski, T.; Stepien, K.; Zuperl, U. The study of the roundness and cylindricity deviations of parts produced with the use of the additive manufacturing. *Int. J. Adv. Manuf. Technol.* **2022**, *121*, 7427–7437. [CrossRef]
9. Pellegrini, G.; Ravasio, C. Study of the Law Motion of the Micro-EDM Drilling Process. *J. Manuf. Mater. Process.* **2023**, *7*, 165. [CrossRef]
10. Liu, F.; Liang, L.; Xu, G.; Hou, C.; Liu, D. Measurement and evaluation of cylindricity deviation in Cartesian coordinates. *Meas. Sci. Technol.* **2020**, *32*, 035018. [CrossRef]
11. Rangappa, R.; Patel, G.C.M.; Chate, G.R.; Lokare, D.; Lakshmikanthan, A.; Giasin, K.; Pimenov, D.Y. Coaxiality error analysis and optimization of cylindrical parts of CNC turning process. *Int. J. Adv. Manuf. Technol.* **2022**, *120*, 6617–6634. [CrossRef]
12. Jiang, K.; Chen, S.; Li, Z.; Ma, T.; Ren, J.; Mao, Z. Cylindrical-cam modulation instantaneous angular speed sensor for reciprocating rotary machine. *Mech. Syst. Signal Process.* **2023**, *205*, 110860. [CrossRef]
13. Maropoulos, P.G.; Ceglarek, D. Design verification and validation in product lifecycle. *CIRP Ann.* **2010**, *59*, 740–759. [CrossRef]
14. Chen, X.; Dai, Y.; Hu, H.; Tie, G.; Guan, C. Research on High Precision and Deterministic Figuring for Shaft Parts Based on Abrasive Belt Polishing. *Materials* **2019**, *12*, 1389. [CrossRef] [PubMed]
15. Çiçek, A.; Kıvak, T.; Ekici, E. Optimization of drilling parameters using Taguchi technique and response surface methodology (RSM) in drilling of AISI 304 steel with cryogenically treated HSS drills. *J. Intell. Manuf.* **2013**, *26*, 295–305. [CrossRef]
16. Zhu, W.-L.; Beaucamp, A. Compliant grinding and polishing: A review. *Int. J. Mach. Tools Manuf.* **2020**, *158*, 103634. [CrossRef]
17. Ma, Y.Z.; Wang, X.H.; Li, H.M.; Dong, X.; Kang, Y.H. A New Capacitive Sensing System for Roundness Measurement. *Adv. Mater. Res.* **2013**, *662*, 754–757. [CrossRef]
18. Duan, B.; Yin, Z.; Chai, N.; Meng, S.; Yao, J. A new single probe scanning method for on-machine measurement of roundness error. *Int. J. Adv. Manuf. Technol.* **2021**, *116*, 2861–2871. [CrossRef]
19. Liang, J.; Sun, Z. Air conditioning compressor crankshaft roundness measurement based on a laser profiler. *Meas. Sci. Technol.* **2024**, *35*, 055010. [CrossRef]
20. Bai, J.; Wang, Y.; Wang, X.; Zhou, Q.; Ni, K.; Li, X. Three-Probe Error Separation with Chromatic Confocal Sensors for Roundness Measurement. *Nanomanuf. Metrol.* **2021**, *4*, 247–255. [CrossRef]
21. Oertel, E.; Manske, E. Radius and roundness measurement of micro spheres based on a set of AFM surface scans. *Meas. Sci. Technol.* **2021**, *32*, 044005. [CrossRef]
22. Okuyama, E.; Fukuda, T. Roundness Profile Measurement Using a Combination Method of Three-Point Method for Roundness Profile Measurement and Integration Method for Straightness Profile Measurement. *Int. J. Autom. Technol.* **2024**, *18*, 77–83. [CrossRef]
23. Küng, A.; Meli, F.; Thalmann, R. Ultraprecision micro-CMM using a low force 3D touch probe. *Meas. Sci. Technol.* **2007**, *18*, 319–327. [CrossRef]
24. Gao, W.; Kim, S.W.; Bosse, H.; Haitjema, H.; Chen, Y.L.; Lu, X.D.; Kunzmann, H. Measurement technologies for precision positioning. *CIRP Ann.* **2015**, *64*, 773–796. [CrossRef]
25. Bai, J.; Li, J.; Wang, X.; Zhou, Q.; Ni, K.; Li, X. A new method to measure spectral reflectance and film thickness using a modified chromatic confocal sensor. *Opt. Lasers Eng.* **2022**, *154*, 107019. [CrossRef]
26. Bai, J.; Li, X.; Wang, X.; Wang, J.; Ni, K.; Zhou, Q. Self-reference dispersion correction for chromatic confocal displacement measurement. *Opt. Lasers Eng.* **2021**, *140*, 106540. [CrossRef]
27. Bai, J.; Li, X.; Wang, X.; Zhou, Q.; Ni, K. Chromatic Confocal Displacement Sensor with Optimized Dispersion Probe and Modified Centroid Peak Extraction Algorithm. *Sensors* **2019**, *19*, 3592. [CrossRef] [PubMed]
28. Bai, J.; Li, X.; Zhou, Q.; Ni, K.; Wang, X. Improved chromatic confocal displacement-sensor based on a spatial-bandpass-filter and an X-shaped fiber-coupler. *Opt. Express* **2019**, *27*, 10961–10973. [CrossRef] [PubMed]
29. Li, Q.; Shimizu, Y.; Saito, T.; Matsukuma, H.; Cai, Y.; Gao, W. Improvement of a Stitching Operation in the Stitching Linear-Scan Method for Measurement of Cylinders in a Small Dimension. *Appl. Sci.* **2021**, *11*, 4705. [CrossRef]

30. Li, Q.; Shimizu, Y.; Saito, T.; Matsukuma, H.; Gao, W. Measurement Uncertainty Analysis of a Stitching Linear-Scan Method for the Evaluation of Roundness of Small Cylinders. *Appl. Sci.* **2020**, *10*, 4750. [CrossRef]
31. Li, Q.; Shimizu, Y.; Wang, X.; Li, X.; Gao, W. High-accuracy roundness measurement of small cylindrical workpieces by a high-frequency filtering method. *Precis. Eng.* **2024**, *85*, 241–246. [CrossRef]
32. Li, Q.; Wang, Y.; Li, J.; Wang, X.; Li, X. Non-contact ultra-precision metrology of superfine cylinders with a developed two-dimensional coordinate measuring device. *Measurement* **2023**, *223*, 113727. [CrossRef]
33. Gao, W.; Haitjema, H.; Fang, F.Z.; Leach, R.K.; Cheung, C.F.; Savio, E.; Linares, J.M. On-machine and in-process surface metrology for precision manufacturing. *CIRP Ann.* **2019**, *68*, 843–866. [CrossRef]
34. Gao, W.; Kiyono, S.; Sugawara, T. High-accuracy roundness measurement by a new error separation method. *Precis. Eng.* **1997**, *21*, 123–133. [CrossRef]

sensors

MDPI

Article

A Binocular Stereo-Imaging-Perception System with a Wide Field-of-View and Infrared- and Visible Light-Dual-Band Fusion

Youpan Zhu [1,2], Dan Zhang [3], Yongkang Zhou [2,4], Weiqi Jin [1,*], Lingling Zhou [3], Guanlin Wu [1] and Yong Li [5]

1 MOE Key Laboratory of Optoelectronic Imaging Technology and System, Beijing Institute of Technology, Beijing 100081, China; zhuypbit@aliyun.com (Y.Z.); 3120220595@bit.edu.cn (G.W.)
2 Kunming Institute of Physics, No. 31, Jiaochang East Road, Wuhua District, Kunming 650223, China; zyk1120102464@163.com
3 Yunnan North Optical & Electronic Instrument Co., Ltd., No. 300, Haikou Town, Xishan District, Kunming 650114, China; zhangdanynoei@163.com (D.Z.); lynzhll@163.com (L.Z.)
4 School of Life Science, Beijing Institute of Technology, 5 South Zhongguancun Street, Haidian District, Beijing 100081, China
5 Chengdu Zhongke Information Technology Co., Ltd., No. 1369 Kezhi Road, Xinglong Street, Tianfu New District, Chengdu 610042, China; scly2012@163.com
* Correspondence: jinwq@bit.edu.cn

Abstract: With the continuous evolution of autonomous driving and unmanned driving systems, traditional limitations such as a limited field-of-view, poor ranging accuracy, and real-time display are becoming inadequate to satisfy the requirements of binocular stereo-perception systems. Firstly, we designed a binocular stereo-imaging-perception system with a wide-field-of-view and infrared- and visible light-dual-band fusion. Secondly we proposed a binocular stereo-perception optical imaging system with a wide field-of-view of 120.3°, which solves the small field-of-view of current binocular stereo-perception systems. Thirdly, For image aberration caused by the wide-field-of-view system design, we propose an ellipsoidal-image-aberration algorithm with a low consumption of memory resources and no loss of field-of-view. This algorithm simultaneously solves visible light and infrared images with an aberration rate of 45% and 47%, respectively. Fourthly, a multi-scale infrared- and visible light-image-fusion algorithm is used, which improves the situational-awareness capabilities of a binocular stereo-sensing system in a scene and enhances image details to improve ranging accuracy. Furthermore, this paper is based on the Taylor model-calibration binocular stereo-sensing system of internal and external parameters for limit correction; the implemented algorithms are integrated into an NVIDIA Jetson TX2 + FPGA hardware framework, enabling near-distance ranging experiments. The fusion-ranging accuracy within 20 m achieved an error of 0.02 m, outperforming both visible light- and infrared-ranging methods. It generates the fusion-ranging-image output with a minimal delay of only 22.31 ms at a frame rate of 50 Hz.

Keywords: binocular stereoscopic-perception system; wide field-of-view; infrared and visible light fusion; image distortion

1. Introduction

In the field of autonomous driving, the primary methods for perception systems currently include millimeter-wave radar perception, LiDAR perception, millimeter-wave radar, and LiDAR fusion with visual perception, monocular vision, and binocular stereo vision. Radar- and vision sensor-fusion methods involve the calibration of multiple sensors in both time and space, transforming their acquired data into a common coordinate system, followed by information processing. The authors in [1] used the radar sensor and visual sensor approaches to investigate the vehicle environment. This approach is more accurate in terms of the information obtained but it suffers from the complexity of calibrating multiple sensors with inconsistent sensing ranges and becomes increasingly expensive as accuracy increases. Monocular stereo vision utilizes a single camera to project the three-dimensional world onto a two-dimensional plane, resulting in a loss of depth information. Monocular stereo vision systems assume static ground and solve for a dynamic camera pose to achieve distance-measurement functionality [2–4]. R. Mur-Artal et al. [5] introduced the ORB-SLAM monocular simultaneous localization and mapping (SLAM) system, which exhibits robustness against significant motion noise and allows for wide baseline-loop closure and relocalization but requires further improvements in depth accuracy. Raul Mur-Artal et al. [6] presented ORB-SLAM2, a monocular SLAM system that employs bundle adjustment to enhance depth accuracy and features a lightweight localization mode for the efficient reutilization of disabled maps. Carlos Campos et al. [7] introduced ORB-SLAM3, which employs comprehensive short-term, medium-term, and long-term data associations to achieve zero drift in mapped areas. Compared to other methods, monocular stereo vision systems offer the advantage of being small and lightweight. However, such systems can only capture two-dimensional images and cannot determine distances to objects, leading to issues with low ranging accuracy and poor environmental adaptability. Binocular stereo vision, on the other hand, emulates the human visual system and leverages the principle of disparity to acquire target features from different positions using detectors. This approach establishes the relationship between corresponding pixel points in the image based on the three-dimensional geometric positions of certain features on the spatial-target surface. This process involves constructing a binocular stereo vision-imaging model to obtain model parameters [8,9]. Binocular stereo vision is also referred to as passive binocular depth sensing and offers improved robustness in depth computation compared to monocular stereo vision. This technology also enables three-dimensional object recognition. Various methods exist for obtaining real-world depth information in binocular stereo perception systems, such as passive stereo [10], active stereo [11], time-of-flight imaging [12], and defocus depth [13]. In this paper, the passive stereo method of binocular stereo systems is employed.

With the advancement of industrial automation and machine vision, binocular stereo perception systems have been widely researched and applied in various fields, including vehicle driving [14], intelligent transportation [15], 3D reconstruction [16], virtual reality [17], surveying, and rapid positioning [18]. In the realm of autonomous and assisted driving [19], the first assisted-driving vehicle based on a binocular stereo vision system was developed in Japan. This system processes images captured by two onboard cameras. Subsequently, in Europe and the United States, autonomous driving vehicles were developed based on binocular stereo vision. Notable examples include the NavLab series developed at Carnegie Mellon University in the United States [20] (NavLab-1, NavLab-5, and NavLab-11). These vehicles utilize binocular stereo vision for road environment detection and focus on addressing challenging visual perception issues in complex environments. Silicon Valley chip company Ambarella has developed binocular ADAS (Advanced Driver Assistance Systems) and autonomous driving chips [21], as well as binocular stereo-specific chips and solutions tailored to binocular vision systems, which act as valuable complements to integrated chips. These specialized chips can handle a portion of the perception tasks at the edge, creating a smaller and more efficient perception-decision loop.

With the evolution of deep learning, many researchers have proposed more efficient and accurate binocular stereo algorithms. Sun et al. [22] introduced a multi-path Viterbi (MPV) multi-scale fast stereo matching algorithm for generating dense disparity information. Jure Bontar and others [23] trained a convolutional neural network (CNN) to predict the matching degree between two image patches and used this CNN for binocular stereo matching. The authors incorporated left–right consistency checks to eliminate errors in occluded areas. However, their method was found to be ineffective in textureless regions. Li et al. [24] introduced Stereo R-CNN, which takes binocular images as network inputs. The authors extended Faster R-CNN to simultaneously detect and associate objects in left and right images. Through the binocular FPN network, the authors predicted object key points and sizes to generate coarse 3D object detection boxes. Finally, the authors used the region-based photometric correspondence method to refine fine-grained 3D detection boxes. Tang et al. [25] used the YOLOv4-tiny model to generate boundary frameworks and employed an adaptive stereo matching approach based on these boundaries. This approach meets real-time detection requirements and exhibits high localization stability and robustness under varying lighting conditions.

Autonomous driving vehicle is a kind of autonomous and automatic unmanned vehicle, the vehicle not only needs to grasp the dynamic situation of nearby vehicles and obstacles in real-time, but also needs to deal with the complex environment of night, haze, and obstacles. The traditional driving system based on visible light is easy to fail in the area of poor lighting conditions, unable to meet the all-weather needs of automatic driving vehicles, infrared images are not subject to the interference of illumination, sun glare, can penetrate smoke, haze, and other characteristics. The perception ability of automatic driving vehicles can be further improved by fusion of visible light and infrared. Current autopilots commonly use small field-of-view cameras, usually with a field-of-view of 40–60, which limits the system's perception of the external environment, and the field-of-view and perception can be further improved to a wide-field-of-view technology with infrared- and visible light fusion [26].

The field of autonomous driving places significant emphasis on important metrics such as the field-of-view, ranging accuracy, and real-time display in binocular stereo perception systems. This paper integrates the characteristics of visible light and infrared sensors to design a wide-field-of-view binocular stereo imaging perception system using infrared- and visible light-fusion . This system combines a wide-field-of-view binocular stereo optical imaging system with a low-storage image distortion correction algorithm to achieve imaging with a wide field-of-view and minimal distortion. Focusing on the features of visible light and infrared images, this study employs a multi-scale fusion approach for infrared- and visible light images. This work explores a straightforward yet precise method for object ranging based on the Taylor model for calibrating the internal and external parameters of the binocular stereo perception system for extreme correction, thereby enhancing the measurement accuracy of the binocular stereo perception system. The algorithms are implemented using the NVIDIA Jetson TX2 + FPGA hardware framework, enabling real-time ranging and display.

2. System Framework

In this paper, a wide-field-of-view binocular stereo sensing system based on infrared and visible light is established; the system block diagram is shown in Figure 1. This system consists of two sets of uncooled long-wave infrared-imaging modules with a field-of-view greater than 120°, visible-high-definition-imaging modules, and digital-video-processing modules. The system's operating temperature ranges from −40 °C to 70 °C. The digital-video-processing module utilizes hardware circuits featuring the NVIDIA Jetson TX2 + FPGA combination and receives digital video signals from the visible light-imaging component and the infrared-imaging component, enabling the module to perform image processing. A hardware diagram of the digital-video-processing module is shown in Figure 2.

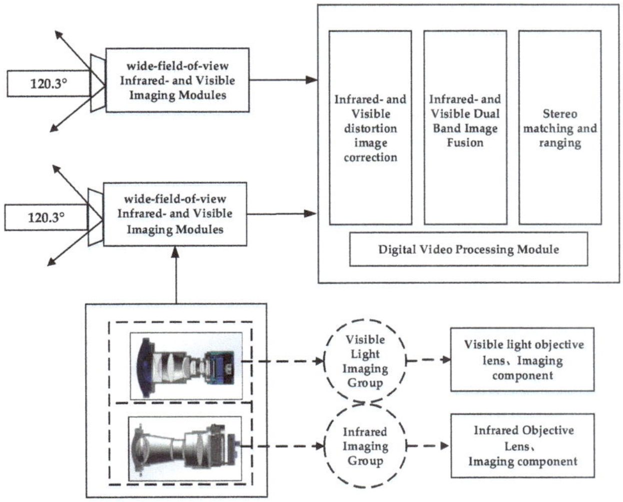

Figure 1. Schematic diagram of the wide-field binocular stereo perception system with both infrared- and visible light components.

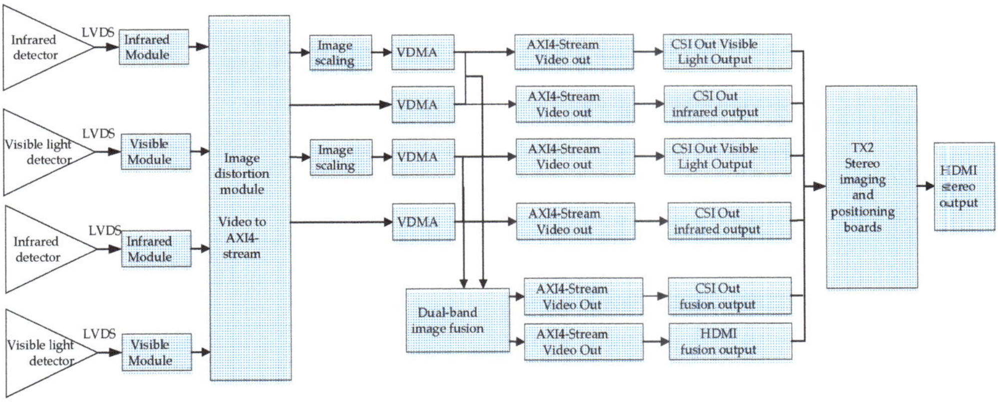

Figure 2. The digital-video-image-processing module.

2.1. Design and Simulation of the Wide-Field Binocular Stereo Perception-Dual-Band-Imaging System

2.1.1. The Principle of Binocular Stereo Imaging

In response to the specific requirements of night-time vehicle driving and rapidly changing work environments, we designed a wide-field-binocular stereo perception system with dual-band-electro-optical imaging. The visiblelight-imaging component incorporates a large-array CMOS device from Rockchip Electronics Co., Ltd. (Jiangsu, China) that has high sensitivity and definition. This system boasts a pixel resolution of 1920 (H) × 1080 (V), with each pixel measuring 13 μm (H) × 13 μm (V), and it operates at a frame rate of 50 Hz. This component can function effectively under low-light conditions down to 10^{-3} lx. Furthermore, the infrared-imaging component employs a non-cooled infrared focal plane detector from Yantai IRay Technology Co., Ltd. (Yantai, China) that offers a pixel resolution of 1024 (H) × 768 (V), with each pixel measuring 14 μm. The frame rate of the infrared component is also 50 Hz.

Binocular stereo sensing systems are mainly divided into two types according to their placement [27]: a parallel model and a convergent model. The parallel model has two camera optical axes parallel to each other. Moreover, the structure is simple and easy to calculate. The advantage of this structure is the presence of only a negative horizontal parallax, with no vertical parallax. Disadvantages include a small common area and a lack of stereoscopic information in the left and right sides of the single viewing area, which will cause a waste of information. The convergence model can adjust the angle between the two optical axes to obtain a larger effective field-of-view, with positive-, negative-, or zero-horizontal parallax; however, the camera body will produce vertical parallax, which causes a certain gradient distortion. In this study, considering the advantages and disadvantages of the parallel model and the convergence model, the visible and infrared components were fused. For this purpose, we selected an optical design based on the parallel model, as shown in Figure 3. To realize binocular stereo vision, the binocular stereo sensing system consisted of two visible light-objective lens groups and two infrared-objective lens groups.

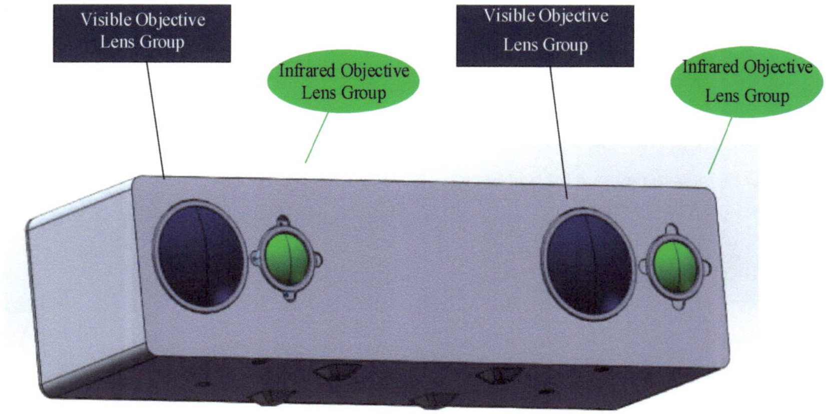

Figure 3. Schematic diagram of the infrared- and visible light-lens structure in the binocular-stereo-perception system.

Binocular-imaging-distance-measurement technology relies on the binocular disparity to establish an ideal model for binocular ranging. In this model, both cameras have identical specifications and parameters, including matching camera models, consistent focal lengths, and the parallel alignment of optical axes. The model is shown in Figure 4.

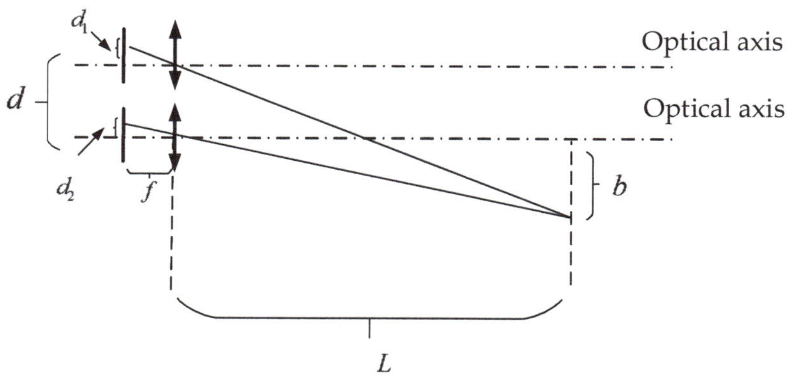

Figure 4. The parallel dual-axis-ranging model with the same focal length.

Here, d represents the camera's inter-image plane spacing (baseline width), d_1 and d_2 are the distance from the image point to the image plane center, f is the camera's focal length, and L is distance to the object. Using the principles of similar triangles, the following relationships can be derived:

$$L = \frac{fd}{d_1 - d_2} = \frac{fd}{c} \tag{1}$$

where c represents the difference in the imaging positions of the object point in the two fields of view, which is commonly referred to as the parallax value.

Under the conditions of parameter determination in a binocular-stereo-perception system (such as focal length and baseline), ranging accuracy is determined by the parameter c. Presently, binocular stereo-matching algorithms can achieve sub-pixel-level matching accuracy, resulting in superior disparity precision. Results can be obtained by differentiating Equation (1):

$$dL = \frac{fd}{c^2}dc = \frac{L^2}{fd}dc. \tag{2}$$

Hence, as long as the parameters of the binocular-stereo-perception system are determined, the distance to the target can be calculated by measuring the disparity. System parameters typically consist of intrinsic and extrinsic parameters. Intrinsic parameters include the focal length, principal point coordinates, and distortion coefficients of the left and right cameras, among others. Extrinsic parameters encompass the relative transformation between the left and right cameras, involving rotation and translation matrices. Due to potential errors during camera installation, such as non-parallel alignment of the lens and imaging plane, it is necessary to recalibrate the camera to obtain updated focal lengths, intrinsic parameters, and extrinsic parameters.

2.1.2. The Simulation and Design of Visible- and Infrared Objective Lenses

A. Simulation design for the visible light objective lens

In the wide-field binocular stereo perception-optical-imaging system, the visible light component employs a high-resolution and low-light CMOS-imaging module from Rockchip Electronics Co., Ltd., which enables imaging in both day and night scenes. This component features large pixels and sensor-imaging areas, imposing stringent requirements on the optical system. We utilized the CODE V(10.2) software for optical system design, implementing a "telephoto-type" optical path structure with 12 lenses. By introducing appropriate non-spherical elements while keeping the total number of lenses, glass thickness, and imaging quality constant, we enhanced the light-gathering capabilities of the visible light objective lens without compromising its transmittance. The system design is depicted in Figure 5. Table 1 presents the optical-design specifications for the visible light objective lens.

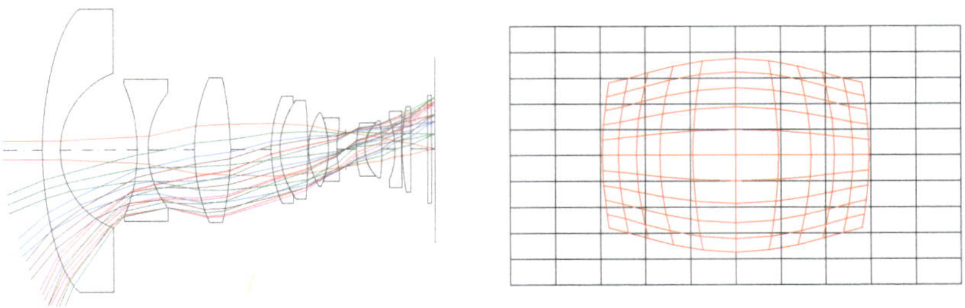

Figure 5. Visible light objective lens-optical system and distortion diagram.

Table 1. Optical-design specifications for the visible light lens.

Parameters	Value	Parameters	Value
Objective focal length (f)	10.96 mm	Diagonal field	133°
F-number	2.1	Distortion (maximum on the diagonal)	−45%
Horizontal field	124°	Maximum working distance	97.16 m
Vertical field	74.4°		

The maximum effective range L of the visible light-optical-imaging system is as follows:

$$L = \frac{f\sqrt{wh}}{8d_{pix}} = 97.16 \text{ m} \quad (3)$$

where w represents human height, h represents shoulder width, and d_{pix} represents pixel size.

At a distance of $L = 15$ m, there is a depth-calculation deviation of $dL \leq 20$ cm. Under the condition of a matching algorithm precision at the 0.1 pixel level, the baseline d can be determined as follows:

$$d \geq \frac{L^2 dC}{f dL} = 13.52 \text{ cm} \quad (4)$$

where L represents the distance from the target to the system, which is referred to as the test distance. dC is set at a 0.1 pixel level, and dL stands for the resolvable distance.

At a distance of $L = 30$ m, the depth-calculation deviation of $\mathrm{d}L \leq 1$ m. When these values are incorporated into Equation (4), the following result is obtained:

$$d \geq \frac{L^2 dC}{f dL} = 10.82 \text{ cm}. \quad (5)$$

It can be seen clearly that $d \geq 13.52$ cm. The formula for visible light stereo acuity $d\gamma$ is

$$d\gamma = \frac{\alpha d}{L^2} dL \quad (6)$$

where α is a constant of 206,265 when converting from radians to arcseconds:

$$d\gamma = \frac{\alpha}{f} dc = 24.47''. \quad (7)$$

Stereopsis is the ability to resolve the smallest horizontal disparity between retinal images from both eyes. The normal value for stereopsis should be less than 60″. A smaller value of stereopsis indicates better stereo vision.

Utilizing CONE V software, we simulated environmental temperature variations and obtained the transfer functions for the visible light-optical system at 20 °C, 50 °C, and −40 °C, as well as the diffuse spots, as shown in Figure 6, Figure 7, and Figure 8, respectively.

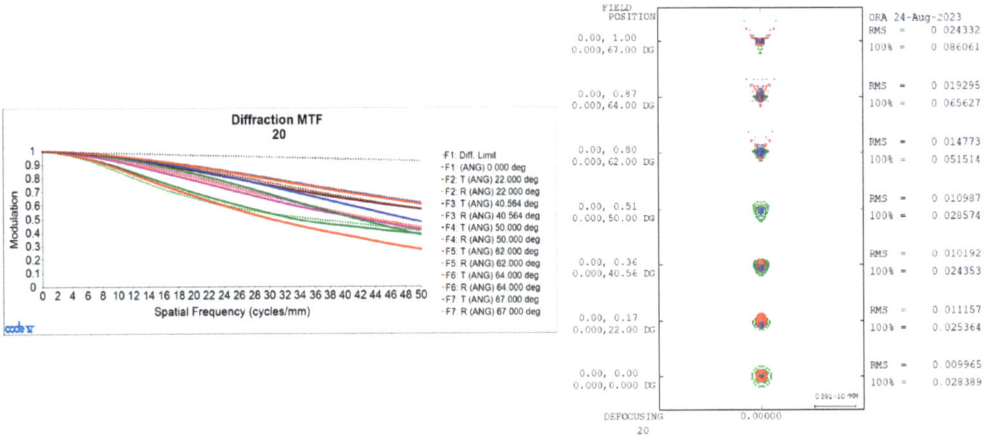

Figure 6. Transfer functions and diffuse spots of visible light-optical systems at 20 °C.

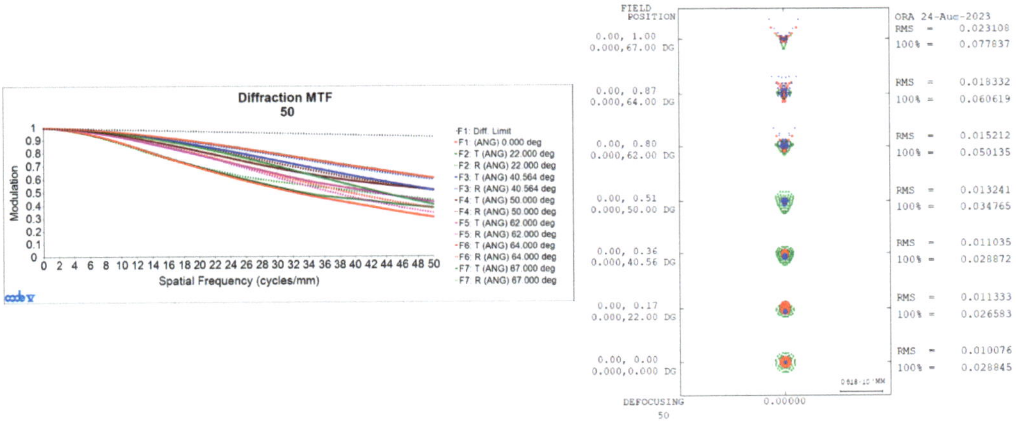

Figure 7. Transfer functions and diffuse spots of visible light-optical systems at 50 °C.

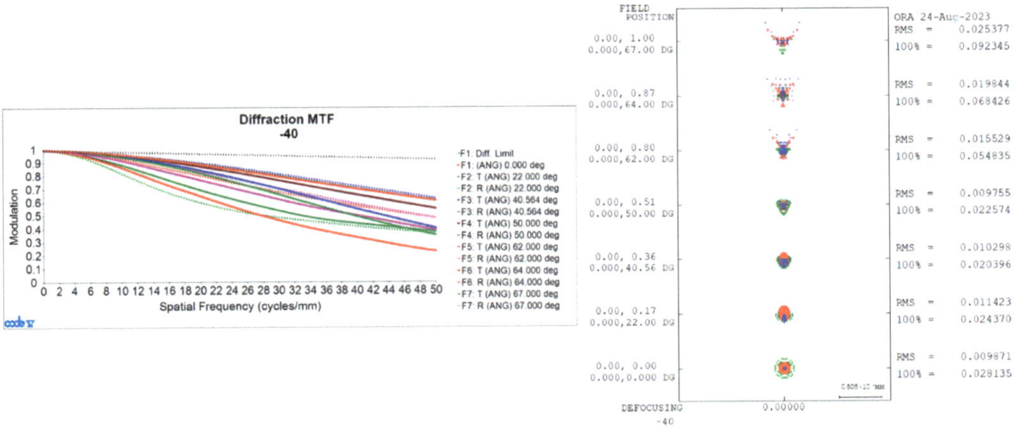

Figure 8. Transfer Functions and Diffuse Spots of Visible Light-Optical Systems at −40 °C.

B. The Simulation Design of an Infrared objective lens

This paper addresses the design of a wide-field, relative aperture long-wave infrared optical system for the 1024 (H) × 768 (V) long-wave infrared detector from Yantai IRay Technology Co., Ltd. The infrared objective lens system, as depicted in Figure 9, operates without active cooling over a wide temperature range. The front surface of the first lens is non-spherical, enabling passive temperature compensation at different temperatures by adjusting the system's back focal length. Table 2 provides the optical-design specifications for the infrared objective lens.

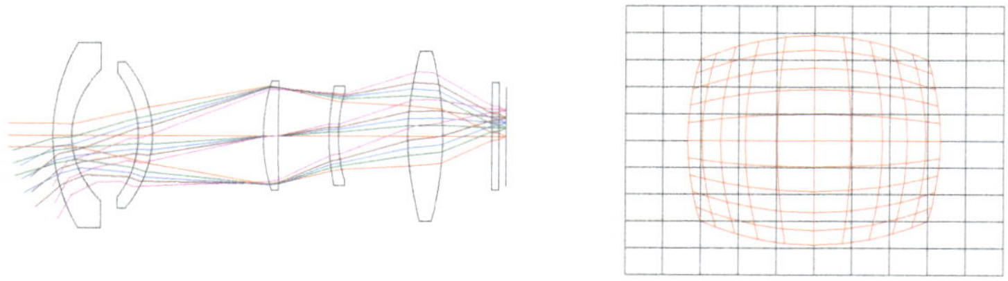

Figure 9. The optical system of the infrared objective lens and a distortion map of the infrared objective lens at 20 °C.

Table 2. The optical-design specifications for the infrared objective lens.

Parameters	Value
objective focal length (f)	6 mm
F-number	2.1
horizontal field	121.6°
vertical field	107°
distortion (maximum on the diagonal)	−47%

The maximum effective range L of the infrared objective lens-optical system is

$$L = \frac{f\sqrt{wh}}{8d_{pix}} = 40.67\ \text{m}. \tag{8}$$

At a distance of $L = 15$ m, with a depth-calculation deviation d$L \leq 30$ cm, the baseline d can be calculated as follows:

$$d \geq \frac{L^2 dC}{f dL} = 21.26\ \text{cm}. \tag{9}$$

At a distance of $L = 30$ m, with a depth-calculation deviation d$L \leq 2$ m, the baseline d can be calculated as follows:

$$d \geq \frac{L^2 dC}{f dL} = 12.75\ \text{cm}. \tag{10}$$

According to Formulas (9) and (10),

$$d \geq 21.26\ \text{cm}. \tag{11}$$

The formula for the stereo acuity $d\gamma$ is as follows:

$$d\gamma = \frac{\alpha d}{L^2} dL \tag{12}$$

where d is the baseline length, dL is the resolvable distance, L is the distance from the target to the system, and dL is the binocular ranging accuracy. The formula is as follows:

$$d\gamma = \frac{a}{f}dc = 58.44'' \tag{13}$$

With a decrease in temperature, the infrared optical system experiences changes in inter-lens spacing, lens thickness, refractive index, and curvature radius. Therefore, temperature variations inevitably lead to defocusing of the system's focal plane, resulting in a degradation of image quality. The transfer functions and diffuse spots of the infrared optical system based on system simulation analysis are shown in Figure 10, Figure 11, and Figure 12, respectively, at temperatures of 20 °C, 50 °C, and −40 °C.

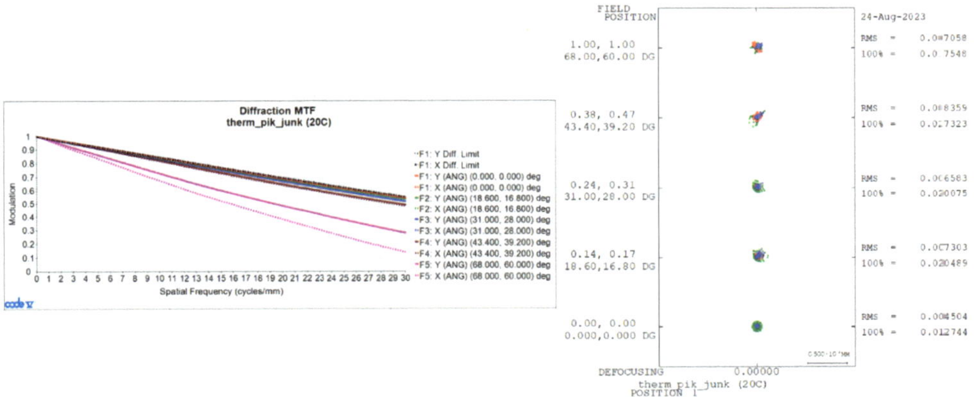

Figure 10. Transfer Functions and Diffuse Spots of Infrared Optical Systems at 20 °C.

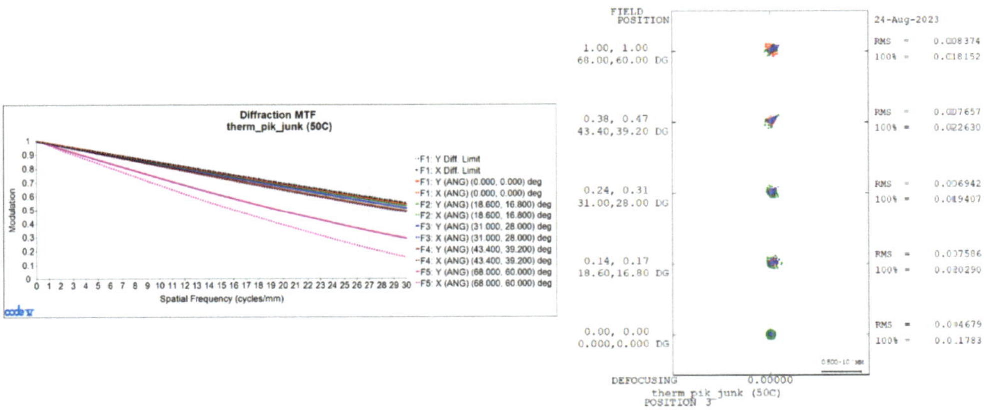

Figure 11. Transfer Functions and Diffuse Spots of Infrared Optical Systems at 50 °C.

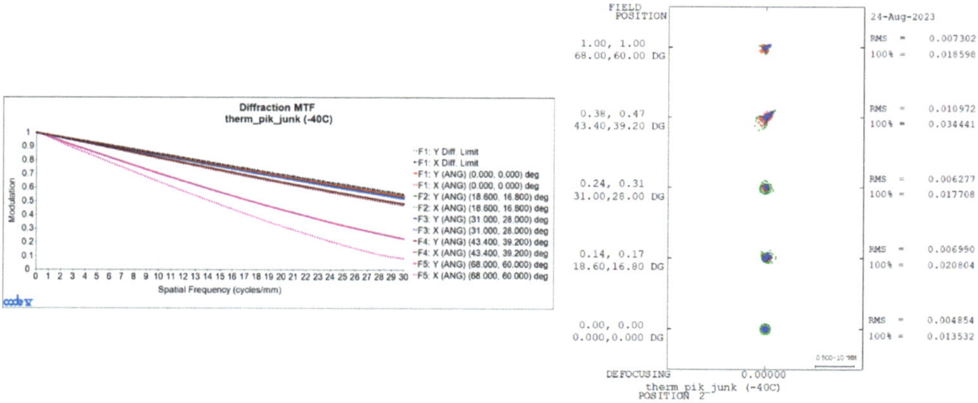

Figure 12. Transfer Functions and Diffuse Spots of Infrared Optical Systems at −40 °C.

In summary, without moving any optical elements, we achieved alignment of the image plane with the detector target surface during changes in environmental temperature. With temperature variations, the lens assembly, employing optical passive thermal compensation, can maintain consistent magnification and requires no active optical components. The relative positions of the optical axes remain relatively unchanged, resulting in high image registration accuracy. Large-field optical systems often exhibit significant distortion to improve the field-of-view. Image distortion is corrected to obtain high-quality fused images. Based on the imaging quality of the fused lens system and image registration effectiveness, further improvements in the system's ranging accuracy were achieved.

Baseline distance $d \geq 21.26$cm; if baseline distance $d = 35$cm, then the ranging accuracy is as follows:

$$dL \geq \frac{L^2 dC}{fd}. \tag{14}$$

Table 3 presents the calculated accuracy of visible and infrared ranging at test distances of 15 m and 30 m.

Table 3. Infrared- and visible light-ranging accuracy.

Serial Number	Measuring Distance (L)	Visible Light-Ranging Accuracy (L)	Infrared-Ranging Accuracy (L)
1	15 m	$dL \geq 7.63$cm	$dL \geq 15.00$cm
2	30 m	$dL \geq 30.50$cm	$dL \geq 60.00$cm

2.2. Real-Time Image Distortion Correction and Simulation

The optical system design model produces aberrations that are unavoidable due to the system's non-coaxial nature, field-of-view, focal length, and other auxiliary factors. In this study, a binocular stereo sensing optical imaging system with a large field-of-view is designed, in which the aberration rate produced by the visible objective lens is −45%, and the aberration rate of the infrared objective lens is −47%, as shown in Section 2.1. Here, we propose a shareable elliptical aberration for real-time correction.

Image distortion is mainly the result of geometric distortion of the pixel positions of the image after imaging. Geometric distortion is further categorized into linear and nonlinear distortion, which refer to a mixture of several distortions that work together [28]. The causes of nonlinear aberrations are mainly categorized as radial aberrations, centrifugal aberrations, and thin prismatic aberrations [29,30]:

$$x_d = x(1 + k_1 r^2 + k_2 r^4 + k_3 r^6) + 2p_1 xy + p_2(r^2 + 2x^2) \tag{15}$$

$$y_d = y(1 + k_1 r^2 + k_2 r^4 + k_3 r^6) + 2p_1 xy + p_2(r^2 + 2y^2) \quad (16)$$

$$x^2 + y^2 = r^2 \quad (17)$$

where x and x_d represent, respectively, the reference image and distortion image in the Xdirection; y and y_d are, respectively, the reference image and distortion image in the Y-direction; k_1, k_2, k_3, p_1, and p_2 are distortion correction parameters. When the aberration coefficient is greater, the aberration correction is less effective. Aberration correction usually considers radial aberrations and ignores the effect of tangential aberrations, which can describe the nonlinear aberrations of the lens.

Because the resolution ratio of the image H: V $\neq$ 1:1, there is a poor edge effect for the image edge distortion correction edge when using the standard concentric circle distortion model. Therefore, an elliptical distortion correction model is proposed using the improved standard concentric circle distortion model:

$$x = x_d \frac{1}{(1 + k_1 r^2)}, y = y_d \frac{1}{1 + k_1 r^2} \quad (18)$$

$$\frac{x^2}{a^2} + \frac{y^2}{b^2} = r^2 \quad (19)$$

where H is the height of the image, V represents the image's width, $a = \mathrm{H}/2$ and $b = \mathrm{V}/2$ are shown in Figure 13.

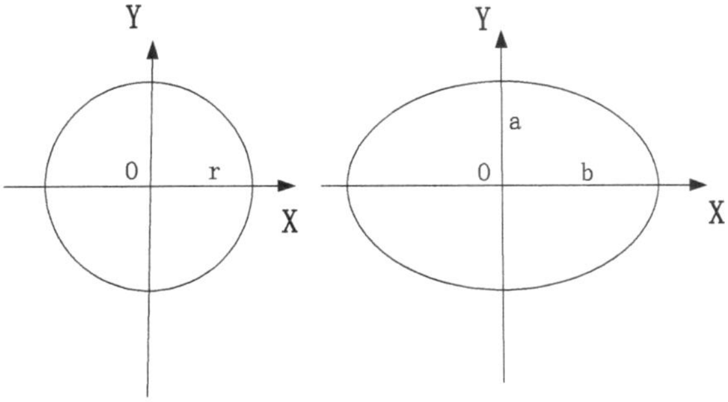

Figure 13. Standard-concentric-aberration model and elliptical-aberration-correction model.

The elliptical distortion correction model cannot cover the image completely ($H \times V$). We further improve the elliptical distortion correction model using an approximate elliptical distortion correction model:

$$x = k_2 \frac{x_d}{1 + k_1 (r - R)^2}, y = k_2 \frac{y_d}{1 + k_1 (r - R)^2} \quad (20)$$

$$\frac{x^2}{a^2} + \frac{y^2}{b^2} = r^2 \quad (21)$$

where k_1 and k_2 represent the distortion coefficients; x and x_d represent the corrected image and distorted image in the X-direction; and y and y_d represent the corrected image and the distorted image in the Y-direction, respectively.

In this paper, real-time corrections are performed based on FPGA. A common approach in FPGA hardware circuits is the correction-mapping table, which first calculates the corresponding correction result of the image and stores it in the hardware circuit and determines the corresponding corrected video output from the correction-mapping table based on the input real-time video. The correction-mapping table is a straightforward computation but suffers from the disadvantage of depositing an approximate elliptic-distortion-correction-mapping table, which requires a large amount of hardware-circuit resources. Thus, we proposed elliptical distortion correction model only needs to store one quarter of the data in the hardware circuitry, as shown in Figure 14.

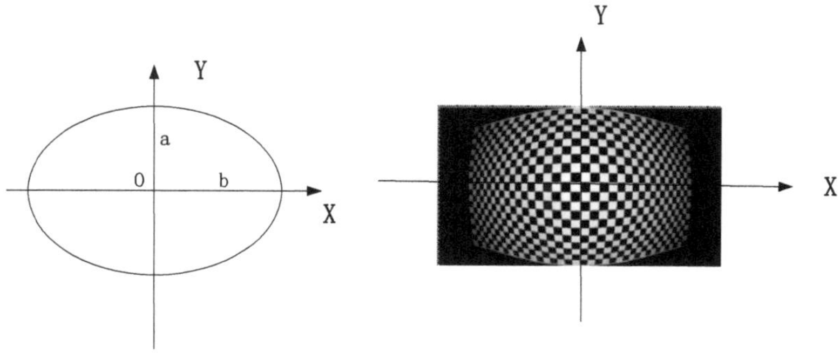

Figure 14. Correction-mapping table for approximate elliptic-distortion correction.

The generated checkerboard grid $H \times V$ images are shown in Figures 15a and 16a and are standard checkerboard grids of 1920 × 1080 and 1024 × 768, respectively. Figures 15b and 16b present, respectively, the aberration simulation images derived using the standard checkerboard grid as the input image through the visible optical system model and those generated using the infrared optical system. Through the myopic elliptic-aberration model proposed in this paper for aberration correction, the aberration-correction map of the visible image and the aberration-correction map of the infrared image are obtained, as shown in Figures 15c and 16c, respectively. The edges of the image are still aberrated, but the field-of-view is increased.

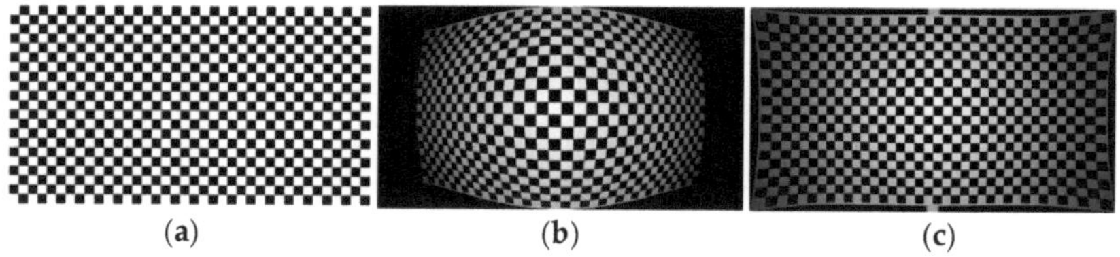

Figure 15. The 1920 × 1080 Visible light distortion effect: (**a**) standard checkerboard grid; (**b**) optical-system-aberration map (aberration rate −45%); and (**c**) aberration-correction-result map.

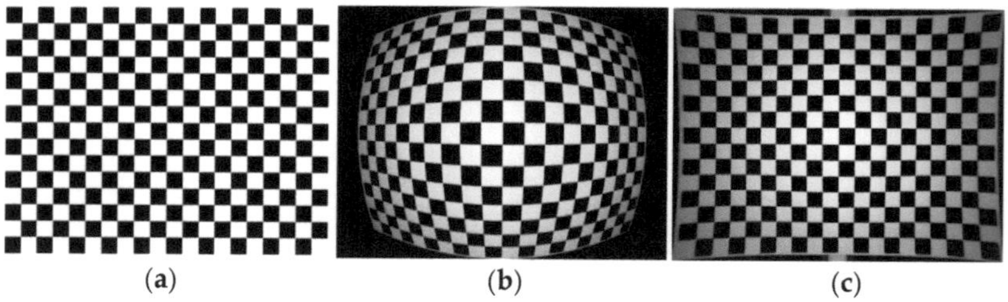

Figure 16. The 1024 × 768 infrared-distortion effect: (**a**) standard checkerboard grid; (**b**) optical-system-aberration map (aberration rate −45%); and (**c**) aberration-correction-result map.

2.3. Infrared- and Visible Light-Fusion Algorithm

The infrared- and visible light images are horizontally aligned after image alignment and limit correction to realize the left and right images, and the common adaptive image-enhancement method [31] is used to enhance the infrared- and visible light images to improve the bright and dark regions in the image to improve the contrast of the image.

Infrared- and visible light-image matching is a type of heterogenous spit matching, and it is difficult to find the same type of corresponding feature points for the two types of images. In this paper, we use the SURF + RANSAC algorithms to realize stereo matching [32,33]. The SURF algorithm has the characteristics of rotation, scale transformation, and brightness invariance; compared to the SIFT algorithm, SURF reduces the complexity of the algorithm, reduces the dimensionality of the feature descriptor from 128 to 65 dimensions, and reduces the computational amount by double [34]. The anomalous data in the matching process is filtered out by using the RANSAC algorithm.

Image fusion involves the use of image information from multiple imaging sensors in a unified scene to increase the perception of the scene and the ability to recognize targets and other objectives. Unlike traditional multi-scale fusion methods, this study adopts a multi-scale infrared- and visible light-fusion method based on the work in [35], which has the unique characteristics of retaining scale-specific information and reducing the edge halo, taking into account the different characteristics of the infrared image and the visible image. This study also adopts the traditional "maximum–absolute" fusion rule. With this optimization, the useful visual details can be better transferred to the fused image while suppressing the noise in the infrared image:

$$V(p) = \left|I_p - I_1\right| + \left|I_p - I_2\right| + \cdots + \left|I_p - I_N\right| \tag{22}$$

where I_p is the intensity value at pixel p in image I, $V(p)$ is the significant value of pixel p, and N is the total number of pixels in image I.

If the two-pixel intensity values are equal, then

$$V(p) = \sum_{j=0}^{L-1} M_j \left|I_P - I_j\right| \tag{23}$$

where j is the pixel point, M_j is the number of pixels with pixel intensity the same as that of point j, and L is the number of gray levels in the image. The infrared image I_{ir} and visible image I_{vi} are the input images. Then, the base layer B_F of the fused image is obtained as

$$B_F = W_b B_1 + (1 - W_b) B_2 \tag{24}$$

$$W_b = 0.5 + \frac{I_{\mathrm{ir}} - I_{vi}}{2} \tag{25}$$

Infrared images usually contain coarse-scale structural information and noise and lack visual details compared with visible light images. The fused detail layer obtained by the "max–absolute" rule is improved to make the image look more natural and more suitable for human visual perception. The weighting factor W^j of "max–absolute" is

$$W^j = \begin{cases} 1, \left|d_1^j\right| > \left|d_2^j\right| (j = 1, 2, \cdots N) \\ \quad 0, otherwise \end{cases}. \tag{26}$$

Next, we apply a Gaussian filtering for denoising:

$$W_d^j = Gaussian(W^j, \delta_s) \tag{27}$$

where setting $\delta_s = 2$, M^j for the fusion details in the j_{th} layer is accomplished according to the "max–absolute" rule, as follows:

$$M^j = W_d^j d_1^j + (1 - W_d^j) d_2^j. \tag{28}$$

The fusion detail layer D^j of the j_{th} layer can be obtained according to the weighted least squares method:

$$D^j = \sum_P \left(\left(D_p^j - M_p^j\right)^2 + \lambda a_p^j \left(D_p^j - \left(d_2^j\right)_p\right)^2 \right) \quad (29)$$

where $a_p^j = \left(\left|\sum_{q \in w_p} (d_1^j)_q\right| + \varepsilon\right)^{-1}$ represents the spatial variation weight; p represents the spatial position of the pixel; ε is a constant equal to 0.0001; and w_p represents a rectangular window centered on pixel point p.

The fused image I_F after combining the base layer B_F and the detail layer $D^1, D^2, \ldots, D^N$ is

$$I_F = B_F + D^1 + D^2 + \cdots + D^N. \quad (30)$$

2.4. Binocular Stereo Ranging Algorithm

Unlike the traditional binocular stereo-ranging method [36], we seek to acquire binocular stereo-ranging information from the image obtained via aberration correction and image fusion. As shown in Figure 17, parallax information is used to restore the depth. Here, Q_L and Q_R are the coordinate origins of the two camera coordinate systems, and A is the baseline length. To find the corresponding point of the world coordinate system in the stereoimage pair for point Q, the corresponding direction vectors in the virtual coordinate system are $\overrightarrow{Q_L Q}$, $\overrightarrow{Q_R Q}$. Q' is the projection of Q in the XO_LZ plane. The angles of the vector $\overrightarrow{O_L Q}$, $\overrightarrow{O_R Q}$ with the plane YO_LZ, YO_RZ are β_1, β_2. If the angles with the O_LZ-axis and O_2Z-axis are both α; then, we apply the following:

$$\theta_1 = \frac{\pi}{2} - \beta_1 \, , \theta_2 = \frac{\pi}{2} - \beta_2. \quad (31)$$

$$\frac{B}{\sin(\pi - \theta_2)} = \frac{A}{\sin\theta} = \frac{A}{\sin(\theta_2 - \theta_1)} \Rightarrow B = \frac{A\sin\theta_2}{\sin(\theta_2 - \theta_1)}. \quad (32)$$

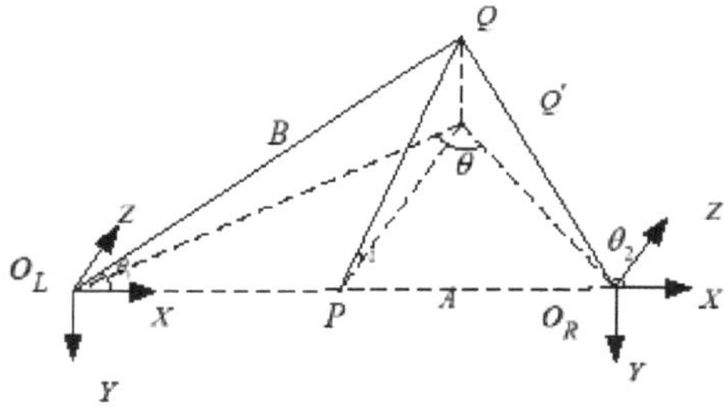

Figure 17. Schematic diagram of binocular stereo-ranging information.

Then, the depth information of Q is

$$Qdepth = Q'P = B\sin\theta_1 \cos a. \quad (33)$$

3. Experiments and Results

3.1. Test Platform

The experiment adopted a wide-field-of-view binocular stereo perception system and a calibration checkerboard grid (ambient and heated state), as shown in Figure 18. The calibration checkerboard grid was based on the Boltier principle design for an active infrared-radiation-calibration checkerboard grid. We used a JY-260 microcomputer temperature controller device from Jiangyin Jinyu Electric Heating Appliance Co., Ltd. (Jiangsu, China) to control the temperature of the calibration checkerboard grid in a range from

−50 °C to −260 °C, the calibration checkerboard grid was 10 × 7, the length of the squares was 100 mm, and the overall size was 1040 mm × 740 mm. Infrared-component calibration was carried out using the power supply. After the infrared component was calibrated, the power supply was used to heat the calibration checkerboard grid, and the temperature was controlled with the JY-260 microcomputer temperature controller at 50 °C, which heated and cooled the white box to produce corners with a large grayscale gradient on the infrared image, which was convenient for corner identification.

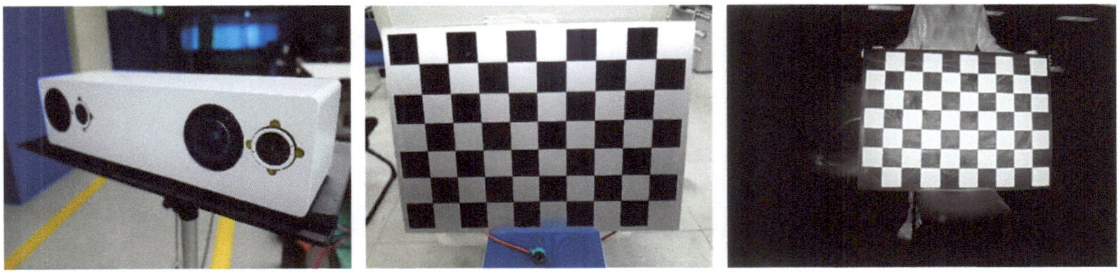

Figure 18. Wide-field-of-view binocular stereo sensing system and calibrated checkerboard grid.

3.2. Calibration Test Results

The binocular stereo sensing system was calibrated using sensors for tessellated grids in a three-dimensional scene. Under the optical imaging model, the relationship between the spatial coordinate system of the object points in the scene and the corresponding image points in the image plane was established, which, in turn, determined the internal and external parameters of the camera. In the experiment, calibration of the internal and external parameters of infrared and visible cameras for binocular stereo sensing systems was based on Taylor model-calibration. The images used in this experiment were acquired by visible light and infrared left and right cameras in different orientations for calibrating the checkerboard grid, and then the coordinates of the corner points were extracted.

We selected 18 images from the collected data for calibration, as shown in the binocular stereo left lens's visible light-camera-calibration diagram in Figure 19, and in the right lens's visible light camera-calibration diagram in Figure 20, and in the right lens's infrared camera-calibration diagram in Figure 21, and in the right lens's infrared camera-calibration diagram in Figure 22. The binocular stereo sensing system's visible camera and its infrared camera inside and outside the parameters of the calibration results are shown in Tables 4 and 5. We perform parameter calibration based on a cal ibrated chessboard, which absorbs heat uniformly after heating, overcoming the problem of difficult to recognize corner points in infrared cameras, and at the same time solving the calibration problems arising from the different imaging characteristics of infrared and visible light.

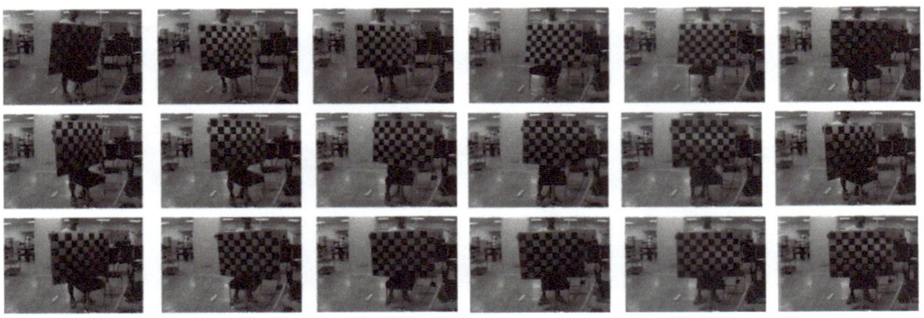

Figure 19. Calibration chart for the visible light camera (left lens).

Figure 20. Calibration chart for the visible light camera (right lens).

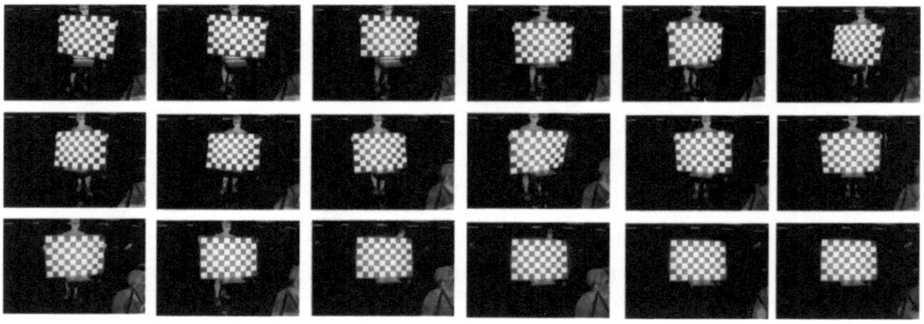

Figure 21. Calibration chart for the infrared camera (left lens).

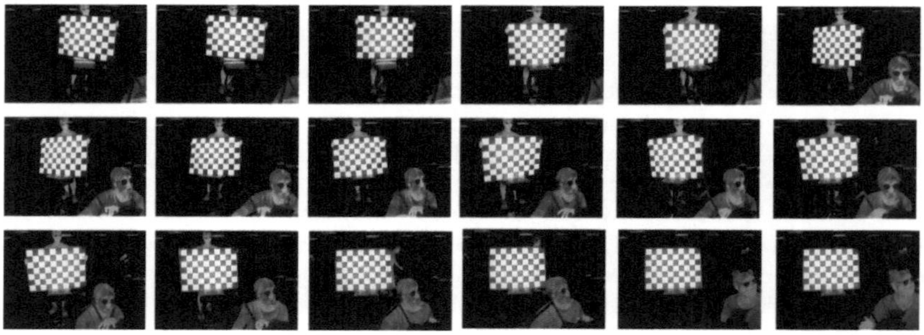

Figure 22. Calibration chart for the infrared camera (right lens).

Table 4. Internal and external parameters of the visible light camera of the binocular stereo sensing system.

Parameter	Calibration Results (Left)	Calibration Results (Right)
(c, d, e)	$(1.0001, -1.2 \times 10^{-5}, 1.7 \times 10^{-5})$	$(1.0001, -2.2 \times 10^{-5}, -2.12 \times 10^{-4})$
(u_0, v_0)	$(487.0978, 964.3319)$	$(464.1441, 949.0153)$
$(a_0, a_1, a_2, a_3, a_4)$	$(-835.384, 0, 3.296 \times 10^{-4}, 3.290 \times 10^{-7}, -3.189 \times 10^{-10})$	$(-819.290, 0, 3.642 \times 10^{-4}, 2.727 \times 10^{-7}, -2.909 \times 10^{-10})$
R	$\begin{pmatrix} 0.9976 & -0.0009 & 0.0684 \\ 0.0007 & 0.9999 & 0.0019 \\ -0.0684 & -0.0018 & 0.9976 \end{pmatrix}$	
T	$\begin{bmatrix} -335.5285 & 9.9994 & -16.8602 \end{bmatrix}$	

Table 5. Internal and external parameters for the infrared camera of the binocular stereo sensing system.

Parameter	Calibration Results (Left)	Calibration Results (Right)
(c, d, e)	$(0.992, -1.47 \times 10^{-3}, 6.2 \times 10^{-5})$	$(0.9998, -9.5 \times 10^{-5}, 1.9 \times 10^{-5})$
(u_0, v_0)	$(479.4945, 959.2869)$	$(449.2642, 932.4781)$
$(a_0, a_1, a_2, a_3, a_4)$	$(-780.385, 0, 2.472 \times 10^{-3}, 1.522 \times 10^{-8}, -7.214 \times 10^{-12})$	$(-796.383, 0, 8.938 \times 10^{-5}, 3.160 \times 10^{-8}, 1.7333 \times 10^{-10})$
R	$\begin{pmatrix} 0.9989 & 0.0371 & 0.0291 \\ -0.0375 & 0.9992 & 0.0133 \\ -0.0286 & -0.0144 & 0.9995 \end{pmatrix}$	
T	$\begin{bmatrix} -355.90 & 11.0089 & 35.05714 \end{bmatrix}$	

3.3. *Aberration-Correction Results*

The aberration maps for this experiment were acquired indoors on a calibrated checkerboard grid using a binocular stereo sensing system. The aberration-correction algorithm in this paper was used to obtain the aberration-correction map (shown in Figures 23 and 24). The aberration rate of the visible image was 45%, indicating that the image aberration caused by the large field-of-view was greatly improved at the edges and that the field-of-view was not lost.

Figure 23. Aberration and correction maps for visible light camera (left lens).

Figure 24. Aberration and correction maps for visible light camera (right lens).

3.4. Ranging Test Results

As shown in Figures 25 and 26, a person was selected as the recognition target and could be recognized by both visible light and infrared cameras. Several ranging measurements were taken at 5 m, 10 m, 15 m, 20 m, 25 m, 30 m, 40 m, and 60 m. By recognizing the target, the target point was framed out and the depth value of the feature point inside the frame was calculated. Sometimes the feature point contained both the foreground and the background feature points. In this study, the analysis was carried out using Equation (34), eliminating the unwanted data. The results of the ranging tests are shown in Table 6.

$$\overline{d_V} \pm \frac{\sum\left|d_i - \overline{d_V}\right|}{n} \tag{34}$$

where d_i is the distance value of the feature point, and $\overline{d_V}$ is the average value of the distance value of the feature point.

(a) (b) (c)

Figure 25. Binocular distance-measurement: (**a**) visible image; (**b**) infrared image; (**c**) and fused image.

(a) (b) (c)

Figure 26. Binocular ranging-scene diagram in the park: (**a**) visible light camera image; (**b**) infrared image; (**c**) and fused image.

Table 6. Binocular distance-measurement results.

Serial Number	Measuring Distance (m)	Visible Light-Ranging		Infrared Ranging		Fusion-Ranging	
		Distance(m)	Difference (m)	Distance (m)	Difference (m)	Distance (m)	Difference (m)
1	5	4.82	0.18	5.05	0.05	5.01	0.01
2	10	10.15	0.15	10.02	0.02	10	0
3	15	14.64	0.36	15.04	0.04	15.02	0.02
4	20	19.01	0.99	19.84	0.16	19.98	0.02
5	25	24.14	0.86	24.53	0.47	24.86	0.14
6	30	31.62	1.62	29.53	0.47	30.37	0.37
7	40	37.99	2.01	39.46	0.54	39.78	0.22
8	60	56.76	3.24	62.55	2.55	60.35	0.35

Table 6 shows the visible ranging error at 0.99 m (distance 20 m), 1.62 m (distance 30 m), and 3.24 m (distance 60 m) and 0.02 m (distance 20 m), 0.47 m (distance 30 m), and 2.55 m (distance 60 m), as well as the fusion-ranging error at 0.16 m (distance 20 m), 0.37 m (distance 30 m), and 0.35 m (distance 60 m). At a close distance, the error of ranging was small, and the error gradually increased with an increase in distance. Comparing the ranging results for visible and infrared cameras, the ranging accuracy of infrared camera was better than that of visible light camera. The reason for this result is that the reprojection error for the calibration of the infrared camera was less than that of the calibration results for the visible light camera. Our proposed fusion-ranging method has greater accuracy than those using only infrared- or visible light. The reason for this result is that the fusion algorithm combines the advantages of visible- light and infrared, and the extracted feature points are more accurate, so its ranging errors were 0.16 m (distance 20 m), 0.37 m (distance 30 m), and 0.35 m (distance 60 m). In addition, its feature points were more accurate, so its ranging results were higher.

In order to improve the performance of this system, this paper improves the running speed of the algorithms by optimizing the related algorithms, parallel processing, and hardware acceleration, as shown in Figure 27. The major algorithms include: an FPGA-based elliptic-aberration correction, which saves FPGA storage space by storing 1/4 of the data, and the processing time of the algorithm is 6 ms; an FPGA-based adaptive image-enhancement algorithm with a processing time of 0.2 ms, lower computational complexity, and faster processing time; FPGA-based multi-scale infrared and visible fusion with a processing time of 0.01 ms; a TX2-based SURF + RANSAC binocular stereo-matching algorithm which uses feature point extraction, feature point descriptor calculation, matching point search and consistency checking, such that the computational complexity is high and the processing time is 16 ms; and, finally, a binocular stereo ranging method that removes the misjudged points to get the final ranging data with a processing time of about 0.1 ms. To summarize, the algorithms in the binocular stereo sensing system have a total processing time of about 22.31 ms.

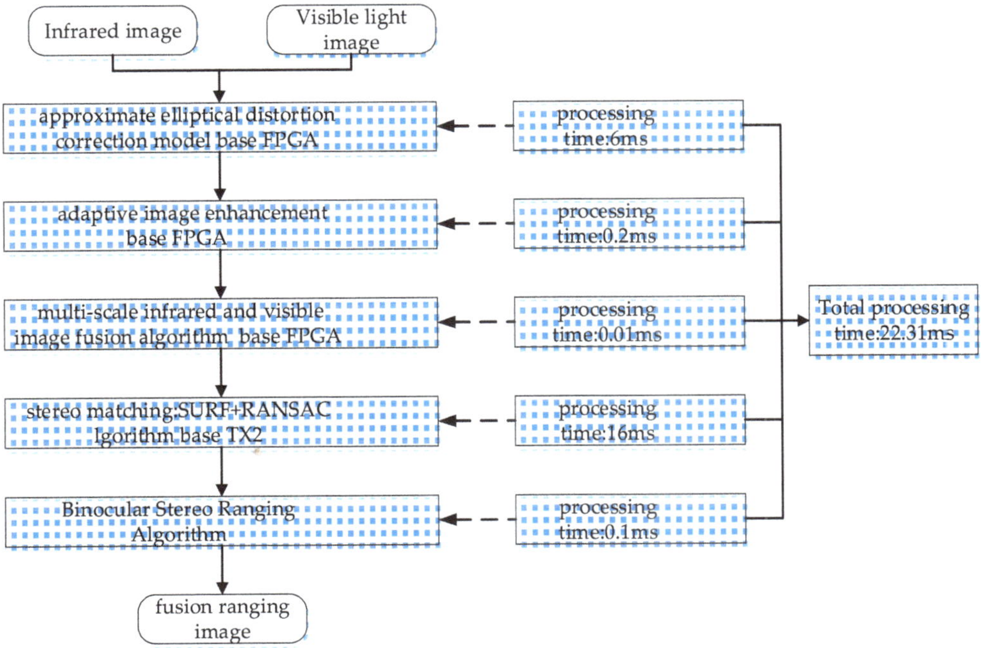

Figure 27. Image-algorithm-processing time for binocular stereo sensing systems.

4. Conclusions

This study proposed a dual-band fusion binocular stereo perception system with a large 120° field-of-view combining infrared- and visible light imaging through two sets of binocular stereo vision. This system enables information acquisition and stereo visual perception within a wide field-of-view, providing enhanced scene understanding and addressing the limitations of traditional binocular stereo vision systems, such as a narrow field-of-view and inaccuracies in short-distance ranging. Aiming to solve the aberration problems of infrared and visible images with a large field-of-view, this paper adopted the elliptic-aberration-correction model, which was corrected in real-time in the hardware circuit in FPGA, thereby saving the storage resources of FPGA and solving the problem of field-of-view loss. By adopting the multi-scale fusion method, the visible and infrared images compensated for each other's differences, fully utilized the texture information of the visible image and the thermal radiation contour information of the infrared image, and significantly improved the ranging accuracy. The simulation and experimental results demonstrated that the fusion-ranging accuracy outperforms the individual ranging accuracies for infrared- and visible light cameras.

The proposed system can maintain high performance under different light and weather conditions and scenarios. Adding image recognition algorithms for target localization of special targets or obstacles, which has application value in many areas such as driverless vehicles, virtual reality, and robot navigation. With the advancement of hardware and algorithm technology, future research will focus on enhancing the real-time performance of binocular stereo perception algorithms and achieving low-power edge computing, engineering the binocular stereo perception system, and further building the binocular stereo perception system with different resolutions to correct aberrations and accurately recognize objects at different distances. By introducing artificial intelligence algorithms, the system can realize adaptive processing and intelligent decision-making for complex scenes.

Author Contributions: Conceptualization, Y.Z. (Youpan Zhu), D.Z. and Y.Z. (Yongkang Zhou); methodology, Y.Z. (Youpan Zhu), D.Z., Y.Z. (Yongkang Zhou) and W.J.; software, Y.Z. (Youpan Zhu), D.Z., Y.Z. (Yongkang Zhou), L.Z. and G.W.; validation, Y.Z. (Youpan Zhu), D.Z., Y.Z. (Yongkang Zhou), L.Z. and Y.L.; formal analysis, G.W. and Y.L.; investigation, Y.Z. (Youpan Zhu), W.J., and G.W.; resources, W.J.; data curation, Y.Z. (Youpan Zhu), D.Z., and Y.Z. (Yongkang Zhou); writing—original draft preparation, Y.Z. (Youpan Zhu), D.Z., Y.Z. (Yongkang Zhou), W.J., L.Z., G.W. and Y.L.; writing—review and editing, Y.Z. (Youpan Zhu), D.Z., Y.Z. (Yongkang Zhou), W.J., L.Z., G.W. and Y.L.; visualization, G.W. and Y.L.; supervision, W.J. and Y.L.; project administration, W.J.; funding acquisition, W.J. All authors have read and agreed to the published version of the manuscript.

Funding: This research was funded by the National Defense Science and Technology Foundation Strengthening Plan (grant number 2021-JCJQ-JJ-1020).

Institutional Review Board Statement: Not applicable.

Informed Consent Statement: Not applicable.

Data Availability Statement: Data are contained within the article.

Acknowledgments: We thank the National Defense Science and Technology Foundation Strengthening Plan for help in identifying collaborators for this work.

Conflicts of Interest: Authors Dan Zhang and Lingling Zhou were employed by the company Yunnan North Optical & Electronic Instrument Co., Ltd. The remaining authors declare that the research was conducted in the absence of any commercial or financial relationships that could be construed as a potential conflict of interest.

References

1. Srivastav, A.; Mandal, S. Radars for Autonomous Driving: A Review of Deep Learning Methods and Challenges. In Proceedings of the Computer Vision and Pattern Recognition, Vancouver, BC, Canada, 7 September 2023.
2. Feng, D.; Haase-Schutz, C.; Rosenbaum, L.; Hertlein, H.; Glaser, C.; Timm, F.; Wiesbeck, W.; Dietmayer, K. Deep Multi-modal Object Detection and Semantic Segmentation for Autonomous Driving: Datasets, Methods, and Challenges. *IEEE Trans. Intell. Transp. Syst.* **2019**, *22*, 1341–1360. [CrossRef]
3. Yan, J.; Jiao, H.; Pu, W.; Shi, C.; Dai, J.; Liu, H. Radar Sensor Network Resource Allocation for Fused Target Tracking: A Brief Review. *Inf. Fusion* **2022**, *86–87*, 104–115. [CrossRef]
4. Fernandez Llorca, D.; Hernandez Martinez, A.; Garcia Daza, I. Vision-based Vehicle Speed Estimation for ITS: A Survey. In Proceedings of the Computer Vision and Pattern Recognition, Nashville, TN, USA, 26 May 2021.

5. Mur-Artal, R.; Montiel, J.M.M.; Tardos, J.D. ORB-SLAM: A Versatile and Accurate Monocular SLAM System. *IEEE Trans. Robot.* **2015**, *31*, 1147–1163. [CrossRef]
6. Mur-Artal, R.; Tardós, J.D. ORB-SLAM2: An Open-Source SLAM System for Monocular, Stereo, and RGB-D Cameras. *IEEE Trans. Robot.* **2017**, *33*, 1255–1262. [CrossRef]
7. Campos, C.; Elvira, R.; Rodríguez, J.J.G.; Montiel, J.M.; Tardós, J.D. ORB-SLAM3: An Accurate Open-Source Library for Visual, Visual–Inertial, and Multimap SLAM. *IEEE Trans. Robot.* **2021**, *6*, 37. [CrossRef]
8. Marr, D. Cooperative computation of stereo disparity. *Science* **1976**, *194*, 283–287. [CrossRef]
9. Marr, D. A Computational Theory of Human Stereo Vision. *R. Soc. B Biol. Sci.* **1979**, *204*, 301–328.
10. Poggi, M.; Tosi, F.; Batsos, K.; Mordohai, P.; Mattoccia, S. On the Synergies between Machine Learning and Binocular Stereo for Depth Estimation from Images: A Survey. *IEEE Trans. Pattern Anal. Mach. Intell.* **2021**, *44*, 5314–5334. [CrossRef]
11. Jin, Y.; Lee, M. Enhancing Binocular Depth Estimation Based on Proactive Perception and Action Cyclic Learning for an Autonomous Developmental Robot. *IEEE Trans. Syst. Man Cybern. Syst.* **2019**, *49*, 169–180. [CrossRef]
12. Gyongy, I.; Dutton, N.A.; Henderson, R.K. Direct Time-of-Flight Single-Photon Imaging. *IEEE Trans. Electron Devices* **2022**, *69*, 2794–2805. [CrossRef]
13. Zhang, J.; Luo, B.; Su, X.; Wang, Y.; Chen, X.; Wang, Y. Depth range enhancement of binary defocusing technique based on multi-frequency phase merging. *Opt. Express* **2019**, *27*, 36717–36730. [CrossRef] [PubMed]
14. Real-Moreno, O.; Rodríguez-Quiñonez, J.C.; Sergiyenko, O.; Flores-Fuentes, W.; Mercorelli, P.; Ramírez-Hernández, L.R. Obtaining Object Information from Stereo Vision System for Autonomous Vehicles. In Proceedings of the 2021 IEEE 30th International Symposium on Industrial Electronics (ISIE), Kyoto, Japan, 20–23 June 2021; pp. 1–6.
15. Yang, F.; Rao, Y. Vision-Based Intelligent Vehicle Road Recognition and Obstacle Detection Method. *Int. J. Pattern Recognit. Artif. Intell.* **2020**, *34*, 2050020. [CrossRef]
16. Montemurro, N.; Scerrati, A.; Ricciardi, L.; Trevisi, G. The Exoscope in Neurosurgery: An Overview of the Current Literature of Intraoperative Use in Brain and Spine Surgery. *J. Clin. Med.* **2021**, *11*, 223. [CrossRef]
17. Yin, Y.; Yu, K.; Yu, C.; Bai, X.; Cai, Z.; Meng, X.; Yang, X. 3D Imaging Using Geometric Light Field: A Review. *Chin. J. Lasers* **2021**, *48*, 1209001.
18. Xu, Y.; Liu, K.; Ni, J.; Li, Q. 3D reconstruction method based on second-order semiglobal stereo matching and fast point positioning Delaunay triangulation. *PLoS ONE* **2023**, *17*, e0260466. [CrossRef]
19. Blake, R.; Wilson, H. Binocular vision. *Vis. Res.* **2011**, *51*, 754–770. [CrossRef]
20. Pomerleau, D. Efficient Training of Artificial Neural Networks for Autonomous Navigation. *Neural Comput.* **2014**, *3*, 88–97. [CrossRef]
21. Shao, N.; Li, H.G.; Liu, L.; Zhang, Z.L. Stereo Vision Robot Obstacle Detection Based on the SIFT. In Proceedings of the 2010 Second WRI Global Congress on Intelligent Systems, Wuhan, China, 16–17 December 2010; pp. 274–277.
22. Sivaraman, S.; Trivedi, M.M. Real-time vehicle detection using parts at intersections. In Proceedings of the IEEE Conference on Intelligent Transportation Systems, Anchorage, AK, USA, 16–19 September 2012; pp. 1519–1524.
23. Bontar, J.; Lecun, Y. Computing the Stereo Matching Cost with a Convolutional Neural Network. In Proceedings of the Conference on Computer Vision and Pattern Recognition (CVPR), Boston, MA, USA, 7–12 June 2015; IEEE: New York, NY, USA, 2014.
24. Li, P.; Chen, X.; Shen, S. Stereo R-CNN based 3D Object Detection for Autonomous Driving. In Proceedings of the IEEE/CVF Conference on Computer Vision and Pattern Recognition (CVPR), Long Beach, CA, USA, 15–20 June 2019; pp. 7636–7644.
25. Tang, Y.; Zhou, H.; Wang, H.; Zhang, Y. Fruit detection and positioning technology for a Camellia oleifera C. Abel orchard based on improved YOLOv4-tiny model and binocular stereo vision. *Expert Syst. Appl.* **2023**, *211*, 118573. [CrossRef]
26. Huang, H.; Yuan, E.; Zhang, D.; Sun, D.; Yang, M.; Zheng, Z.; Zhang, Z.; Gao, L.; Panezai, S.; Qiu, K. Free Field of View Infrared Digital Holography for Mineral Crystallization. *Cryst. Growth Des.* **2023**, *23*, 7992–8008. [CrossRef]
27. Ma, K.; Zhou, H.; Li, J.; Liu, H. Design of Binocular Stereo Vision System with Parallel Optical Axesand Image 3D Reconstruction. In Proceedings of the 2019 China-Qatar International Workshop on Artificial Intelligence and Applications to Intelligent Manufacturing (AIAIM), Doha, Qatar, 1–4 January 2019; pp. 59–62.
28. Hu, Y.; Chen, Q.; Feng, S.; Tao, T.; Asundi, A.; Zuo, C. A new microscopic telecentric stereo vision system—Calibration, rectification, and three-dimensional reconstruction. *Opt. Lasers Eng.* **2019**, *113*, 14–22. [CrossRef]
29. Peng, Y.T.; Cao, K.; Cosman, P.C. Generalization of the Dark Channel Prior for Single Image Restoration. *IEEE Trans. Image Process* **2018**, *27*, 2856–2868. [CrossRef]
30. Wu, J.; Liu, G. Noniterative calibration of a camera lens with radial distortion. *Meas. Sci. Technol.* **2012**, *23*, 105013. [CrossRef]
31. Zhou, Y.; Li, X.; Zhou, J.; Tang, X. A low delay highly dynamic range infrared imaging system for complex scenes based on FPGA. *Measurement* **2023**, *223*, 113745. [CrossRef]
32. Zhu, Y.; Li, H.; Li, L.; Jin, W.; Song, J.; Zhou, Y. A stereo vision depth estimation method of binocular wide-field infrared camera. In Proceedings of the Third International Computing Imaging Conference, Sydney, Australia, 1–3 June 2023; Volume 2921, pp. 252–264.
33. Bay, H.; Tuytelaars, T.; Van Gool, L. SURF: Speeded up robust features. *Comput. Vis. Image Underst.* **2008**, *110*, 346–359. [CrossRef]
34. Bansal, M.; Kumar, M.; Kumar, M. 2D object recognition: A comparative analysis of SIFT, SURF, and ORB feature descriptors. *Multimed. Tools Appl.* **2021**, *80*, 18839–18857. [CrossRef]

35. Ma, J.; Zhou, Z.; Wang, B.; Zong, H. Infrared and visible image fusion based on visual saliency map and weighted least square optimization. *Infrared Phys. Technol.* **2017**, *82*, 8–17. [CrossRef]
36. Mu, Q.; Wei, J.; Yuan, Z.; Yin, Y. Research on Target Ranging Method Based on Binocular Stereo Vision. In Proceedings of the 2021 International Conference on Intelligent Computing, Automation and Applications (ICAA), Nanjing, China, 25–27 June 2021 pp. 81–85.

sensors

MDPI

Article

High-Range and High-Linearity 2D Angle Measurement System for a Fast Steering Mirror

Boshi Du, Yong Lv *, Lishuang Liu and Yang Liu

Instrumental Science and Optoelectronic Engineering Faculty, Beijing Information Science & Technology University, Beijing 100192, China; boshidu1997@gmail.com (B.D.); liulishuang@bistu.edu.cn (L.L.); 20141891@bistu.edu.cn (Y.L.)

* Correspondence: lvyong@bistu.edu.cn

Abstract: In order to solve the problem of the insufficient range of the traditional fast mirror (FSM) angle measurement system in practical applications, a 2D large-angle FSM photoelectric angle measurement system based on the principle of diffuse reflection is proposed. A mathematical model of the angle measurement system is established by combining the physical properties of the diffuse reflecting plate, such as the rotation angle, rotation center, rotation radius, reflection coefficient and the radius of the diffuse reflecting surface. This paper proposes a method that optimizes the degree of nonlinearity based on this mathematical model. The system is designed and tested. The experimental results show that changing the diffuse reflection surface area can improve the nonlinearity of the angle measurement system effectively. When the radius of the diffuse reflection surface is 3.3 mm, the range is ±20°, the non-linearity is 0.74%, and the resolution can reach up to 2.3″. The system's body is simple and compact. It is also capable of measuring a wider range of angles while linearity is guaranteed.

Keywords: photoelectric sensors; fast steering mirror; large angle measurement system; diffuse reflection

1. Introduction

Fast steering mirrors (FSMs) are often used to improve the performance of photoelectric imaging systems, such as airborne imaging systems and space-based optical surveillance systems [1–3]. It is necessary to measure the large deflection angle of the FSM to achieve a wider range of image acquisition capabilities. However, due to its compact spatial structure and fast response characteristics, the common large-angle angle measurement method and contact angle measurement method are not suitable for the structure of the FSM. Currently, common FSM angle measurement systems include capacitive sensor angle measurement systems, eddy current sensor angle measurement systems, digital encoder angle measurement systems, and photoelectric angle measurement systems. It is difficult for the above systems to achieve two-dimensional large-angle measurements of more than ±10°. Therefore, how to achieve a large-scale and high-precision angle measurement has become a research question for many scholars [4–14].

Yue Fan [15] demonstrated an angle measurement method based on an eddy current sensor which achieved an angle measurement accuracy of 1.1 arcsec in the range of ±1.5°. Lin D. [16] introduced a two-axis measurement system based on the principle of a capacitance sensor, with a measurement range of ±5°. OIM101 Inch [17] achieved a nonlinearity of 1% and a resolution of 1″ in the range of ±1.5°. Xin Wu [18] achieved a nonlinearity of 1% in the range of ±1.25°. YingXue Ni [19] achieved a measurement range of ±2° and a nonlinearity of 0.72% through specular reflection and QD sensors. ShaoKang Hung [20] established a spot center positional algorithm and realized a measurement of ±10° and a resolution of 0.0265° by collecting the spot position through QPD. Mingguang Shan [21] proposed a differential angle measurement system based on optical fiber devices which can achieve a complete linear measurement in the range of ±6.1°. Based on the above research, the current two-dimensional FSM angle measurement system can only achieve high resolution and high linearity in a small range, and it is difficult to achieve large angle measurements. Therefore, achieving large angle measurements while ensuring the measurements' accuracy has become a problem.

In this paper, based on the principle of diffuse reflection, a two-dimensional large-angle FSM angle measurement system based on differential optical power was proposed. By analyzing the distribution of diffuse reflection light on the surfaces of different photodiode diffuse reflectors, a mathematical model was established to study the change in the differential optical power with the radius of the diffuse reflection surface. Then, an experimental model was built to test the static performance of the angle measurement system. Later, the correctness of the mathematical model was verified, a large angle measurement in the range of ±20° was realized, and the nonlinearity of the angle measurement system was optimized.

2. Deflection Angle Detection Principle

Figure 1 shows the optical structure of the two-dimensional large-angle angle measurement system, which is divided into upper and lower parts. The upper part comprises the mirror, mirror support, and the diffuse reflection surface of the FSM. The lower part comprises the optical signal transmitting and receiving circuit, whose structure is shown in Figure 2. The acquisition circuit is fixed on a small PCB board. The LED light source in the middle directly faces the center of the diffuse reflector in a spatial position. Four photodiodes are evenly distributed around the light source and are divided into two groups, A and D and B and C. The two groups are placed orthogonally.

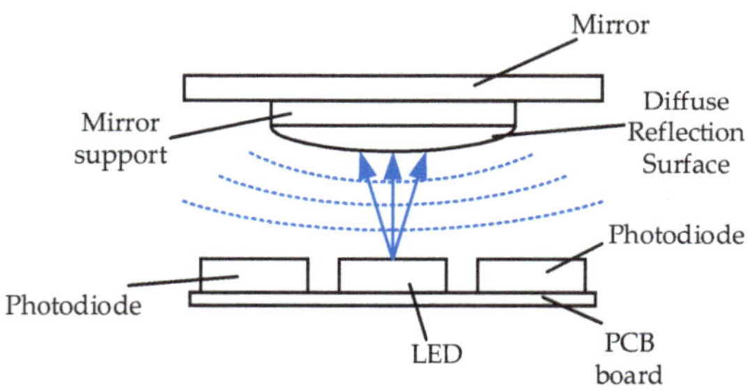

Figure 1. Optical structure design of 2D large angle measurement system.

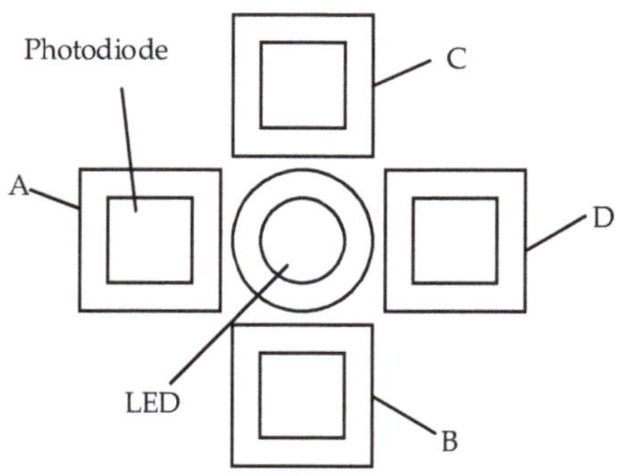

Figure 2. Structural design of optical signal transmitting and receiving Circuit.

When the LED light source irradiates the diffuse reflector, the light is reflected through the diffuse reflection surface. Then, the light is reflected to the optical signal acquisition plate in an irradiation mode similar to the Lambert light source. After the photodiode on the surface of the optical signal acquisition plate is illuminated, different photocurrents are generated according to the change in the surface optical power. When the angle of the mirror of the FSM changes, the angle of the diffuse reflection surface will also change, which will affect the distribution of the reflected light intensity on the photoelectric acquisition plate so that the surface light power of the four photodiodes is different, and the change in the angle of the FSM will be converted into the output change in the photocurrent generated by the photodiode. For example, when the FSM mirror swings in the direction of photodiode A, the diffuse reflection surface will swing in the direction of photodiode D. The optical power of the reflected light of the diffuse reflection surface falling on photodiode D will become larger, the optical power falling on photodiodes A, B, and C will become smaller, and the optical power on the B and C surfaces will change similarly. The relationship between the angle value and the differential optical power can be obtained from the difference in the optical power on the surfaces of the two pairs of photodiodes.

3. Establishment of a Theoretical Model for the Angle Measurement System

3.1. Diffuse Reflection Theory

The angle measurement system realizes the test function by using the diffuse reflection theory. In the theoretical model, it is assumed that the diffuse reflector used in the experiment is an ideal diffuse reflector, namely, Lambert's reflector, and its surface reflects light with equal intensity in all directions. According to the properties of Lambert's reflector, when an arbitrary ray of light irradiates Lambert's reflector, the intensity of the diffuse light is proportional to the cosine of the angle between the direction of the incident light and the normal direction of the incident point surface:

$$I_{ldiff} = k_d \, I_l \cos\theta \tag{1}$$

where k_d ($0 < k_d < 1$) represents the reflection coefficient of the material to the ambient light; Il represents the intensity of the point light source; θ represents the angle between the direction of the incident light and the normal of the vertex, which is called the incident angle ($0^\circ \leq \theta \leq 90^\circ$); and I_{ldiff} represents the light intensity reflected by the interaction between the diffuse reflector and any light. Figure 3 is a plane map of the angle measurement system. In practice, in order to make the reflected light closer to the reflected light of an ideal Lambertian body, a material with a reflection coefficient close to 1 should be used, and the material used in this paper is polytetrafluoroethylene, with a reflection coefficient of 0.983.

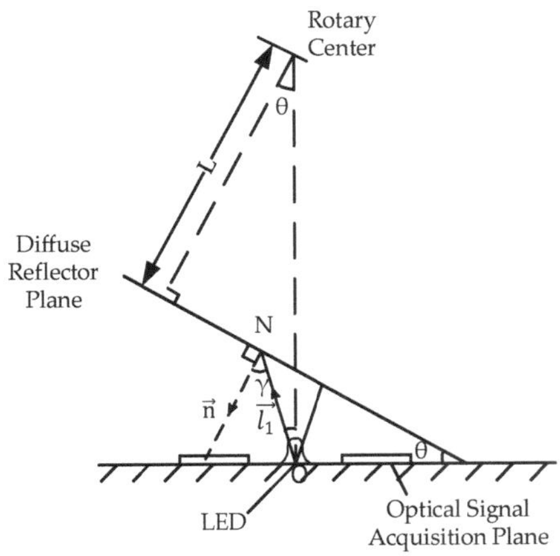

Figure 3. Plane diagram of the angle measurement system.

Where θ represents the tilt angle of the diffuse reflection plane; γ represents the angle between the direction vector of the light source $\vec{l_1}$ and the normal vector of the diffuse reflection surface $\vec{n}$; and N point represents any irradiated point on the diffuse reflection surface. In this way, the light intensity reflected by any point on the diffuse reflection surface can be expressed as follows:

$$I_{ldiff} = k_d\, I_l \cos\gamma = k_d\, I_l \frac{\left|\vec{n} \cdot \vec{l_1}\right|}{\left|\vec{n}\right| \cdot \left|\vec{l_1}\right|} \tag{2}$$

Due to the diffuse reflection of the light, the reflection point N on the diffuse reflection surface can be considered a new point light source, which is a spherical wave with a light intensity of I_{ldiff}, as shown in Figure 4.

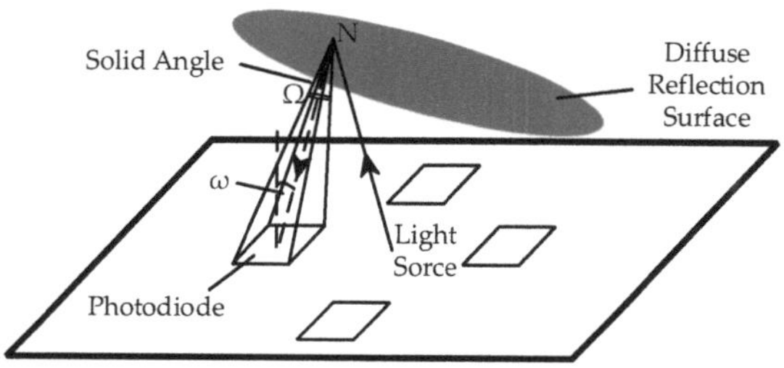

Figure 4. Diagram of a spatial solid angle.

The diffuse reflection surface reflects light to the surface of the photodiode, and the optical power per unit surface of the photodiode can be expressed as follows:

$$dE = I\, d\Omega \tag{3}$$

where dE represents the unit surface light power; I represents the light intensity; and dΩ represents the solid angle corresponding to the unit surface area, which can be expressed as

$$d\Omega = \frac{dS \cdot \cos\omega}{l^2} \tag{4}$$

where dS represents the unit surface area of the observation surface; ω represents the included angle between the normal of the observation surface and the incident light; and l represents the distance from the point light source to the center of the unit surface.

That is, after determining the solid angle Ω of the photodiode's surface to the point light source N, the optical power of the light emitted by any point in the point light source on the photodiode's surface can be acquired.

As shown in Figure 4, when integrating dΩ, ω changes. Since the surface area of the photodiode is very small, for any point N on the diffuse reflection surface, when integrating the diffuse reflection surface, the actual change range of ω is less than 7°. Therefore, ω can be simplified, that is, cosω in all unit area sources is considered approximately equal, and S can be defined as the surface area of the photodiode as follows:

$$\Omega = \frac{S \cdot \cos\omega}{l^2} \tag{5}$$

Inserting Equations (2) and (4) into Equation (3), the surface optical power E of the photodiode reflected by any point on the diffuse reflection surface can be obtained.

$$E = k_d\, I_l \frac{\left|\vec{n_1} \cdot \vec{l_1}\right|}{\left|\vec{n_1}\right| \cdot \left|\vec{l_1}\right|} \frac{S \cdot \cos\omega}{l^2} \tag{6}$$

3.2. Diffuse Reflection Model of Angle Measurement System

Based on the above theory, a spatial coordinate system is established with the light source as the origin. The light intensity IL of the light source is determined by the light distribution curve of the light source, and the light distribution curve is expressed as the function F (x, y, z). When the FSM mirror rotates, the diffuse reflection surface also rotates around the rotation center of the FSM mirror. The coordinates of the rotation center are M (0, 0, M), the radius of the diffuse reflection surface is L, and the diffuse reflection surface is a circular surface with a radius of r. The coordinates of the four photodiode centers are A (0, −h, 0), B (h, 0, 0), C (−h, 0, 0), and D (0, h, 0), and the photodiode is a square with a side length of d. The distances from any point N′ (x, y, z) on the diffuse reflection surface to the surfaces of the four photodiodes are l_A, l_B, l_C, and l_D, respectively, as shown in Figure 5.

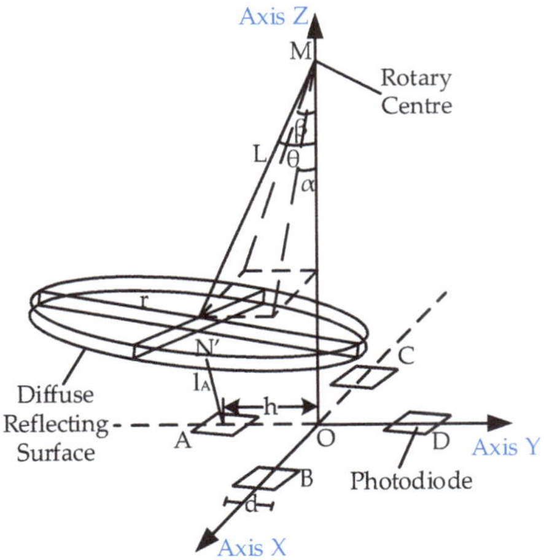

Figure 5. Space model of angle measurement system.

Here, α represents the included angle between the turning radius and the xoz plane; β represents the included angle between the turning radius and the yoz plane; and θ represents the spatial angle of the turning radius, as follows:

$$\tan\theta = \sqrt{\tan^2\alpha + \tan^2\beta} \tag{7}$$

The normal vector of the photodiode is the same as that of the xoy plane $\vec{n_0} = (0,\ 0,\ 1)$. The light vector of the light source to the diffuse reflection surface is defined as $\vec{n_1} = (x,\ y,\ z)$, the normal vector of the diffuse reflection surface is defined as $\vec{n_2} = (\tan\alpha,\ \tan\beta,\ -1)$, and the light vector of any point N′ on the diffuse reflection surface to the center point of photodiode A is defined as $\vec{n_3} = (x,\ y - h,\ z)$.

The vector in Equation (2) is extended from a plane vector to a space vector to obtain the diffuse reflection intensity of any point N′ (x, y, z) on the diffuse reflection surface, as follows:

$$I'_{ldiff} = k_d I_l \frac{\left|-\vec{n_1} \cdot \vec{n_2}\right|}{\left|\vec{n_1}\right| \cdot \left|\vec{n_2}\right|} = k_d\, F(x,\ y,\ z) \frac{x\tan\alpha + y\tan\beta\ -\ z}{\sec\theta \cdot \sqrt{x^2+y^2+z^2}} \tag{8}$$

The distance between any point on the diffuse reflection surface and the center point of photodiode A is

$$l_A = \sqrt{x^2 + (y\ -\ h)^2 + z^2} \tag{9}$$

where cosω can be expressed as

$$\cos\omega = \frac{\left|-\vec{n_3} \cdot \vec{n_0}\right|}{\left|\vec{n_3}\right| \cdot \left|\vec{n_0}\right|} = \frac{z}{\sqrt{x^2 + (y\ -\ h)^2 + z^2}} \tag{10}$$

Inserting Equations (8)–(10) into Equation (6), the optical power from any point on the diffuse flat plate reflector to the surface of photodiode A can be obtained as follows:

$$\Delta E_A = \frac{k_d\, F(x,\ y,\ z)}{\sec\theta} \frac{x\tan\alpha + y\tan\beta\ -\ z}{\sqrt{x^2+y^2+\ z^2}} \frac{d^2 \cdot z}{\sqrt{x^2 + (y\ -\ h)^2 + z^2}^{\,3}} \tag{11}$$

By integrating (11) on the diffuse reflection surface, the optical power of all diffuse reflection rays on the surface of the photodiode can be obtained. The method of obtaining the integral interval is as follows.

When the diffuse reflector rotates around the center of rotation, the space equation of the diffuse reflector plane is

$$-\,x\tan\alpha\ -\ y\tan\beta + z + L\sec\theta\ -\ M = 0 \tag{12}$$

The circular diffuse reflection surface can be expressed as a plane formed by the intersection lines between the diffuse reflection surface plane and the spatial cylindrical surface. The equation of the spatial cylindrical surface is

$$(-x\tan\beta + y\tan\alpha)^2 + [x + \tan\alpha(z\ -\ M)]^2 + [y + \tan\beta(z -\ M)]^2 = r^2\sec^2\theta \tag{13}$$

Equation (14) of the projection surface P of the space plane on the xoy plane is obtained by combining function (12) with function (13), and the projection surface P is the integral interval.

$$P: \begin{cases} x^2 + y^2 + [-x\tan\alpha\ - y\tan\beta + L\sec\theta]^2 < r^2 + L^2 \\ z = 0 \end{cases} \tag{14}$$

By integrating optical power on the interval P, the optical power value of all diffuse reflection rays reflected on the surface of photodiode A can be obtained.

$$E_A = k_d\, d^2\,(M\ -\ L\sec\theta) \iint_P \frac{F(x,\ y,\ z)}{\sqrt{x^2+y^2+z^2}} \frac{z}{\sqrt{x^2 + (y\ -\ h)^2 + z^2}^{\,3}} dxdy \tag{15}$$

Supposing that the light distribution curve L obeys second-order normal distribution; then, the light source opening angle is 17°, r = 5.0, L = 7.0, M = 10.0, d = 1.4, D = 3.5, k_d = 0.983, β = 0, and θ = α. When α changes from −20° to 20°, the changes in the surface light intensity of a single photodiode in the α and β directions are shown in Figure 6.

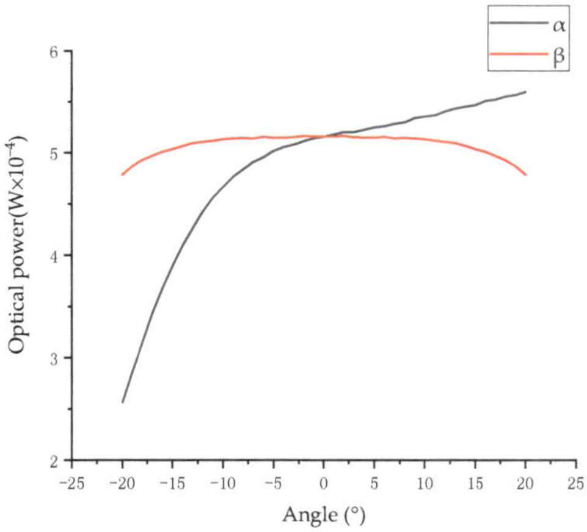

Figure 6. Surface optical power of a single photodiode.

Figure 6 shows changes in the light intensity on the surface of a single photodiode. In principle, the angle can be measured by using a photodiode in both directions. However, as shown in Figure 6, within a large angle range, the nonlinearity of the angle measured by a single sensor in the α direction is 38.32%, and it is non-monotonic. For a non-monotonic system with poor linearity, the linearity can be optimized by difference. Therefore, the difference model is obtained by subtracting the optical power of the surfaces of the photodiodes A and D:

$$E_{A-D} = k_d(M - L\sec\theta)d^2\iint_{\pi} \frac{F(x, y, z)}{\sqrt{x^2+y^2+z^2}}\left(\frac{z}{\sqrt{x^2+(y+h)^2+z^2}^{3}} - \frac{z}{\sqrt{x^2+(y-h)^2+z^2}^{3}}\right)dxdy \quad (16)$$

For the same reason, the difference model is obtained by subtracting the optical power of the surfaces of the photodiodes B and C:

$$E_{B-C} = k_d(M - L\sec\theta)d^2\iint_{P} \frac{F(x, y, z)}{\sqrt{x^2+y^2+z^2}}\left(\frac{z}{\sqrt{(x-h)^2+y^2+z^2}^{3}} - \frac{z}{\sqrt{(x+h)^2+y^2+z^2}^{3}}\right)dxdy \quad (17)$$

Putting $r = 5.0$, $L = 7.0$, $M = 10.0$, $d = 1.4$, $D = 3.5$, $k_d = 0.983$, $\beta = 0$, and $\theta = \alpha[-20°, 20°]$ into (16) and (17), the differential optical power of the two groups of photodiode surfaces can be obtained, as shown in Figure 7.

In Figure 7, the nonlinearity of the differential optical power is 11.20%. It can be seen that when the optical power is used as the reference standard for the angle change, the differential form can effectively reduce the nonlinearity of the measurement results and improve the measurement accuracy.

In this way, the parameters in Equation (16) can be adjusted to reduce the nonlinearity. Among the five parameters r, L, M, d, and D, the radius of the diffuse reflection surface r most significantly influences the differential optical power ΔE. Keeping the other parameters unchanged, different r values are substituted into the equation to obtain the relationship between the light intensity difference and the radius of the diffuse reflection plate, as shown in Figure 8.

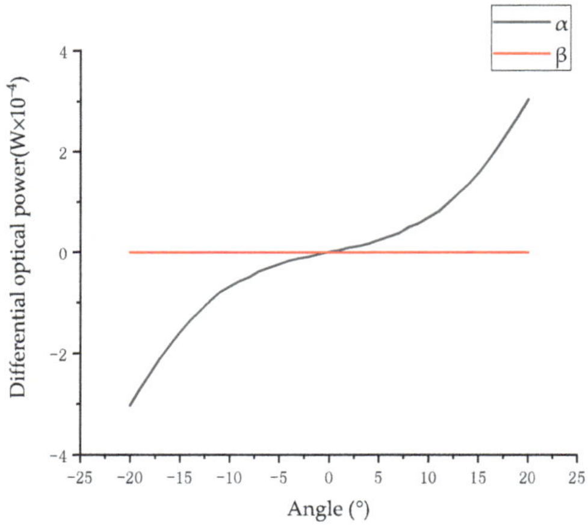

Figure 7. Relationship between the surface differential optical power and the angle of two pairs of photodiodes.

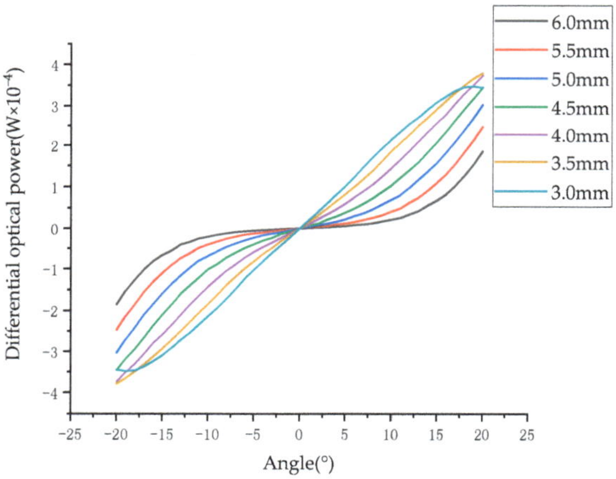

Figure 8. Relationship between the surface differential optical power and the angle at different radii.

When the differential optical power changes with the angle, the greater the nonlinearity of the differential optical power, the greater the difference in the resolution of the angle measurement at different angles. As shown in Figure 8, when the radius of the diffuse reflection surface is 6 mm, the differential optical power does not obviously change within the angle range of $-10°$to $10°$. As a result, the angle measurement system has poor angular resolution within the range of $-10°$to $10°$. When the nonlinearity of the system is small, the system can clearly distinguish the changes in the differential splitting power with angle values between $-20°$and $20°$.

Starting from the diffuse reflection surface radius of 6 mm, as the diffuse reflection surface radius decreases, the differential optical power changes approximately linearly. According to the above simulation data, compared with the other radii shown in Figure 8, when the diffuse reflection surface radius is 3.5, the minimum nonlinearity of the differential optical power is 1.8%. When the radius of the diffuse reflector is less than or equal to 3.0 mm, there is obvious non-monotonicity within the range of $-20°$ to $20°$, which makes a diffuse reflector with a radius of less than 3 mm unsuitable for angle measurement.

4. Experiments and Results

4.1. Experimental Structure of the Angle Measurement System

To verify the validity of the model, a test mechanism was designed based on the principle of the simulation results, as shown in Figure 9.

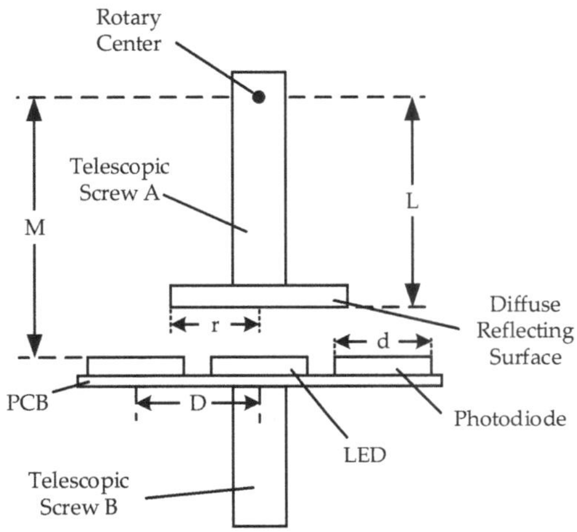

Figure 9. Schematic diagram of the testing apparatus structure.

The rotation radius L can be adjusted by adjusting the length of telescopic screw A, and the distance M between the LED light source and the rotation center can be adjusted by adjusting telescopic screw B. The LED and the photodiode are installed on the PCB plate according to the structure shown in Figure 2 to form an optical signal processing plate. The PCB is fixed on telescopic screw B, and the adjustment of the photodiode spacing D can be realized by replacing the PCB plate.

The test device was built according to the test structure. The light source of the test system adopted an SFH4441 LED with a wavelength of 940 nm, and the opening angle of the light source was 17°. For the material of the diffuse reflecting surface, the reflection coefficient of a 940 nm wavelength is 0.983. The model of the four photodiodes was PD15-22B/TR8, and the surface of each photodiode was a rectangular surface of 1.4 mm × 1.4 mm. Retractable screw A was fixed on the rotating platform through an adaptor, and the rotating platform was controlled by a stepper motor. The motor model was K431-60, with an accuracy of 0.004°. Figure 10 shows the test system.

When the rotating platform rotates, telescopic screw rod A rotates around the center of the rotating platform by following the adapter, driving the diffuse reflector to rotate around the rotating center. When the light source irradiates the diffuse reflector, the light intensity on the surfaces of the four photodiodes changes with the angle of the diffuse reflector, and the light intensity difference of the pair of photodiodes changes relatively. The differential voltage of the angle measuring circuit changes, and the analog circuit is converted into a digital signal via ADC, that is, the AD value corresponds to the light intensity difference. In this experiment, the deflection angle of the angle motion table is used as the theoretical value, and the AD value is the actual value. The AD value corresponds to the angle value, and the floating AD value corresponds to the measurement accuracy value. The relationship can be verified according to the corresponding relationship between the AD value and the angle.

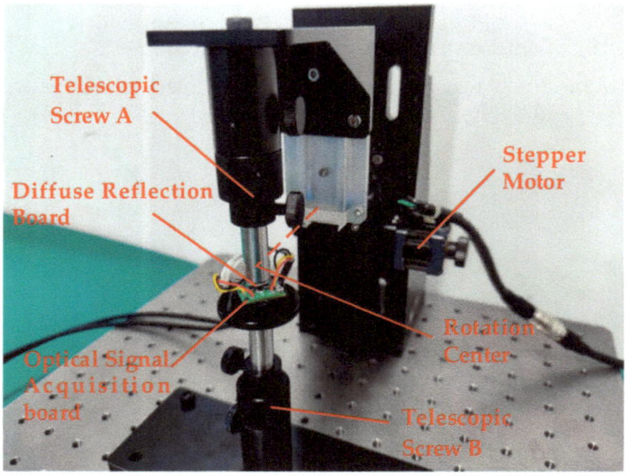

Figure 10. Static photoelectricity angle measurement system.

4.2. Results and Analysis of the Deflection Angle Measurement System

The test system was set according to the following parameters. The distance from the center of the four photodiodes to the light source was 3.5 mm, the distance from the center of the diffuse reflection surface to the light source was 3.0 mm, and the distance from the diffuse reflection surface to the rotation center was 7.0 mm. The rotation angle of the diffuse reflection surface changed in the range of −20° to + 20°, and the radii of the diffuse reflection plate were 3.0 mm, 3.5 mm, 4.0 mm, 4.5 mm, 5.0 mm, 5.5 mm, and 6.0 mm, respectively. The rotation angle of the rotation center was adjusted by the electronically controlled rotating platform to obtain the experimental results, as shown in Figure 11.

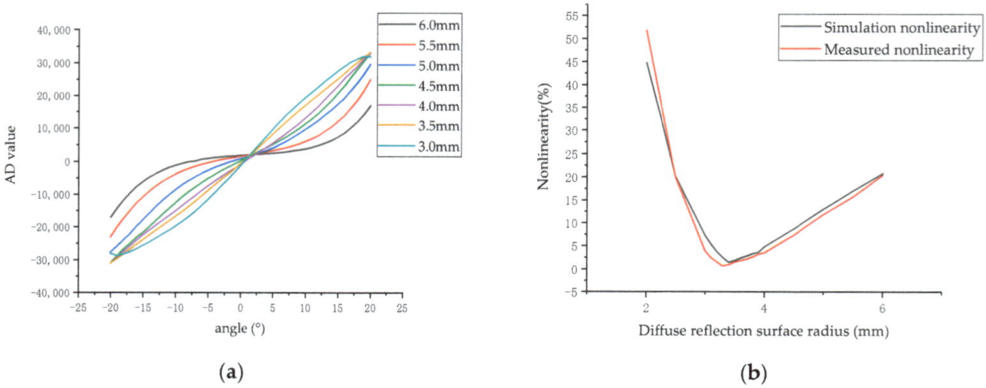

Figure 11. (**a**) The relationship between the angle value and the differential optical power for different diffuse surface radii; (**b**) the relationship between the diffuse reflection surface radius and the nonlinearity of the differential optical power.

As Figure 11a shows evidently, the measurement value changes as the angle value changes. When the radius of the diffuse reflection surface is less than 3.0 mm, the angle measurement system is non-monotonic within ±20°. When the radius of the diffuse reflection surface is more than 4.0 mm, the angle measurement system shows an evident nonlinearity trend. As the radius of the diffuse reflection surface increases, the variation in the differential optical power with the angle value increases. When the radius of the diffuse reflection surface is between 3.0 mm and 4.0 mm, there is a point of minimum nonlinearity. Figure 11b illustrates nonlinearity changes for different radii of the diffuse reflection surface.

The simulated calculation results show that the system's minimum nonlinearity is 0.81% when the radius of the diffuse reflection surface is 3.38 mm. The experimental tests were conducted around 3.38 mm. When the radius is 3.3 mm, nonlinearity is 0.74%; when the radius is 3.4 mm, the nonlinearity is 0.92%. Both Figures 8 and 11a show that the experimental results are consistent with the simulation results.

In summary, the mathematical models (16) and (17) prove the correctness and feasibility of this angle measurement method based on the photodiodes' differential optical power through simulated and experimental results. The experimental results are highly consistent with the simulation results. Therefore, this scheme is effective for the design of the angle measurement system of an FSM.

The diffuse reflection surface with a radius of 3.3 mm was analyzed. The AD value was tested in the range of $-20°$ to $20°$, and linear fitting was performed to obtain the measurement line and the fitting line, as shown in Figure 12.

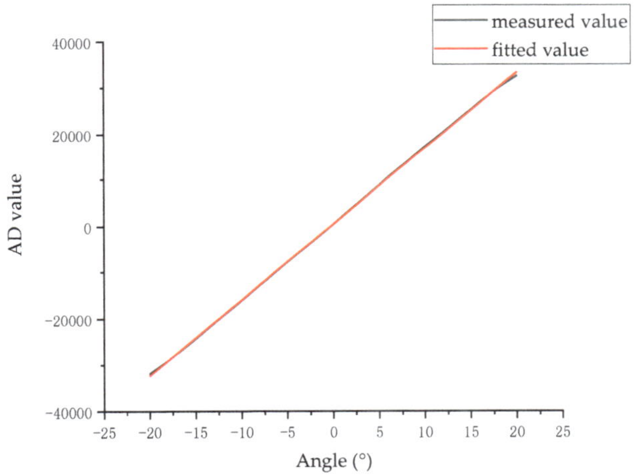

Figure 12. Measured values and fitting curves.

The fitting function of the angle value and the differential AD value in the angle range of $\pm 20°$ is

$$F(\theta) = 1641 \cdot \theta + 482.7$$

The resolution of the angle measurement system is 2.3″. Combined with Figure 12, it can be seen that the fitting function is approximately coincident with the measured value, and the R2 value is 0.9999.

5. Conclusions

This paper investigates the working principle of diffuse reflection and analyzes the advantages and disadvantages of an angle measurement system for an FSM. It also proposes a solution for achieving large-angle measurement using diffuse reflection. A differential optical power angle measurement model which is applicable to various FSMs is established. The structural variables that actually affect the measurement results are clarified in the model. It also proposes an optimized method for nonlinearity in angle measurement. Through experiments and simulations, the validity of the model is proved. The model plays an auxiliary role in the structural design of an angle measurement system of a large-angle FSM. We also designed a testing apparatus to test the range, nonlinearity, and accuracy of the angle measurement system. The testing results demonstrate that this proposed system can achieve measurements within a $\pm 20°$ angle range, a resolution of 2.3″, and an optimized nonlinearity of 0.74%. Compared to other 2D FSM angle measurement systems, it reaches larger measurement ranges, smaller nonlinearity, and is suitable for various FSM applications.

Author Contributions: Conceptualization, B.D. and Y.L. (Yang Liu); methodology, B.D.; software, L.L.; validation, B.D.; formal analysis, B.D.; investigation, B.D.; resources, B.D.; data curation, B.D.; writing—original draft preparation, B.D.; writing—review and editing, Y.L. (Yong Lv); visualization, Y.L. (Yong Lv); supervision, Y.L. (Yong Lv); project administration, Y.L. (Yong Lv); funding acquisition, Y.L. (Yong Lv). All authors have read and agreed to the published version of the manuscript.

Funding: This research was funded by the National Natural Science Foundation of China, grant number 51775034.

Institutional Review Board Statement: Not applicable.

Informed Consent Statement: Not applicable.

Data Availability Statement: No new data were created or analyzed in this study. Data sharing is not applicable to this article.

Acknowledgments: The authors would like to thank Ruikongxin Technology Co., Ltd., Beijing, for providing the testing apparatus. Thanks for the help of my girlfriend.

Conflicts of Interest: The authors declare no conflict of interest.

References

1. Shinshi, T.; Shimizu, D.; Kodeki, K.; Fukushima, K. A Fast Steering Mirror Driven by Voice Coil Motors and Supported by Magnetic Suspension. In Proceedings of the 23rd International Conference on Mechatronics Technology (ICMT), Salerno, Italy, 23–26 October 2019.
2. Wang, L.; Liu, X.L.; Wang, C.Y. Modeling and design of fast steering mirror in image motion compensation for backscanning step and stare imaging systems. *Opt. Eng.* **2019**, *58*, 9. [CrossRef]
3. Xiantao, L.; Li, B.T.; Wang, L.; Li, J.Z.; Zhou, Z.M.; Li, C.; Zhang, B. Negative Effect Compensation of Disturbance for High-Precision Voice Coil Fast Steering Mirror. *Math. Probl. Eng.* **2022**, *2022*, 13. [CrossRef]
4. Bohkman, E.; Burnashev, M.; Filatov, Y.; Pavlov, P. Implementation of the dynamic laser goniometer for noncontact measurement of angular movement. *Opt. Eng.* **2016**, *55*, 5. [CrossRef]
5. Bokhman, E.; Burnashev, M.; Filatov, Y.; Pavlov, P. Results of implementation of the dynamic laser goniometer for non-contact measurement of angular movement. In Proceedings of the Conference on Electro-Optical Remote Sensing, Photonic Technologies, and Applications IX, Toulouse, France, 21–22 September 2015.
6. Chen, W.; Chen, S.H.; Luo, D. Design and experimental investigations of a two-dimensional laser scanner based on piezoelectric actuators. *Opt. Eng.* **2015**, *54*, 6. [CrossRef]
7. El-Gohary, M.; McNames, J. Human Joint Angle Estimation with Inertial Sensors and Validation with A Robot Arm. *IEEE Trans. Biomed. Eng.* **2015**, *62*, 1759–1767. [CrossRef] [PubMed]
8. Fan, M.W.; Huang, L.H.; Li, M.; Rao, C.H. A fast high voltage driver for the piezoelectric fast steering mirror. In Proceedings of the 20th International Symposium on High Power Laser Systems and Applications (HPLS and A), Chengdu, China, 25–29 August 2014.
9. Jia, X.D.; Wan, Q.H.; Yu, H.; Lu, X.R.; Zhao, C.H. Small-sized visual angular displacement measurement technology. *Measurement* **2019**, *135*, 406–412. [CrossRef]
10. Konyakhin, I.; Hoang, V.P.; Artemenko, Y.; Li, R.P.; Smekhov, A. Optic-electronic system for measuring the three-dimensional angular deformation of pipe sections at large constructions. In Proceedings of the Conference on Optical Measurement Systems for Industrial Inspection IX, Munich, Germany, 22–25 June 2015.
11. Padma, S.; Umesh, S.; Pant, S.; Srinivas, T.; Asokan, S. Fiber Bragg Grating based Tunable Sensitivity Goniometer. In Proceedings of the Conference on Photonic Instrumentation Engineering III, San Francisco, CA, USA, 17–18 February 2016.
12. Rezende, A.; Alves, C.; Marques, I.; Silva, M.A.; Naves, E. Polymer Optical Fiber Goniometer: A New Portable, Low Cost and Reliable Sensor for Joint Analysis. *Sensors* **2018**, *18*, 4293. [CrossRef] [PubMed]
13. Vikas, V.; Crane, C.D. Joint Angle Measurement Using Strategically Placed Accelerometers and Gyroscope. *J. Mech. Robot.* **2016**, *8*, 7. [CrossRef]
14. Yu, H.; Jia, X.D.; Wan, Q.H.; Zhao, C.H.; Sun, Y. High-resolution angular measurement arithmetic based on pixel interpolations. *Measurement* **2020**, *149*, 5. [CrossRef]
15. Fan, Y.; Ma, W.L.; Jiang, P.; Huang, J.L.; Chen, K.W.; Pan, N.A. Improving Angular Accuracy of a Scanning Mirror Based on Error Modeling and Correction. *Sensors* **2019**, *19*, 367. [CrossRef]
16. Lin, D.; Wu, Y.M.; Zhu, F. Research on Precision Tracking on Fast Steering Mirror and Control Strategy. In Proceedings of the International Conference on Power and Energy Engineering (ICPEE), Ottawa, ON, Canada, 13–15 September 2017.

17. OIM101-One Inch FSM. Available online: https://www.opticsinmotion.net/OIM101%20INFO%20Update%2010-08-10.pdf (accessed on 10 August 2010).
18. Wu, X.; Chen, S.; Chen, W.; Yang, M.; Fu, W. Large angle and high linearity two-dimensional laser scanner based on voice coil actuators. *Rev. Sci. Instrum.* **2011**, *82*, 105103. [CrossRef] [PubMed]
19. Ni, Y.X.; Wu, J.B.; San, X.G.; Gao, S.J.; Ding, S.H.; Wang, J.; Wang, T. Deflection angle detecting system for the large-angle and high-linearity fast steering mirror using quadrant detector. *Opt. Eng.* **2018**, *57*, 9. [CrossRef]
20. Hung, S.K.; Chung, Y.H.; Chen, C.L.; Chang, K.H. Optoelectronic Angular Displacement Measurement Technology for 2-Dimensional Mirror Galvanometer. *Sensors* **2022**, *22*, 872. [CrossRef]
21. Shan, M.G.; Min, R.; Zhong, Z.; Wang, Y.; Zhang, Y.B. Differential reflective fiber-optic angular displacement sensor. *Opt. Laser Technol.* **2015**, *68*, 124–128. [CrossRef]

sensors

MDPI

Article

A Low-Delay Dynamic Range Compression and Contrast Enhancement Algorithm Based on an Uncooled Infrared Sensor with Local Optimal Contrast

Youpan Zhu [1,2], Yongkang Zhou [2,3], Weiqi Jin [1,*], Li Zhang [2], Guanlin Wu [1] and Yiping Shao [4]

[1] School of Optics and Photonics, Beijing Institute of Technology, 5 South Zhongguancun Street, Haidian District, Beijing 100081, China; zhuypbit@aliyun.com (Y.Z.)
[2] Kunming Institute of Physics, No. 31, Jiaochang East Road, Wuhua District, Kunming 650221, China; zyk1120102464@163.com (Y.Z.)
[3] School of Life Science, Beijing Institute of Technology, 5 South Zhongguancun Street, Haidian District, Beijing 100081, China
[4] College of Mechanical Engineering, Zhejiang University of Technology, Hangzhou 310023, China
* Correspondence: jinwq@bit.edu.cn

Abstract: Real-time compression of images with a high dynamic range into those with a low dynamic range while preserving the maximum amount of detail is still a critical technology in infrared image processing. We propose a dynamic range compression and enhancement algorithm for infrared images with local optimal contrast (DRCE-LOC). The algorithm has four steps. The first involves blocking the original image to determine the optimal stretching coefficient by using the information of the local block. In the second, the algorithm combines the original image with a low-pass filter to create the background and detailed layers, compressing the background layer with a dynamic range of adaptive gain, and enhancing the detailed layer for the visual characteristics of the human eye. Third, the original image was used as input, the compressed background layer was used as a brightness-guided image, and the local optimal stretching coefficient was used for dynamic range compression. Fourth, an 8-bit image was created (from typical 14-bit input) by merging the enhanced details and the compressed background. Implemented on FPGA, it used 2.2554 Mb of Block RAM, five dividers, and a root calculator with a total image delay of 0.018 s. The study analyzed mainstream algorithms in various scenarios (rich scenes, small targets, and indoor scenes), confirming the proposed algorithm's superiority in real-time processing, resource utilization, preservation of the image's details, and visual effects.

Keywords: infrared image; dynamic range compression; local optimal contrast; real-time imaging

1. Introduction

Because of its advantages for passive imaging and night imaging, infrared imaging systems have been widely used in military, aerospace, security, and other fields. It has always been a goal to obtain high-quality imaging results. Infrared imaging systems typically use 14-bit or higher ADC (analog to digital converter) acquisition circuits to gather information from an infrared scene with a high dynamic range, while the majority of image display devices are 8-bit displays with a low dynamic range. This may result in some details being lost during the display process. To combine the application scenarios of infrared imaging devices in order to better adapt them to the visual characteristics of the human eye, we need to perform high-dynamic-range compression of the image. Of course, the loss of information from the image in this process is inevitable, but different processing methods will exhibit different details, and the effect of the processing method depends mostly on human vision. In addition, in recent years, uncooled infrared sensor technology has been significantly improved, and the applications have become increasingly widespread, especially in civilian products. Therefore, how to compress images with a high dynamic range into those with a low dynamic range while preserving the majority of the detailed information is one of the critical technologies for infrared systems; at the same time, low-delay image processing is also essential for back-end target detection and tracking.

Contrast refers to the measurement of different brightness levels between the brightest white and the darkest black in the light and dark areas of an image. The greater the difference, the greater the contrast, and the smaller the difference, the lower the contrast. It can be global or local (i.e., concentrated in a small area).

Appropriate local contrast can make the image present a concave and convex three-dimensional sense, which can improve the visual effect of the human eye. In addition, the enhancement of local contrast can minimize the halo effect, making the picture clearer.

To address the problem of dynamic range compression, many researchers have developed algorithms for displaying high dynamic ranges and enhancing details [1]. Overall, high-dynamic-range image compression algorithms can be classified into traditional mapping-based algorithms, gradient domain-based compression algorithms, and image layering-based compression algorithms. Besides these, there are also a few methods [2,3] that cannot be classified into any of these three classes. Different algorithms are applicable for various application contexts and vary in their complexity and performance.

Related Work

Mapping-based high-dynamic-range IR (infrared radiation) image enhancement algorithms are the simplest and most widely used, and the visual effect of the human eye is obviously improved. These algorithms include self-gain-based linear mapping, Gamma curve correction, histogram projection, etc. In order to make the distribution of the histogram as uniform as possible, the earliest method of histogram equalization redistributed the image's grayscale using a cumulative distribution function. A uniform probability density distribution can be obtained using this method, but there are drawbacks as well, including overenhancement, higher noise levels, the loss of some details, and fading. To solve these issues, a threshold-based plateau histogram equalization (PHE) algorithm was proposed in the literature [4]. The plateau histogram modifies the ordinary histogram algorithms by adding a threshold. When the density of each gray value exceeds the threshold, it will be processed, thereby improving the global contrast and avoiding a small number of outliers affecting the global distribution of the gradient. The literature [5] has proposed an adaptive histogram equalization algorithm (AHE), which computes a histogram function based on a local window and can enhance the local contrast of an image while further highlighting its finer details. However, the AHE algorithm is prone to creating a great deal of noise. In light of this, a contrast-limited adaptive histogram equalization algorithm (CLAHE) was proposed in the literature [6], which reduces the amplitude of the noise with a clipping threshold. Some researchers [7,8] have also made some improvements to the CLAHE algorithm. Generally, the self-gain-based equalization algorithms perform noticeably worse than the histogram-based equalization algorithms. Nevertheless, the histogram-based algorithms lack flexibility in some particular applications and are also susceptible to overenhancement, block effects, etc., because they are entirely dependent on the histogram's statistics.

A gradient dynamic range compression algorithm compresses the large-scale gradient and keeps the low-amplitude information. Fattal [9] proposed an algorithm as the framework of the gradient-domain-based high-dynamic-range image compression algorithm (GDHDRC). Based on this, a detail-preserving algorithm (GDHDRC-DPS) was proposed, and further research [10–13] refined the GDHDRC-DPS algorithms even more. On most occasions, these algorithms produce rather acceptable outcomes; however, there are significant limitations in certain occasions. On the one hand, the factors used to smooth the data items and gradient items need to be properly designed in order to prevent oversaturation; on the other hand, the details of the output images are still not sharp enough to meet some of the requirements of displaying infrared images with a HDR (high dynamic range).

The algorithms based on image layering [14–27] decompose the original image into various components, such as the background layer of the image and the detailed layer of the image, and then process each component separately; afterwards, the compressed background layer image and the stretched detailed layer image are reintegrated into an image in which the dynamic range is compressed and the detailed layer image is enhanced. The basic framework is shown in Figure 1, where the original image is decomposed into a background layer image and a detailed layer image after being processed by a low-pass filter; the background layer image undergoes dynamic range compression, and the detailed layer image is enhanced. A final image with a low dynamic range is obtained after re-fusing. Although the choice of algorithms can differ for low-pass filtering, the background layer, the detailed layer, and image fusion, they are basically improved and promoted in this framework.

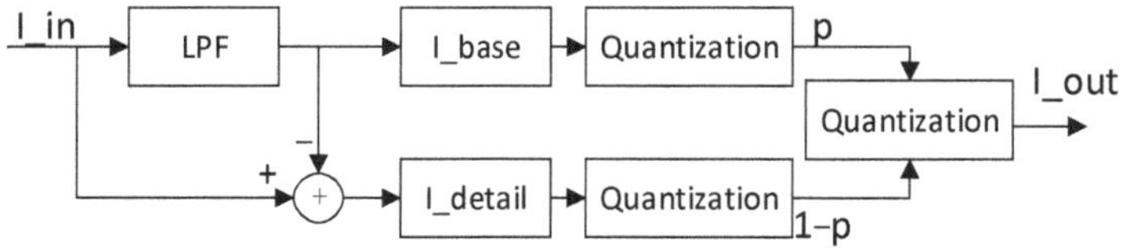

Figure 1. Framework of the dynamic range compression algorithm based on image layering. LPF, low-pass filter; I_base, the background layer image; I_detail, the detailed layer image; p, parameter.

Specifically, an infrared image enhancement algorithm based on bilateral filtering (bilateral filter and dynamic range partitioning; BF-DRP) was proposed in the literature [14]. The algorithm divides the original image into a background layer image and a detailed layer image using bilateral filtering, then performs Gamma curve processing (compression or expansion, respectively) on the background layer image and the detailed layer image, and finally reintegrates the background layer and the detailed layer image to obtain the detail-enhanced image. The bilateral filter is a nonlinear filter that takes both the spatial proximity and the pixels' grayscale difference into account, smoothing out uniform areas of the input image such as the sky and sea while preserving strong edges such as floors. However, the drawbacks of the BF-DRP algorithm are also obvious: (1) several parameters of the algorithm require fine-tuning to achieve better enhancement; (2) the original image minus the image after bilateral filtering is used to obtain a detailed layer image, the edges of which will be sharper than the edges of the original image or even have gradient inversion, while the highlighted noise appears in flat areas [15,16].

Based on this, an improved version of the BF-DRP algorithm, namely the bilateral filter and digital detail enhancement (BF-DDE) algorithm, was proposed in the literature [17,18], with the goal of obtaining a corrected background layer image that was closer to the original image. In addition, research on the human visual system [17] showed that human eyes are more sensitive to noise in uniform regions than in complex regions. The literature [19] first described the noise mask function, and the literature [20] used a noise mask function based on local variance, both of which clearly improved the image's overall contrast and improved the information on the target and the detail. The algorithms based on a bilateral filter have two problems: (1) the detailed layer image obtained with the bilateral filter is prone to gradient flipping because the mean filter method based on Gaussian weights is unstable when one pixel has a large difference from its adjacent pixels; (2) the computational complexity of bilateral filter is O(Nr2), and as the filter window increases, the computational time will increase quadratically. To solve these problems, a linear filter, i.e., a guided filter (GF), for the detail-enhancement algorithm was proposed in the literature [21], while the studies of both [22,23] introduced an algorithm for enhancing images with a high dynamic range (GF-DDE) based on a guided filter. The guided filter was used instead of the bilateral filter in the GF-DDE method, which significantly decreased the algorithm's computing complexity but worsened its edge retention. An algorithm for enhancing images with a high dynamic range (LEPF-DDE) based on a local edge-preserving filter was proposed by [24]. The local edge-preserving filter (LEPF) was used to separate the original image into a background layer image and one or more detailed layer images. Next, the multi-scale background layer image undergoes maximum entropy-based Gamma curve correction, and the detailed layer undergoes of elimination of the artifacts and amplification of the detail. Finally, the detailed layer image and the background layer image are re-bounded. An edge-preserving filter algorithm called LEPF has been proposed in the literature [24,25], which aimed to filter low-amplitude noise while maintaining strong edges.

On the whole, the traditional image mapping method has a general effect on the retention of detail and image contrast, in which the gradient compression algorithm retains the details better and the contrast is enhanced, but the parameters need to be carefully adjusted for different scenes. The algorithm based on image layering results in greater improvement in the retention of detail and noise suppression, but traditional methods are still used in the background layer, and the contrast enhancement effect is general and can easily cause halos and other problems.

The DRCE-LOC algorithm proposed in this study focused on improving image delay, storage resources, FPGA-based implementation, and the trade-off between global contrast and local contrast while applying the framework of dynamic range compression based on image layering (Figure 1).

At the same time, the algorithm was optimized and implemented on an FPGA (field-programmable gate array). For the low-delay image processing requirements of offline applications [26], it is often necessary to design a hardware processing method based on an FPGA. This is different from many CPU (central processing unit)-based algorithms. Multiplication can be implemented on a parallel pipelined FPGA with only one clock cycle, but exponential calculation that is simple on a CPU is difficult to implement on an FPGA. Therefore, it is necessary to effectively utilize the characteristics of the FPGA to study the design and improvement of algorithms. We conducted experiments on an infrared imager equipped with the FPGA and uncooled infrared sensors and evaluated the effect of the algorithm.

2. Methods

2.1. DRCE-LOC

An uncooled infrared sensor does not need a refrigeration device and can work at room temperature. It has many advantages, such as fast start-up, low power consumption, small size, light weight, long life, and low cost. In recent years, such sensors have been widely used in military and civil night vision products. The proposed algorithm was applied to an infrared imager equipped with 1024×768 uncooled infrared sensors, and all experiments were carried out on this system.

The framework of the DRCE-LOC algorithm is shown in Figure 2. Its basic principle is as follows. Firstly, the brightness value and contrast value are re-corrected in each local block to achieve the best possible local contrast. Secondly, the Gaussian model [28] is used to smooth the contrast value of the local block to eliminate the block effect. Then, a brightness adjustment algorithm guided by the global brightness is applied. The algorithm uses the background layer obtained by the guided filter to obtain the brightness-guided image after passing through the global image. This step maintains the brightness relationship of the original image as much as possible. Finally, the detailed layer obtained via the guided filter and the image obtained after brightness and contrast correction are re-fused to obtain the final image. The process of implementing the algorithm includes the following eight steps.

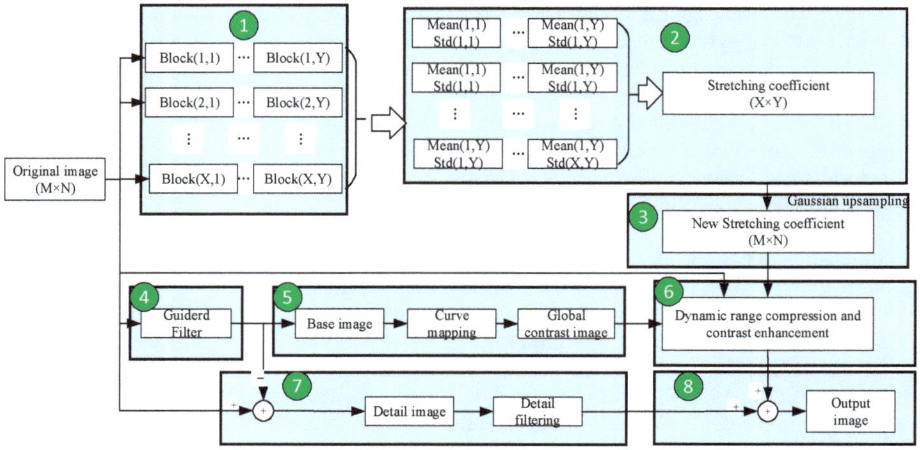

Figure 2. Framework of the DRCE-LOC algorithm.

- Step 1: Image segmentation.

To reduce the amount of storage, the original image I_{in} (with a size of $M \times N$) is divided into $X \times Y$ local blocks. The size of each local block is $\frac{M}{X} \times \frac{N}{Y}$. The mean value ($Block_{mean}$) and its standard deviation are calculated. $Block_{mean}$ reflects the level of brightness of local blocks in the original image, and $Block_{std}$ reflects the level of richness of the details of local blocks in the original image. The sizes of $Block_{mean}$ and $Block_{std}$ are both $X \times Y$.

- Step 2: Calculation of the local block's stretching coefficient.

The stretching coefficient of each local block is determined by the standard deviation of each local block, as shown in Equation (1)

$$Stretch_{para} = \frac{255}{Block_{std} + Bias} \tag{1}$$

where $Block_{str}$ is the stretching coefficient of the local block and $Bias$ is a constant set to limit the coefficient of gain from being too large. After processing, the size of $Block_{str}$ is $X \times Y$.

- Step 3: Upsampling of the parameters of the local block.

$Stretch_{para}$ and $Block_{str}$ are upsampled into $M \times N$ dimensions, the basic principle of which is shown in Figure 3. The stretching parameter and the mean value of each local block are projected to the center of the local block of the original image, and the stretching parameter and mean parameter are calculated at the position of each pixel in the original image. They are calculated from the parameters of each local block and the Gaussian weights of the distance to the center of each local block.

$$NStretch_{para}(i,j) = \frac{\sum_{x=1}^{X}\sum_{y=1}^{Y} Stretch_{para}(x,y) * \exp(-\frac{{d_{(i,j)(x,y)}}^2}{\delta^2})}{\sum_{x=1}^{X}\sum_{y=1}^{Y} \exp(-\frac{{d_{(i,j)(x,y)}}^2}{\delta^2})} \tag{2}$$

$$NMean_{para}(i,j) = \frac{\sum_{x=1}^{X}\sum_{y=1}^{Y} Block_{mean}(x,y) * \exp(-\frac{{d_{(i,j)(x,y)}}^2}{\delta^2})}{\sum_{x=1}^{X}\sum_{y=1}^{Y} \exp(-\frac{{d_{(i,j)(x,y)}}^2}{\delta^2})} \tag{3}$$

where $NStretch_{para}$ and $NMean_{para}$ are the stretching parameter and the mean parameter after upsampling, respectively, and their sizes are both $M \times N$; δ^2 is a constant used to control the Gaussian weights; and $d_{(i,j)(x,y)}$ represents the distance from the coordinates of the original image (i,j) to the coordinates of the local block's center (x,y).

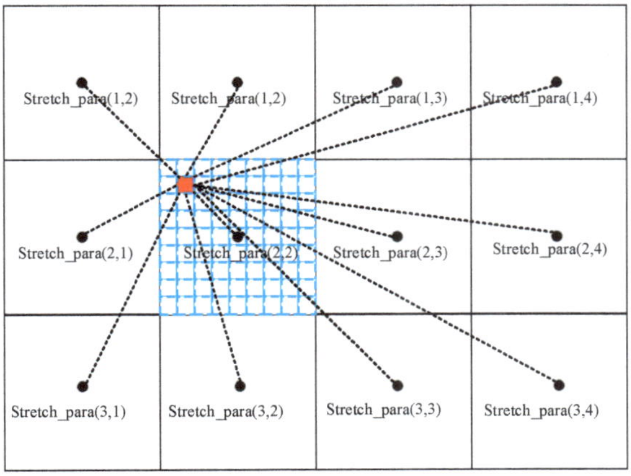

Figure 3. Schematic of the image upsampling process.

- Step 4: Layering of the guided filter.

Guided filters [19] are edge-preserving linear filters that have been widely used in image enhancement. Here, they are used to divide the original image into a background layer image and a detailed layer image. We assumed that the input image is p, the output image is q, and the guide image is I. The fundamental idea is to combine an original image with a guide image to produce a filtered image that resembles the guide image. The equation is displayed in Equation (4)

$$\begin{cases} q = a_k I + b_k, \forall i \in w_k \\ a_k = \frac{\overline{p_k I_k} - \overline{p_k}\,\overline{I_k}}{\overline{I_k^2} - \overline{I_k}^2 + \varepsilon} = \frac{\frac{1}{|w|}\sum_{i \in w_k} I_i p_i - u_k \overline{p_k}}{\sigma_k^2 + \varepsilon} \\ b_k = \overline{p_k} - a_k u_k \end{cases} \tag{4}$$

where w_k is a square window with a diameter of w with k as the center; $|w|$ represents the number of the pixels in the window; a_k and b_k are linear coefficients and are constants in the local window w_k; and i denotes the number of pixels in w_k. It should be ensured that the linear coefficient satisfies the formula above so that the difference between q and p is minimized.

- Step 5: Calculation of the brightness guide map.

In order to increase the computational speed of the FPGA and maintain the light–dark relationship of the original image, a linear mapping method is used. Gamma curves, histogram equalization, and other methods can be used in some specialized applications. The mapping method is shown in Equation (5)

$$I_{gc}(i,j) = \frac{255 * (I_{in}(i,j) - \overline{I_{in}})}{Std(I_{in}) + \lambda} + Bright \tag{5}$$

where $Std(I_{in})$ is the standard deviation of the original image, $Bright$ is the desired brightness of the image, λ is a constant to avoid excessive gain, and $I_{gc}(i,j)$ is the mapped 8-bit global brightness guide.

- Step 6: Dynamic range compression and contrast enhancement.

Four factors are needed for dynamic range compression and contrast enhancement: I_{in}, I_{gc} , $NMean_{para}$, and $NStretch_{para}$. The corresponding equation is

$$Base_{out}(i,j) = k1 \times NStretch_{para} \times \left[I_{in}(i,j) - NMean_{para}(i,j)\right] + k2 \times I_{gc} \tag{6}$$

where $Base_{out}(i,j)$ is the output image after contrast enhancement, $k1$ is the parameter that controls the local contrast, $k2$ is the parameter that controls the global contrast, I_{gc} is the brightness-guided image calculated in Step 5, and $NMean_{para}$ and $NStretch_{para}$ are the local mean parameters and stretching parameters calculated in Step 3, respectively.

- Step 7: Filtering of the detailed layer.

The detailed layer obtained through the previous steps contains noise. A noise mask function based on human vision [17] is utilized to reduce noise. The basic principle of the function is that the human eye is sensitive to noise in a uniform scene but is insensitive to noise in areas with large amounts of detail.

$$Detail_{out} = Detail_{image} \times |a_k| * [g_L + (g_H - g_L)] \tag{7}$$

where $Detail_{image}$ is the detailed layer image obtained after layering of the guided filter and a_k is the parameter that was calculated during the process of layering the guided filter, which reflects the quantity of local information.

- Step 8: Image enhancement.

The final image is obtained by re-integrating the contrast-enhanced image and the detailed layer image.

$$I_{out} = Base_{out} + DDE \times Detail_{out} \tag{8}$$

where I_{out} is the final output image and DDE is the detail enhancement factor.

2.2. Noise Suppression

The algorithm includes two sources of noise: noise in the detailed layer and noise in the background layer. The detailed layer's noise adopts the principle of [17]. The over-enhancement of the local contrast, which is primarily the source of the noise in the background layer, is suppressed by *Bias* in Equation (1). In order to analyze the influence of stretching the background layer on local noise, we set $I_{gc} = 128$. At this point, the stretching of the image in terms of global contrast is 0, and all the noise in the image comes from the stretching of local contrast. The algorithm next divides the 1024 × 768 infrared images into 4 × 3 squares and set δ^2 to 3600 (the selection of this parameter is introduced in the next section). Then, the effect of *Base_out* imaging under different values of bias is shown in Figure 4.

AGC (adaptive gain control) is a method of compressing high dynamic ranges that maps the dynamic range of an input image to a specified dynamic range. Since a 14-bit image cannot be visualized on most display devices, we considered the image processed by AGC as the input image and compared it with images processed by other algorithms.

Figure 4 illustrates how the final image's noise level in terms of contrast varied significantly with various biases. The image in the uniform area of the scene was overamplified when $Bias = 0$, producing a significant amount of noise. The noise in the uniform region steadily reduced with an increase in the bias, while the local contrast of the image diminished with the noise. The image's noise was smaller and the local contrast performed better when the bias was around 256.

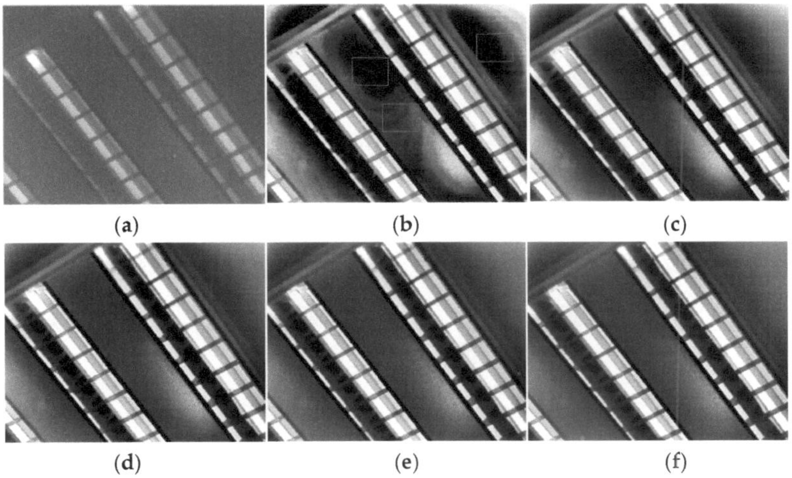

Figure 4. Effects of local noise suppression. (**a**) AGC; (**b**) I_{gc} = 128, bias = 0; (**c**) I_{gc} = 128, bias = 128; (**d**) I_{gc} = 128, bias = 256; (**e**) I_{gc} = 128, bias = 384; (**f**) I_{gc} = 128, bias = 512.

2.3. Halo Suppression and Contrast Control

A halo is a virtual shadow that extends from the edge of the image, which is a problem that can easily happen during image blocking. Suppressing halos can keep the image in its original state. In this study, we combined local and global information, but the halo problem could also exist in the case of improper parameter selection. In Equation (6), $k2$ was used to control the halo problem and the global contrast. When $k2 = 0$, the algorithm became a stretching algorithm with pure local contrast; as $k2$ increased, the weight of global contrast in the final image gradually increased. In particular, in Equation (6), the mean component of the local stretched image was eliminated rather than the algorithm just adding the local stretched image and the global image. Therefore, the global image plays a role in guiding the global contrast in Equation (6). The final image's global contrast will be closer to the global compressed image as $k2$ increases, and the halo will be weaker; however, better local contrast can be achieved by controlling $k1$. Figure 5 shows the effect of the algorithm under different values of $k2$ ($k1 = 1$; the basis for selecting this parameter is presented in the next section). The red rectangles in the images facilitated a comparison of the halo suppression effect under different values of k. It can be seen from Figure 5 that when $k2$ is large, the contrast of the final image was excessively high and some local information was even obscured, as shown in Figure 5a. As $k2$ increased, the overall contrast of the image steadily declined, and when $k2 = 0.1$, the brightness of the local blocks of the image was nearly the same, losing the overall brightness and the blackness of the original image. When $k2 = 0.1$, if a continuous video sequence enters the algorithm module proposed in this study, a transformed halo will develop as a result of the alteration of the local information. It can be seen from Figure 5d that when $k2 = 0.5$ or so, the final image is nearly identical to the global contrast image, and the halo phenomenon almost disappears, whereas improved global contrast and local contrast were achieved.

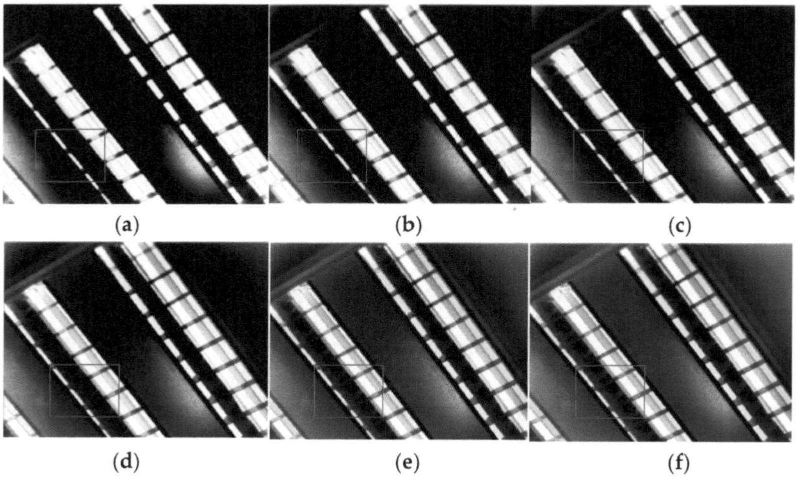

Figure 5. The halo suppression effect under different values of $k2$. (**a**) $k2 = 1.5$; (**b**) $k2 = 1$; (**c**) $k2 = 0.8$; (**d**) $k2 = 0.5$; (**e**) $k2 = 0.2$; (**f**) $k2 = 0.1$.

2.4. Parameter Settings

The choice of parameters determines how well the algorithm performs. The selection of the parameters in the algorithm was as follows:

$\frac{M}{X} \times \frac{N}{Y}$ is the local block size of the image. In general, the smaller the local block size is, the better the local contrast will be. The computational effort will rise as the local block size falls, since more local blocks are required. Tiny squares also have a tendency to bring about localized halo alterations because of the movement of items in real-world settings. For continuous video, a value of $\frac{M}{X} \times \frac{N}{Y} = 64 \times 64$ achieves better results.

Bias is used to suppress the local overenhancement phenomenon, which is described in Section 2.2, and generally achieves better results when it is between 128 and 384.

The parameter δ^2 is used to control the Gaussian kernels' curvature during downsampling and upsampling of the image. The selection of δ^2 is strongly influenced by the size of the image block. Better results are achieved when $\delta^2 \geq \frac{M}{X} \times \frac{N}{Y}$, which eliminates the obvious block effect that may be present in the image.

The parameter λ is set to prevent the global contrast from being overenhanced. It performs the same function as bias, with the exception that bias controls the local contrast while the global contrast is controlled when the parameter is modified. Similarly, better results are achieved when λ is between 128 and 384.

The parameter $k1$ is used to control the local contrast. Since Equation (2) already places a limit on the range of the image, higher results can be achieved when $k1 = 1$.

The parameter $k2$ is used to suppress halos and control the global contrast. A detailed description is given in Section 2.3. Better results can be achieved when $k2$ is in the range of 0.2 and 0.7.

2.5. Complexity of the Algorithm

The algorithm proposed in this study is built on a framework based on guided filter layering, which does not make the original guided filter any more computationally complex. Similarly, the computational complexity of the algorithm proposed in this study is O(n) and depends solely on the number of pixels. However, the following improvements can be applied during the construction of the algorithm, particularly for real-time videos, to further reduce the method's resource consumption:

- Calculating the local information of the image local blocks, which can be achieved by using the local information of the data of the previous image of consecutive frames, thus achieving an algorithm with a delay of less than one frame while storing the complete image frame differently.
- In the calculations of upsampling and downsampling of the image, the Gaussian kernel can be saved in advance as a parameter to avoid exponential calculations during the execution of the algorithm.

3. Results and Discussion

3.1. Quantitative Assessment

To analyze and evaluate the effect of the algorithm, four image indicators, namely the root mean square difference [29] (root mean square, RMS), image entropy [30], structural similarity [31] (SSIM), and the Tenengrad clarity index [32], and the algorithm's running time were used. For calculating the running time, the experiment was carried out in the same environment as follows: operating system, Windows 11; CPU, 12th Gen Intel (R) Core (TM) i9-12900HX 2.30 GHz; RAM, 32.0 GB.

RMS is defined as

$$RMS = \sqrt{\frac{1}{MN}\sum_{i=1}^{M}\sum_{j=1}^{N}\left(I(i,j) - \overline{I}\right)^2} \tag{9}$$

where I denotes the image to be evaluated, $\overline{I}$ represents the mean value of the image, M is the width of the image, and N is the height of the image. The larger the RMS, the better the overall contrast of the image.

Image $entropy$ is defined as

$$Entropy = -\sum_{i=0}^{255} p_i \log p_i \tag{10}$$

where p_i represents the probability of each grayscale value. Greater entropy indicates that the grayscale has been stretched further.

$SSIM$ is defined as

$$SSIM(I, ref) = \frac{(2u_I u_{ref} + C_1)(2\sigma_{ref} + C_2)}{(u_I{}^2 + u_{ref}{}^2 + C_1)(\sigma_I{}^2 + \sigma_{ref}{}^2 + C_2)} \tag{11}$$

where ref represents the reference image, I is the image obtained by stretching the original image with adaptive gain, u represents the mean value, σ represents the standard deviation, and C_1 and C_2 are very small numbers to prevent the denominator from becoming 0. A larger $SSIM$ indicates that the two images are more similar in structure.

Tenengrad is a gradient-based function that extracts the gradient values in horizontal and vertical directions through the Sobel operator [33]. It is defined as

$$S(i,j) = \sqrt{G_x * I(i,j) + G_y * I(i,j)} \tag{12}$$

$$Tenengrad = \frac{1}{MN}\sum_{i=1}^{M}\sum_{j=1}^{N} S(i,j)^2 \tag{13}$$

$$G_x = \frac{1}{4}\begin{bmatrix} -1 & 0 & 1 \\ -2 & 0 & 2 \\ -1 & 0 & 1 \end{bmatrix}, G_y = \frac{1}{4}\begin{bmatrix} 1 & 2 & 1 \\ 0 & 0 & 0 \\ -1 & -2 & -1 \end{bmatrix} \tag{14}$$

where $S(i,j)$ represents the gradient of Image I at the point (i,j); G_x and G_y represent the Sobel convolution kernels in the horizontal direction and vertical direction, respectively. A larger Tenengrad means a clearer image.

To evaluate the processing effect of the DRCE-LOC algorithm in this study, we used the typical AGC, HE, CLAHE, GF-DDE, and BF-DRP algorithms as a comparison in three typical scenarios: a rich scene, a scene with a small target, and an indoor scene. The following can be seen from Figures 6–8.

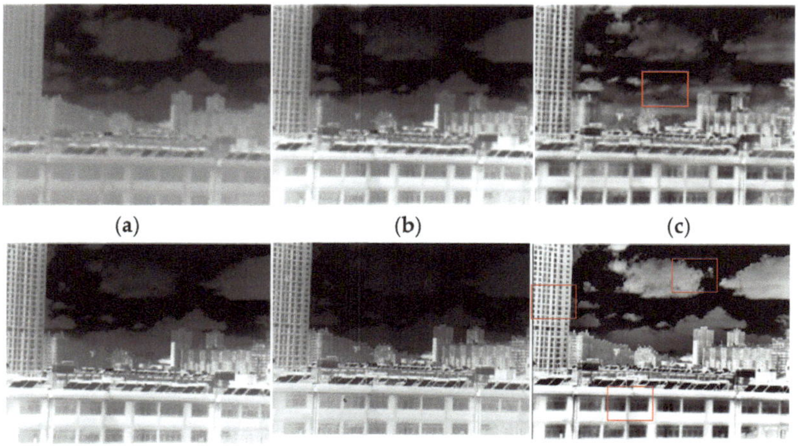

(a) (b) (c)

(d) (e) (f)

Figure 6. Comparison of the effects for the rich scene (Scene 1). (**a**) AGC; (**b**) HE; (**c**) CLAHE; (**d**) GF-DDE; (**e**) BF-DRP; (**f**) proposed method.

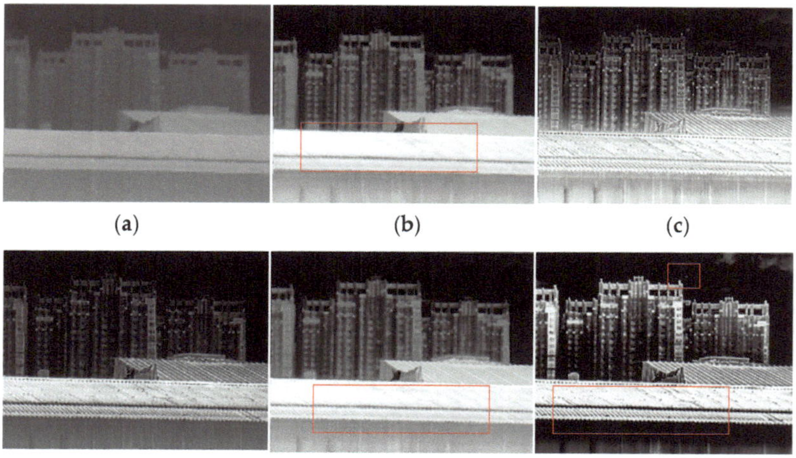

(a) (b) (c)

(d) (e) (f)

Figure 7. Comparison of the effects for the scene with a small target scene (Scene 2). (**a**) AGC; (**b**) HE; (**c**) CLAHE; (**d**) GF-DDE; (**e**) BF-DRP; (**f**) proposed method.

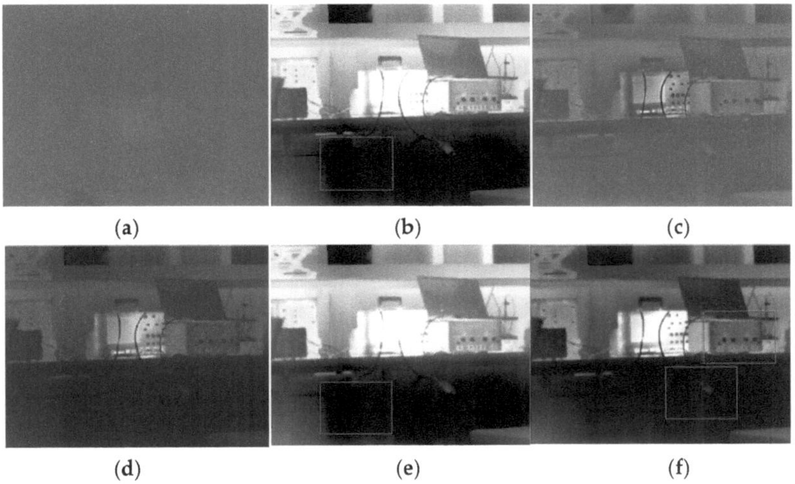

Figure 8. Comparison of the effects for the indoor scene (Scene 3). (a) AGC; (b) HE; (c) CLAHE; (d) GF-DDE; (e) BF-DRP; (f) proposed method.

(1) The CLAHE algorithm and the proposed algorithm had the best results in terms of image local contrast. CLAHE had two problems. First, it easily produced an overenhancement phenomenon, leading to the noise being amplified, as seen in Figure 6c, where the red rectangle has a significant amount of noise. Second, the scene's energy was weak when the overall contrast was low, as shown in Figure 7c, where the thermal radiation of the indoor scene was weak and the overall contrast of the whole image was low.

(2) The proposed algorithm had the best results in terms of retaining small targets and details, as shown in Figure 6, where the two small dots at the top right are aircraft targets. The proposed algorithm could highlight the aircraft targets and other detailed information without overexposure.

(3) The proposed algorithm had the best global contrast, as shown in Figures 6 and 8. As shown in Figure 6, each algorithm maintained the details well, but the proposed algorithm was more transparent and more informative. In Figure 8, HE, BF-HDR, and the proposed algorithm all had good global contrast, but HE and BF-HDR both showed overenhancement. The proposed algorithm achieved better global contrast while suppressing overenhancement.

Table 1 shows the results of the evaluation of the three scenarios shown in Figures 6–8. It can be seen from the table that for the *RMS* index, no algorithm showed particularly good superiority; that is, several algorithms had no significant difference in the global contrast. For entropy, the HE algorithm performed best in all three scenarios; that is, the distribution of gray in the image after HE processing was the most uniform, which was determined by the principle of the HE algorithm. However, the HE-based algorithm is prone to excessive stretching, which can also be seen in Figures 6–8. For the *SSIM* index, the closer it is to 1, the greater the similarity to the original image. It can be seen from the index that the performance of several algorithms was not very good. This is because the original image was a 14-bit image, and the image's structure underwent changes after compression. *Tenengrad* represents the clarity of the image. The proposed algorithm showed very obvious superiority in all three scenarios, indicating that the proposed algorithm retained the most detailed information and had the best image clarity after dynamic range compression.

Table 1. Evaluation results of the image quality of different algorithms.

	Index	AGC	HE	CLAHE	GF-DDE	BF-DRP	Proposed Method
Scene 1	RMS	31	49	42	**51**	50	48
	Entropy	6.989	**7.991**	7.799	7.711	7.929	7.744
	SSIM	1	**0.8061**	0.561	0.6092	0.6980	0.6168
	Tenengrad	3.257	7.581	19.642	14.813	11.8181	**28.561**
	Time (s)	**0.0020**	0.6856	4.6780	0.9384	24.8654	0.3168
Scene 2	RMS	1	31	16	22	32	**33**
	Entropy	2.972	**7.990**	7.322	7.609	7.888	7.393
	SSIM	1	**0.4653**	0.344	0.408	0.386	0.454
	Tenengrad	0.283	15.371	29.906	23.515	18.754	43.940
	Time (s)	**0.0024**	0.6670	4.6160	0.9418	25.3805	0.3651
Scene 3	RMS	3	31	8	28	**32**	17
	Entropy	4.587	**7.953**	6.278	7.142	7.924	7.378
	SSIM	1	0.556	**0.774**	0.483	0.656	0.547
	Tenengrad	1.411	7.625	6.042	7.878	5.489	**9.661**
	Time (s)	**0.0018**	0.6843	4.6874	0.9400	23.3256	0.3358

The results of running the software on a computer and its implementation on the FPGA were consistent.

In terms of the operation time, the algorithm based on AGC had the least delay, followed by the algorithm proposed in this study. The AGC algorithm had the worst enhancement effects for all image indicators. Therefore, the algorithm proposed in this study has obvious advantages in terms of the operation time.

3.2. Implementation of the Algorithm on an FPGA

The proposed algorithm's FPGA implementation is depicted in Figure 9. This is made up of five parts, each of which is indicated by a green circle. The algorithm is applied to a 1024 × 768 infrared imager with an image frame rate of 25 Hz, a pixel clock of 30 MHz, and 200 pixel clocks of line fading. The algorithm was implemented on the Xilinx (San Jose, CA, USA) A7 FPGA chip and was written in VHDL. Each component's implementation and resource utilization were examined in turn.

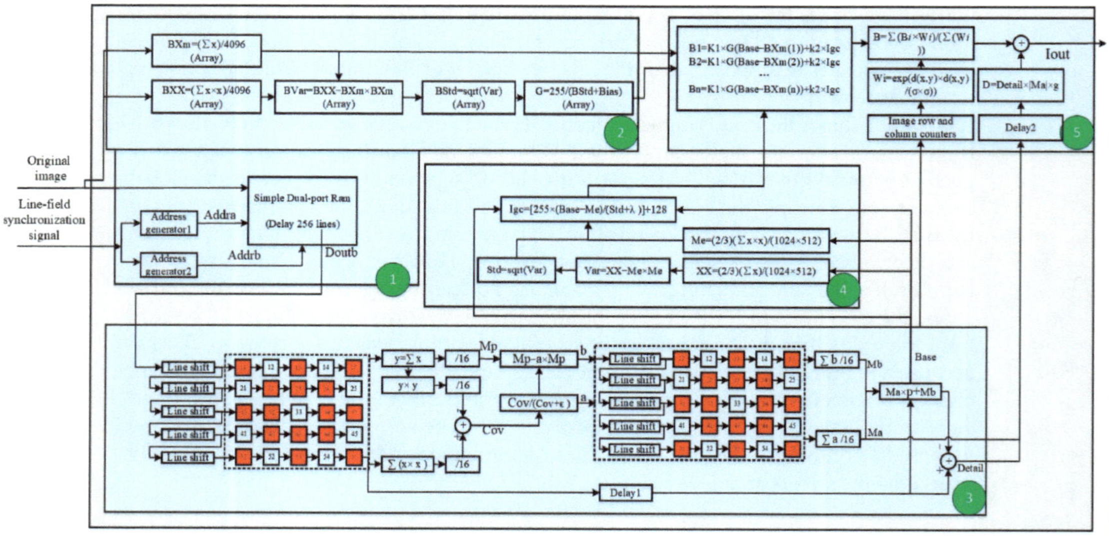

Figure 9. Diagram of FPGA-based implementation of the algorithm.

Part 1: The first part was to finish the process of image buffering in 256 lines so that the calculation of local image stretching and the bias factor needed to wait for 256 lines (in a local block of 64, where $\delta^2 = 25$). This module's operation was straightforward and only required 1 Mb of Block RAM, but it consumed the most resources and had the longest image delay, with a total of 2.1875 Mb of Block RAM and an image delay of 313,344 pixel clocks.

Part 2: The second part was to calculate the pull-up and bias coefficients of the image's local blocks. Each local block had 4096 pixels and was 64×64 pixels in size. There were 192 local blocks in a single image frame. To store the values of data accumulation, square accumulation, and the gain coefficient within the local block, three arrays of 192 bits each were built by the computation module. The data were 26 bits wide. The image was shifted 12 bits to the right after all the data in the local block had been calculated to produce the image's mean value for the local block (BXm) and the mean image variance (BXX). The module accumulated the image data according to the row and column numbers corresponding to the relevant positions in the arrays; furthermore, the local block variance Bstd was computed by calculating the difference between the square of BXm and BXX, and the local standard deviation was generated by taking the square root of Bstd. Finally, the gain coefficient of the local block G could be calculated by Equation (1). A divider was needed for this stage. Overall, this part used resources of 0.0142 Mb of Block RAM, a divider, and some other auxiliary computing resources. The local block calculation required a delay of 64 rows of data, and the calculation of the mean, variance, standard deviation, gain coefficient, and algorithm required 1, 1, 34, and 26 pixel clocks, respectively, with a total delay of 78,398 pixel clocks or about 0.0026 s. Moreover, since Part 1 and Part 2 can run simultaneously, the system was not subjected to any additional delay.

Part 3: The third part was image layering based on a guided filter, which used a 5×5 window, and the data stream was be divided into two main steps:

(1) Calculate the coefficients a,b. To create a real-time image window of 5×5, the original image was successively entered into shift registers of 5 lines after a 256-line delay, and the data from each line were successively entered into 5-pixel buffer registers. The mean and variance of the image were calculated independently for the numbers in the window, and the method of calculation was the same as in Part 2. Specifically, to reduce the computational volume and image delay, the standard deviation and mean values of the 16 domain pixels highlighted in red were used to decrease the computational volume and image delay. The mean value of the 16 data points was computed by simply shifting the sum value 4 bits to the right instead of using a divider, and the local values of a and b were calculated by Equation (4). This step used pixel shift registers of 5 lines + 25 pix and one divider; the image's delay was 3 lines + 36 pix.

(2) Image layering. To create a 5×5 real-time image window, a and b were entered into 5-line shift registers, and the data from each line were entered into 5-pixel buffer registers. To calculate a and b and arrive at the mean values of Ma and Mb, the same 16 neighborhood values were chosen. Finally, the background layer image *Base* was calculated by applying Equation (4); the detailed layer image was $Detail = q - Base$. To match the alignment of a and b, this step specifically called for a delay of 3 lines + 8 pix of the original image. The image delay in this stage was 3 lines + 8 pixel clocks or 3680 clocks, and it used pixel shift registers of 13 line + 58 pix, as well as various other auxiliary computing resources.

All of Part 3 required pixel shift registers of 15 lines + 75 pix, or 0.0147 Mb of Block RAM, a divider, and several auxiliary calculation resources; the image's delay was 6 lines + 44 pixel clocks or 7388 clocks.

Part 4: The fourth part involved the calculation of the global brightness guide map. It involved the standard computation of the global image as described in Part 2. To reduce the amount of calculation, downsampling and summing were performed in the process of image summation, following the method of discarding one out of every three pixels. Meanwhile, the standard deviation of the previous frame was used as a parameter to calculate the guide image of the current frame in the process of continuous video processing. Finally, Equation (5) was used to calculate the global brightness guide image of the frame in real time. A divider, a root calculator, and some more processing resources were needed for this section, and the image's delay, which was caused by the divider, was only 26 clocks.

Part 5: The fifth part was to calculate the final output image, which included three main steps:

(1) Determining the brightness value of the original image under different local block stretching factors, while 192 local blocks were calculated simultaneously to yield B1, B2, ..., Bn. The calculation used Equation (6), and the delay was only one clock.

(2) Determining the weight value using the formula in Equation (2). The different local block weights were based on the distance of the actual pixel from each local block, and the distance could be obtained by extrapolating the row counter. In this step, exponential calculation was involved, so optimization was carried out as follows

$$\exp(-\frac{{d_{(i,j)(x,y)}}^2}{\delta^2}) = \exp(-\frac{{d_{(i)(x)}}^2 + {d_{(j)(y)}}^2}{\delta^2}) = \exp(-\frac{{d_{(i)(x)}}^2}{\delta^2}) \times \exp(-\frac{{d_{(j)(y)}}^2}{\delta^2}) = ex \times ey \quad (15)$$

where $d_{(i,j)(x,y)}$ denotes the Euclidean distance from a pixel with the coordinates (i,j) to the center of the local block with the coordinates (x,y), $d_{(i)(x)}$ denotes the horizontal distance from a pixel with the coordinates $(i,:)$ to the center of the local block with the coordinates $(x,:)$, and $d_{(j)(y)}$ denotes the vertical distance from a pixel with the coordinates $(:,j)$ to the center of the local block with the coordinates $(:,y)$. According to δ^2, when $d_{(i)(x)} = 256$ or $d_{(j)(y)} = 256$, ex ($ex = \exp(-\frac{{d_{(i)(x)}}^2}{\delta^2})$) and ey ($ey = \exp(-\frac{{d_{(j)(y)}}^2}{\delta^2})$) almost decay to zero, so they can be ignored. Therefore, ey and ey have at most 256 cases, or 256 values at $x, y = 0, 1, \ldots 255$. Therefore, to avoid exponential calculations in the FPGA, the process calculated the value at $x, y = 0, 1, \ldots 255$ and saved it in the Block RAM. During the real-time computation, only the lookup table was needed to obtain ex and ey, and only a simple multiplication operation needed performing to obtain the weights. The process of reading the lookup table only delays this step by one clock but requires a Block RAM with a depth of 256 and a width of 16 bits.

(3) The weighted average calculated the final B. The final B was obtained by weighting B1, B2, ..., Bn. This step involved a divider, delayed by 26 pixel clocks.

(4) The last step was to calculate the final image. The detailed layer image was adaptively enhanced and delayed before being added to the background layer image to obtain the final image. The total delay was 55 clocks, with the detailed layer requiring a delay of 53 clocks.

Therefore, Part 5 requires two dividers, 0.0039 Mb of Block RAM, and 55 clocks of image delay.

Overall, the entire process used 2.2554 Mb of Block RAM, five dividers, a root calculator, and some other auxiliary computational resources, with a total image delay of 0.018 s. The specific resource utilization of the FPGA is shown in Table 2.

Table 2. Utilization report.

Slice LUTs (total: 134,600)	39,045 (29%)
Slice registers (total: 269,200)	38,919 (14.46%)
Block RAM (total: 365)	105 (28.77%)
DSPs (total: 740)	319 (43.1%)

The proposed algorithm can be widely used in infrared imaging modules, which can be applied to various military and civilian thermal imagers and scientific research equipment. The infrared imager used in this study was small in size, as shown in Figure 10a, and the effect is shown in Figure 10b. It achieved good imaging results for different scenes. Figure 10c is a schematic diagram of the imaging module mounted on the optomechanical system.

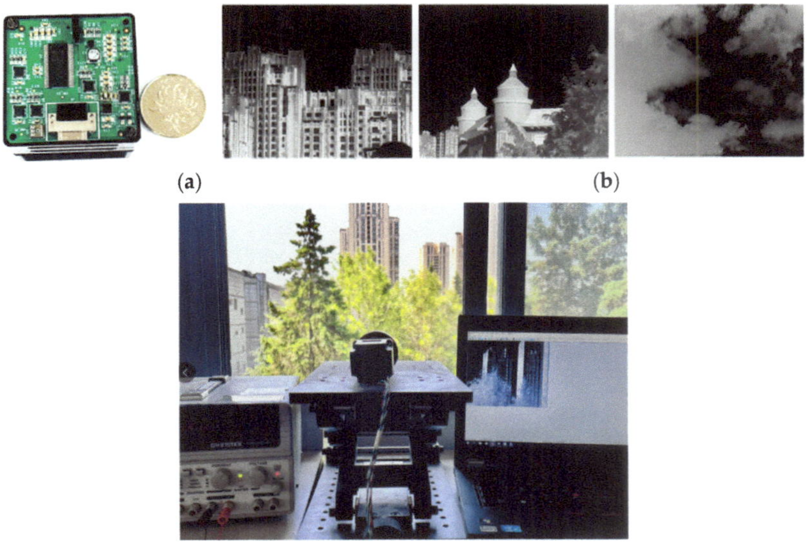

(a) (b)

(c)

Figure 10. Infrared imager and intended display effects. (**a**) The infrared imager; (**b**) The different display effects in different scenes; (**c**) experimental environment.

4. Conclusions

In this study, an algorithm for enhancing the contrast and compressing the dynamic range of infrared images with local optimal contrast was proposed. The focus was to strengthen and improve the following aspects:

(1) Low-delay image processing technology. Most of the algorithms used for conventional background layer image processing are Gamma correction algorithms or histogram-based equalization algorithms. Among them, Gamma-based correction algorithms are generally effective for comparing images, while histogram-based or improved algorithms must count one frame before outputting the final image, which means that at least one whole image will be delayed and there may be phenomena such as image overenhancement. In particular, in continuous video sequences, we found in experiments that when the histogram statistics of the previous frame are used as the current mapping curve, the brightness of the continuous image will flicker due to the mismatch between the real image and the mapping curve. Therefore, an image processing algorithm with a low delay (less than one frame) focuses on the compression of the background layer, so this study proposed a local contrast enhancement framework based on guided global compression of the image.

(2) Minimal storage resources. One of the most crucial components of FPGA-based image processing algorithms is the Block RAM. It is necessary to keep more than one frame of image data when an image processing algorithm has a delay of more than one frame. The block-based local optimal contrast stretching algorithm proposed in this study uses fewer storage resources and only needs two parameters for each local block.

(3) The design of the algorithm structure of the pipeline and the efficient implementation of the algorithm based on FPGA. The pipeline of the algorithm's structure is conducive to the implementation of FPGA; however, exponentiation, division, and other calculations require a significant amount of the FPGA's resources; therefore, this study designed the entire algorithm in accordance with the pipeline and optimized the calculations in the algorithm to facilitate the efficient operation of the FPGA.

(4) Consideration of both global and local contrast. Previous algorithms based on local histograms are prone to block effects, and global class-based algorithms have poor local contrast. Based on this, this study improved the local contrast while ensuring that the

global contrast closely resembled the original image. Specifically, the global background layer's compressed image is used as the guide image to achieve the optimal enhancement of local contrast.

In general, in order to effectively suppress local noise and the halo effect while maintaining both global and local contrast, the proposed algorithm combines local information with global information and uses the global contrast compression map as the guide image. Meanwhile, this study used infrared image with a resolution of 1024 × 768 as an example for optimization and implementation of the FPGA-based algorithm, and the entire algorithm used 2.2554 Mb of Block RAM, five dividers, one root calculator, and some other auxiliary computing resources, with a total image delay of 0.018 s. Finally, the image processing results of different scenes showed that the algorithm proposed in this study had good results in rich scenes, scenes with small targets, and indoor scenes. At the same time, the algorithm has low complexity and low delay, which is beneficial for its application to scenes with high requirements in real time.

Author Contributions: Conceptualization, Y.Z. (Youpan Zhu), W.J. and Y.Z. (Yongkang Zhou); methodology, Y.Z. (Youpan Zhu), Y.Z. (Yongkang Zhou) and W.J.; software, Y.Z. (Youpan Zhu), Y.Z. (Yongkang Zhou), L.Z. and G.W.; validation, Y.Z. (Youpan Zhu), Y.Z. (Yongkang Zhou), L.Z. and Y.S.; formal analysis, G.W. and Y.S.; investigation, Y.Z. (Youpan Zhu), W.J. and G.W.; resources, W.J.; data curation, Y.Z. (Youpan Zhu) and Y.Z. (Yongkang Zhou); writing—original draft preparation, Y.Z. (Youpan Zhu), Y.Z. (Yongkang Zhou), W.J., L.Z., G.W. and Y.S.; writing—review and editing, Y.Z. (Youpan Zhu), Y.Z. (Yongkang Zhou), W.J., L.Z., G.W. and Y.S.; visualization, G.W. and Y.S.; supervision, W.J. and Y.S.; project administration, W.J.; funding acquisition, W.J. All authors have read and agreed to the published version of the manuscript.

Funding: This research was funded by the National Defense Science and Technology Foundation Strengthening Plan (grant number 2021-JCJQ-JJ-1020).

Institutional Review Board Statement: Not applicable.

Informed Consent Statement: Not applicable.

Data Availability Statement: Not applicable.

Acknowledgments: We thank the National Defense Science and Technology Foundation Strengthening Plan for help in identifying collaborators for this work.

Conflicts of Interest: The authors declare no conflict of interest.

Abbreviations

Abbreviations	Full Name
ADC	Analog to digital converter
IR	Infrared radiation
HDR	High dynamic range
PHE	Plateau histogram equalization
CLAHE	Contrast-limited adaptive histogram equalization
GDHDRC	Gradient domain-based high dynamic range image compression
FPGA	Field-programmable gate array
CPU	Central processing unit
DRCE	Dynamic range compression and enhancement algorithm
LOC	Local optimal contrast
BF-DRP	Bilateral filter and dynamic range partitioning
LPF	Low-pass filter
GF	Guided filter
LEPF	Local edge-preserving filter
DDE	Digital detail enhancement
pix	Pixel
AGC	Adaptive gain control

References

1. Jamrozik, W.; Górka, J.; Batalha, G.F. Dynamic Range Compression of Thermograms for Assessment of Welded Joint Face Quality. *Sensors* **2023**, *23*, 1995. [CrossRef] [PubMed]
2. Lv, Z.; Li, J.; Li, X.; Wang, H.; Wang, P.; Li, L.; Shu, L.; Li, X. Two adaptive enhancement algorithms for high gray-scale RAW infrared images based on multi-scale fusion and chromatographic remapping. *Infrared Phys. Technol.* **2023**, *133*, 104774. [CrossRef]
3. Lang, Y.Z.; Qian, Y.S.; Wang, H.G.; Kong, X.Y.; Wu, S. A real-time high dynamic range intensified complementary metal oxide semiconductor camera based on FPGA. *Opt. Quantum Electron.* **2022**, *54*, 304. [CrossRef]
4. Vickers, V.E. Plateau equalization algorithm for real-time display of high-quality infrared imagery. *Opt. Eng.* **1996**, *35*, 1921–1926. [CrossRef]
5. Pizer, S.M.; Amburn, E.P.; Austin, J.D.; Cromartie, R.; Geselowitz, A.; Greer, T.; ter Haar Romeny, B.; Zimmerman, J.B.; Zuiderveld, K. Adaptive histogram equalization and its variations. *Comput. Vis. Graph. Image Process.* **1987**, *39*, 355–368. [CrossRef]
6. Zuiderveld, K. Contrast limited adaptive histogram equalization. In *Graphics Gems*, 4th ed.; Heckbert, P., Ed.; Elsevier Inc.: Amsterdam, The Netherlands, 1994; pp. 474–485.
7. Schatz, V. Low-latency histogram equalization for infrared image sequences: A hardware implementation. *J. Real-Time Image Process.* **2013**, *8*, 193–206. [CrossRef]
8. Ashiba, H.I.; Mansour, H.M.; Ahmed, H.M.; Dessouky, M.I.; El-Kordy, M.F.; Zahran, O.; Abd El-Samie, F.E. Enhancement of IR images using histogram processing and the Undecimated additive wavelet transform. *Multimed. Tools Appl.* **2019**, *78*, 11277–11290. [CrossRef]
9. Fattal, R.; Lischinski, D.; Werman, M. Gradient domain high dynamic range compression. In Proceedings of the 29th Annual Conference on Computer Graphics and Interactive Techniques, San Antonio, TX, USA, 12 July 2002.
10. Kim, J.H.; Kim, J.H.; Jung, S.W.; Noh, C.K.; Ko, S.J. Novel contrast enhancement scheme for infrared image using detail-preserving stretching. *Opt. Eng.* **2011**, *50*, 077002. [CrossRef]
11. Zhang, F.; Xie, W.; Ma, G.; Qin, Q. High dynamic range compression and detail enhancement of infrared images in the gradient domain. *Infrared Phys. Technol.* **2014**, *67*, 441–454. [CrossRef]
12. Wang, Z.; Sun, S.; Li, Y.; Yue, Z.; Ding, Y. Distributed Compressive Sensing for Wireless Signal Transmission in Structural Health Monitoring: An Adaptive Hierarchical Bayesian Model-Based Approach. *Sensors* **2023**, *23*, 5661. [CrossRef]
13. Zhu, J.; Jin, W.; Li, L.; Han, Z.; Wang, X. Multiscale infrared and visible image fusion using gradient domain guided image filtering. *Infrared Phys. Technol.* **2018**, *89*, 8–19. [CrossRef]
14. Branchitta, F.; Diani, M.; Corsini, G.; Romagnoli, M. New technique for the visualization of high dynamic range infrared images. *Opt. Eng.* **2009**, *48*, 096401. [CrossRef]
15. Durand, F.; Dorsey, J. Fast bilateral filtering for the display of high-dynamic-range images. In Proceedings of the 29th Annual Conference on Computer Graphics and Interactive Techniques, San Antonio, TX, USA, 12 July 2002.
16. Bae, S.; Paris, S.; Durand, F. Two-scale tone management for photographic look. *ACM Trans. Graph.* **2006**, *25*, 637–645. [CrossRef]
17. Zuo, C.; Chen, Q.; Liu, N.; Ren, J.; Sui, X. Display and detail enhancement for high-dynamic-range infrared images. *Opt. Eng.* **2011**, *50*, 127401. [CrossRef]
18. Liu, N.; Chen, X. Infrared image detail enhancement approach based on improved joint bilateral filter. *Infrared Phys. Technol.* **2016**, *77*, 405–413. [CrossRef]
19. Anderson, G.L.; Netravali, A.N. Image restoration based on a subjective criterion. *IEEE Trans. Syst. Man Cybern.* **1976**, *6*, 845–853. [CrossRef]
20. Katsaggelos, A.K.; Biemond, J.; Schafer, R.W.; Mersereau, R.M. A regularized iterative image restoration algorithm. *IEEE Trans. Signal Process.* **1991**, *39*, 914–929. [CrossRef]
21. He, K.; Sun, J.; Tang, X. Guided image filtering. *IEEE Trans. Pattern Anal. Mach. Intell.* **2012**, *35*, 1397–1409. [CrossRef]
22. Liu, N.; Zhao, D. Detail enhancement for high-dynamic-range infrared images based on guided image filter. *Infrared Phys. Technol.* **2014**, *67*, 138–147. [CrossRef]
23. Song, Q.; Wang, Y.; Bai, K. High dynamic range infrared images detail enhancement based on local edge preserving filter. *Infrared Phys. Technol.* **2016**, *77*, 464–473. [CrossRef]
24. Farbman, Z.; Fattal, R.; Lischinski, D.; Szeliski, R. Edge-preserving decompositions for multi-scale tone and detail manipulation. *ACM Trans. Graph.* **2008**, *27*, 1–10. [CrossRef]
25. Gu, B.; Li, W.; Zhu, M.; Wang, M. Local edge-preserving multiscale decomposition for high dynamic range image tone mapping. *IEEE Trans. Image Process.* **2012**, *22*, 70–79.
26. Kwan, C.; Chou, B.; Yang, J.; Rangamani, A.; Tran, T.; Zhang, J.; Etienne-Cummings, R. Deep Learning-Based Target Tracking and Classification for Low Quality Videos Using Coded Aperture Cameras. *Sensors* **2019**, *19*, 3702. [CrossRef] [PubMed]
27. Shao, Y.; Xu, F.; Chen, J.; Lu, J.; Du, S. Engineering surface topography analysis using an extended discrete modal decomposition. *J. Manuf. Process.* **2023**, *90*, 367–390. [CrossRef]
28. Shao, Y.; Du, S.; Tang, H. An extended bi-dimensional empirical wavelet transform based filtering approach for engineering surface separation using high definition metrology. *Measurement* **2021**, *178*, 109259. [CrossRef]
29. Peli, E. Contrast in complex images. *JOSA A* **1990**, *7*, 2032–2040. [CrossRef]
30. Mittal, A.; Soundararajan, R.; Bovik, A.C. Making a "completely blind" image quality analyzer. *IEEE Signal Process. Lett.* **2012**, *20*, 209–212. [CrossRef]

31. Zhou, W. Image quality assessment: From error measurement to structural similarity. *IEEE Trans. Image Process.* **2004**, *13*, 600–613
32. Tenenbaum, J.M. *Accommodation in Computer Vision*; Stanford University: Stanford, CA, USA, 1971.
33. Pratt, W.K. *Digital Image Processing*, 3rd ed.; John Wiley & Sons, Inc.: New York, NY, USA, 2001.

sensors

MDPI

Article

Simulation of Laser Profilometer Measurements in the Presence of Speckle Using Perlin Noise

Sara Roos-Hoefgeest [1,*], Mario Roos-Hoefgeest [2], Ignacio Álvarez [1] and Rafael C. González [1]

[1] Department of Electrical, Computer Electronics and Systems Engineering, University of Oviedo, 33003 Oviedo, Spain; ialvarez@uniovi.es (I.Á); rcgonzalez@uniovi.es (R.C.G.)
[2] Desarrollo de Soluciones Integrales Plus S.L., 33211 Gijón, Spain; mroos@cinsystems.es
* Correspondence: roossara@uniovi.es

Abstract: In the manufacturing industry, inspection systems play a crucial role in ensuring product quality. High-resolution profilometric sensors have become increasingly popular for inspection due to their ability to provide detailed surface information. However, the development and testing of inspection systems can be costly and time-consuming. This paper presents the development of a simulation of an inspection system using a high-resolution profilometric sensor. A geometrical and noise model is proposed to simulate the readings of any actual profilometric sensor. The model replicates the sensor's movement on the CAD model of the inspected part. The model incorporates the physical properties of the sensor and combines noise sources from sensor uncertainty and speckle noise induced by the roughness of the material. Our contribution lies in noise modeling. This work proposes a combination of Perlin noise to simulate the speckle noise and Gaussian noise for the uncertainty-related noise. Perlin noise is generated based on the surface roughness parameters of the inspected part. The accuracy of the simulation system is evaluated by comparing the simulated scans with real scans. The results highlight the ability to simulate real scans of different parts, using commercial sensor specifications and the CAD model of the inspected part.

Keywords: laser triangulation profilometry; Perlin noise; speckle; surface roughness; simulation; computer vision

1. Introduction

As increasingly higher requirements are demanded from manufactured products, industries rely on inspection systems to guarantee that their products achieve the expected quality level. This is especially relevant in some industrial sectors, such as the aerospace and automotive sectors or precision parts manufacturing. These sectors demand a full inspection of their production that implies the acquisition of metrological information as well as the detection of other possible defects, such as cracks or irregularities on the product surface. This kind of superficial inspection may be carried out at an intermediate stage of the process to discard defective parts as soon as possible.

However, developing inspection systems can be costly and time-consuming, especially when dealing with high-precision sensors and complex geometries. Therefore, there is a need for CAD tools that can facilitate the development of industrial inspection applications [1,2]. These tools can optimize the design and reduce costs by providing initial data for algorithm development and evaluation. These tools allow for a simulation of the inspection process before actually implementing it on the production line. From the simulation data, it is possible to identify potential issues and refine the inspection process, ultimately improving product quality. To be useful, the simulator must accurately replicate the behavior of the inspection system using models that can be easily related to real hardware. Designers will use that tool to test its performance under different conditions to obtain an accurate and cost-effective solution while reducing the development time and costs.

One of the most common methods for product inspection in industrial applications is laser triangulation. Sensors using this technology employ a laser beam and a camera to measure distances and surface profiles. It works by projecting the laser onto a target surface and analyzing the position of the reflected light in the camera image to calculate the distance. The resolution of triangulation laser sensors refers to their ability to detect and measure small changes or variations in distance. It determines the smallest measurable distance increment that the sensor can reliably capture. The resolution is limited by the uncertainties in the measuring process arising from the different sources of noise in the measuring chain. Among these sources, one of the most important is speckle [3].

Speckle is an interference effect caused by the micro topology of the inspected surface, due to the spatial coherence of the illuminating source [3]. In [4], the influence of speckle in the measurements is analyzed. When lasers are used as light sources on surfaces that are optically rough, speckle can occur due to diffraction. Optically rough surfaces have spatial variations that cause reflected light rays to travel different distances, resulting in phase shifts. These phase-shifted rays can either reinforce or cancel each other out, leading to a pattern of bright and dark areas in the reflected laser light.

Although previous studies have investigated the simulation of laser triangulation sensors for scanning parts, focusing on the geometrical model of the sensor, the impact of the roughness of the scanned material is not usually considered. For example, the work in [5] presents the simulation of scanning laser triangulation sensors for design and evaluation purposes. Based on the CAD files of the parts, the simulation allows for the virtual recreation of the camera acquisition, enabling collision and occlusion detection for complete and effective digitization. In [6], Abu-Nabah et al. proposed a method to validate new laser triangulation designs or commercial ones before their use in the oil and gas industry. They used the commercial animation software Autodesk® 3ds Max® 2017 to create a virtual model of the sensor and generate a synthetic image that is analyzed later on to obtain the depth profile. Using this environment, they were able to check whether the sensor would achieve the required precision and measurement ranges that are needed to retrieve the geometry of weld seams and from that model predict the expected life of the welded product.

Simulation using CAD models can generate measurements that may differ from actual sensor readings because they often overlook the effect of speckle. Just considering the geometrical configuration of the system may produce an accurate ideal measurement, but it will discard the main source of uncertainty, thus limiting the advantages of simulations. It is especially important to consider this discrepancy in applications like surface inspection and defect detection. In this kind of application, the main objective is to assess if differences in surface measurements are relevant and represent a defect or not, for example, an impact mark or a superficial crack. To effectively train algorithms for defect detection and classification, in particular, those used in machine learning or artificial intelligence, it is imperative to include the effects of the main noise sources affecting sensor readings in the simulation models. Ignoring the effect of speckle in simulations can introduce biases and inaccuracies that will compromise the reliability of algorithms in real-world situations.

Speckle simulation has been addressed in some previous works. In [7], Mohammadikaji et al. analyzed the different sources of uncertainty associated with the process of 3D measuring using laser triangulation profilometry. One of the sources of uncertainty that they considered was associated with the localization of the laser line in the image. The accuracy of these values is affected by the presence of speckle combined with quantization effects, sensor noise, sampling, and interpolation methods. They proposed modeling all these effects as two independent zero-mean Gaussian random variables that sum up to the x and y coordinates of the laser line position.

The same research group proposed in [1] a method to obtain realistic simulated images produced by a laser triangulation profilometer. Their objective was to develop a tool for the evaluation and planning of such measurement systems during the development phase. Their model included the simulation of the optical components as well as wave optics effects. They modeled speckle as a diffraction effect caused by a rough surface capable of creating all phase differences. They computed the intensity due to coherent light sources in the Fourier domain, as the product of the estimated amplitude transfer function and the phase of the wave field. As an input to their simulations, they used the CAD model of the object to be measured as well as its bidirectional reflectance distribution function with the corresponding roughness profile. A main drawback was the computational burden of the process. In their work, they stated that rendering an image took about 10 h, using eight cores running at 3.7 GHz.

Csencsics et al. [8] proposed a different approach to simulate speckle. In their simulation, they define all the different elements that form the measuring system, including the geometry, the laser source characteristics, the measured surface, and the imaging system. Then, they approximate the laser spot by a certain number of point sources within the spot diameter, randomly distributed in the z-direction according to the expected surface roughness. For each point source, a given number of rays evenly distributed over a preset solid angle were ray-traced. The solid angle is selected to match the size of the imaging lens. To compute the intensity at each detector point, they combined the different rays arriving at a point according to their phase at that point. They were able to obtain realistic speckle patterns, and they used these results to propose methods to reduce speckle-induced uncertainty.

Other authors have used complex ray tracing methods to model laser triangulation profilometers. In [9], Beermann et al. proposed a simulation model for laser triangulation measurement in an inhomogeneous refractive index field. The simulation setup is performed with a virtual camera and a multistep ray tracing optimization to model a complete triangulation process. The authors explore the benefits of the optical inspection of hot workpieces in multistage forming and the challenges posed by inhomogeneous refractive index fields. The study presents a simulation approach that predicts measurement deviations in such conditions and suggests the design of compensation routines to improve measurement accuracy.

Physically based simulation methods can be very useful when the objective is to design and validate new sensors. As more steps are physically modeled, it is possible to gain more insight into how the sensor performs and the bottlenecks to achieve the desired performance. It may be possible to measure the impact of each component in the final result. Nevertheless, such insight information is useless when the designers want to use an off-the-shelf sensor. In this case, the sensor is a black box that provides depth information. The designer will receive from the camera an array of depth values, where each row corresponds to a single line scan. This information must be processed to extract higher-level information, such as a geometric model or a decision about the presence of defects in the inspected part.

In this work, a proposal to directly simulate the depth measures produced by an off-the-self laser triangulation profilometric sensor is presented. A simple geometrical model provides a base depth value and allows for the simulation of a complete scanning of a part described by its CAD model and its relative position with respect to the sensor during the process. A Gaussian noise model is added to the geometrical depth value to take into account the sensor's depth resolution and precision. The main contribution of this work is to propose the use of Perlin noise [10,11] to model the effect of all the different physical noise sources present in the scanning process, especially the uncertainty due to speckle. This enables the development of a realistic simulation of the sensor that captures its characteristics and the effect of noise sources. This enhances the effectiveness of the simulation for developing, assessing, and optimizing the algorithms that must analyze this information.

Perlin noise is a type of procedural lattice gradient noise originally used in computer graphics to create procedural textures. It has been used both in films to add realism to computer-generated images and in video games to create terrains procedurally. In the computer graphics community, the term noise is used to refer to a random number generator, while the term procedural refers to the fact that the random value at a given point is computed using an algorithm. The speckle effect is to add a random value to the ground truth one. Instead of simulating the physical process that produces speckle and that will affect the image captured by the sensor camera, we use this well-known procedural noise to compute a random value to be added to any point in the depth image returned by an ideal sensor. A method to tune the parameters governing the Perlin noise algorithm is also described. A good survey of procedural noise functions used in computer graphics can be found in [12,13]

There are some works describing the use of Perlin noise to simulate surfaces. Li et al. [14] proposed a method to simulate the reflection and refraction of water surfaces using Perlin noise and ray tracing. It uses multioctave Perlin noise to construct a random height field of the water surface. Acosta et al. [2] applied Perlin noise to simulate oxide textures for developing an image-processing model for rust detection. They simulate 2D color images with a wide range of possible rusted textures, using the Perlin Noise to simulate only the visual aspect of the surface. Other related works are the one in [15], where the authors use Perlin noise to model the material micro-structure, or the one in [16], which proposes a method to enhance the photogrammetric 3D reconstruction performance of featureless surfaces using noise-function-based patterns. In this work, Perlin noise is one of the noise functions used.

As the final objective is to simulate an off-the-shelf sensor, the information provided by the manufacturer is used to build a the geometrical model. The surface geometry of the part to be inspected is modeled using an STL (Standard Tessellation Language) CAD model. STL is a file format commonly used in the manufacturing industry to provide a digital representation of a 3D object surface. To complete the simulation, the trajectory followed by the sensor with respect to the part surface is used to complete the scanning process. Once the simulation is finished, a set of simulated profiles, arranged as a bidimensional array, is obtained. This is the most typical format in which commercial sensors deliver their measurements to the final user.

2. Materials and Methods

Our goal is to achieve a realistic simulation of the readings provided by an off-the-shelf laser profilometric sensor. This should include the effect of speckle, as it is the most relevant source of noise affecting the measurement process. It is a distinct form of noise that occurs as a result of light scattering interference on rough surfaces. In particular, laser triangulation profilometric sensors experience speckle noise due to the irregularities in the surface roughness of the object being scanned. These irregularities introduce fluctuations in the phase and amplitude of the reflected light. The result is the recognizable speckle pattern in the raw image of the laser line. It presents a granular pattern that causes an error in the estimation of the position of its center point. This error is therefore propagated to the estimation of the distance to the object.

This work presents a simulation model that comprises three components: a geometric scene model and two noise models. One accounts for the influence of speckle and the other consider the sensor's accuracy:

$$\hat{r}(x,y) = d(x,y) + PN(x,y) + P_G(0,\sigma) \tag{1}$$

where $\hat{r}(x,y)$ represents the global simulated reading, $d(x,y)$ is the distance provided by a geometric model of the scanning, $PN(x,y)$ represents the effect of speckle in the reading, $P_G(0,\sigma)$ represents the error due to the sensor's accuracy limits, x represents the position along the laser line, and y represents the scan line.

2.1. Geometrical Model

Laser triangulation is a widely used technique in machine vision and metrology for the 3D scanning of surfaces. This non-contact method involves the use of a laser beam and a camera that is directed toward the inspection target. By adopting a known angular offset between the laser source and the camera sensor, it is possible to measure the distance between the surface of the object and the camera using trigonometric principles.

The goal is to simulate the readings provided by a commercial laser triangulation profilometric sensor, which provides high-accuracy depth measurements over a line. In this work, a simple geometric model of the sensor is proposed. The model is defined by a set of parameters extracted from the sensor datasheet that are shown in Figure 1. The first parameter is the working distance (W_d), which accounts for the distance from the camera to the center of the depth measurement range. The working distance is the ideal operating point, where the laser is at its sharpest focal point and the reflected spot is in the center of the detector. Then, we consider the Z-Range (Z_R) value. It defines the difference between the maximum and minimum measurement distances of the sensor. Therefore, the scanned surface must be within the range $W_d \pm \frac{Z_R}{2}$ to obtain a valid measurement.

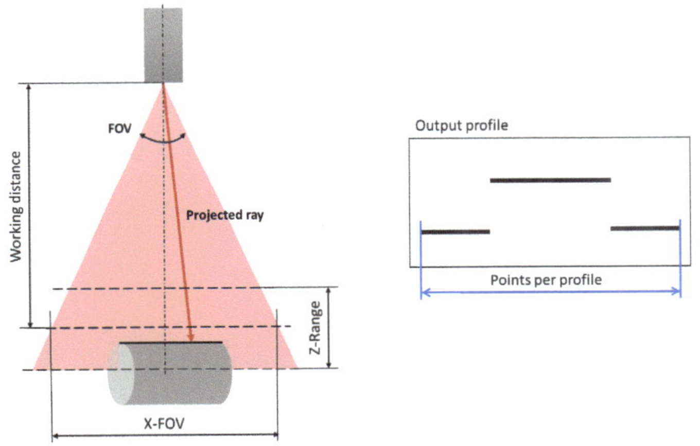

Figure 1. Laser triangulation sensor: key parameters scheme.

The third parameter used is the field of view (FOV) of the camera. The FOV is the laser beam opening angle and it represents the maximum aperture of the sensor and thus the angular limits of the projected beam. It is computed from the X-FOV and the W_d. The X-FOV is the total length of the measurement line at the working distance. Therefore, a simple trigonometric computation leads to the following:

$$FOV = 2\arctan\left(\frac{\text{X-FOV}}{2W_d}\right) \tag{2}$$

The parameter points per profile determine the number of points to be measured in each profile, thus defining the number of rays to be projected in each acquisition. Those rays are evenly distributed within the FOV.

A single profile is not enough to analyze the quality of a surface. It is necessary to scan several profiles and join them to obtain a 3D image of the same. To complete the simulation, the relative trajectory followed by the sensor with respect to the object is replicated. The trajectory is defined by a sequence of sensor pose values and the sensor speed between two consecutive poses. The intermediate position for each individual profile simulation is determined using the maximum profile speed in Hz. This parameter is also available in the sensor datasheet and represents the number of full 3D profiles per second provided by the sensor. Its inverse is the time between two consecutive acquisitions. By employing this temporal interval and the sensor's travel speed, each intermediary pose is computed through linear interpolation.

The 3D model of the inspected piece is loaded as a triangular mesh, which describes a surface by the unit normal and vertices of the triangles. The profile acquired in each measurement corresponds to the intersections between the projected rays and the triangulated 3D model. It is calculated as a line–triangle intersection case, using the Möller–Trumbore ray–triangle intersection algorithm [17].

The output of the simulation system provides the results of the scan in both a 3D point cloud form and a 2D image, where each row of the image corresponds to a profile of the sensor acquisition, so the size of the image is determined by the number of points per profile and the total number of profiles scanned during the acquisition. Each pixel of the image corresponds to the distance measured by the profilometric sensor.

2.2. Speckle Noise Model: Perlin Noise

The inspected surface will always present a certain degree of optical roughness that will cause speckle when is illuminated with a coherent light source. This effect is the main source of error in the measurement. We model this effect by adding a Perlin noise term to the value produced by the geometric model.

2.2.1. Perlin Noise

Perlin noise [10] is a type of gradient noise originally used in computer graphics to create procedural textures with a pseudo-random appearance, an example of Perlin noise can be seen in Figure 2. It is defined by the composition of multiple scaled versions of the same noise function, with different frequencies (f_m) and amplitudes (A_m), known as octaves. It is described according to Equation (3), where m is the m-th noise function being added and $n_{octaves}$ is the number of octaves.

$$PN(x,y) = \sum_{m=1}^{n_{octaves}} A_m N_m(f_m x, f_m y) \quad (3)$$

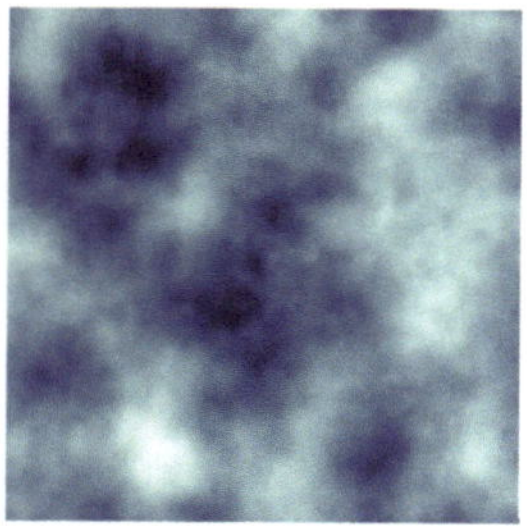

Figure 2. Perlin noise image.

The level of detail and complexity in the generated texture or pattern is determined by the frequency and amplitude of each octave. The frequency refers to the number of cycles of the noise function within a given unit of space, and higher frequency values result in more rapid variations, producing a more intricate pattern. The frequency of each octave is typically set by multiplying the base frequency (f_0) by a factor of 2 for each successive octave, expressed as $f_m = f_0 \cdot 2^{m-1}$.

The amplitude determines the range of values that the noise function can output. Higher amplitude values produce more pronounced peaks and valleys, resulting in a more dynamic pattern. The amplitude of each octave is usually determined by a persistence factor p, expressed as $A_m = A_0 \cdot p^{m-1}$.

To implement a 2D Perlin noise function, the initial step involves defining a 2D grid of control points. For each control point, a fixed pseudo-random gradient vector of unit length is generated. To compute the value at a point $P(x,y)$, it is necessary to determine the control points lying on the corners of the cell where the point falls $C(x_i, y_j)$, where indices (i,j) denote the 4 corners of the cell, and i and j are either the floor or the ceil of x and

y, respectively, as shown in Figure 3. Then, the position vector of the point $P(x, y)$ with respect to each corner, ($\vec{d}(x_i, y_j)$), is computed. Equation (4) shows this calculation.

$$\begin{aligned} \vec{d}(x_i, y_j) &= C(x_i, y_j) - P(x, y) \\ i &\in \{\lfloor x \rfloor, \lceil x \rceil\} \\ j &\in \{\lfloor y \rfloor, \lceil y \rceil\} \end{aligned} \tag{4}$$

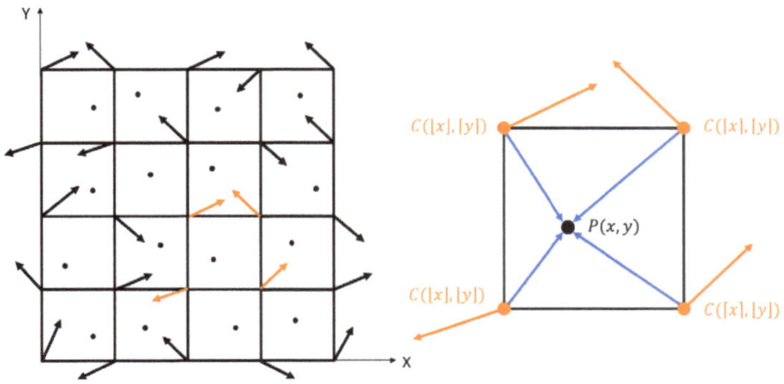

Figure 3. Two-dimensional Perlin noise generation process. Left image shows the 2D grid with the pseudo-random gradient vectors for each control point. Right image represents the process in one input coordinate $P(x, y)$. $C(x_i, y_j)$ are the corners of the grid, blue arrows represent the displacement vectors $\vec{d}(x_i, y_j)$ and orange arrows represent the gradient vectors.

For each corner, the dot product between its gradient and displacement vectors is calculated to obtain the influence values $n(x_i, y_j)$ for point $P(x, y)$.

$$n_{ij} = \vec{g}(x_i, y_j) \cdot \vec{d}(x_i, y_i) \tag{5}$$

The calculation of the 2D Perlin noise function at point $P(x, y)$ is performed according to Equation (7). An interpolation between the 4 dot products is performed to estimate the blending of the noise contribution from the four corners, where n_0 and n_1 are the interpolations between n_{00}, n_{10} and n_{01}, n_{11}, respectively (see Equation (6)).

Interpolation is performed using the improved version of the general form of the *smoothstep* function [11], as shown in Equation (8). This function has zero first and second derivatives in the edges, and this ensures a smooth transition between neighboring cells, giving the Perlin noise its characteristic look.

$$\begin{aligned} n_0 &= n_{00} + smoothstep(P_x - C_{x_0}) \cdot (n_{10} - n_{00}) \\ n_1 &= n_{01} + smoothstep(P_x - C_{x_0}) \cdot (n_{11} - n_{01}) \end{aligned} \tag{6}$$

$$N(x, y) = n_0 + smoothstep(P_y - Cy_0) \cdot (n_1 - n_0) \tag{7}$$

$$smoothstep(x) = \begin{cases} 0 & x \leq 0 \\ 6x^5 - 15x^4 + 10x^3 & 0 \leq x \leq 1 \\ 1 & x \geq 1 \end{cases} \tag{8}$$

2.2.2. Perlin Noise Modeling with Roughness Parameters

Based on the properties of Perlin noise, it can be modeled in a way that, by adjusting its parameters appropriately, allows for the simulation of the speckle noise found in real measurements These parameters are the size and the number of control points of the 2D grid; the signal amplitude; the number of octaves; and the persistence factor. In this subsection, we will describe how to select an appropriate set of values to achieve a realistic simulation. Through this approach, a 2D Perlin noise function is estimated and subsequently incorporated into the full scan originating from the geometric model.

Given the variability in the noise in each real measurement, the aim is not to reproduce point-by-point the exact noise patterns but to capture the fundamental characteristics of the noise to allow its simulation. In the presence of speckle, the real readings provided by the laser profilometric sensor look like a rough surface. Although they do not necessarily reflect the true surface roughness of the object, these readings can be adequately characterized using the same parameters that are commonly used to assess the roughness of materials. Roughness parameter values can be obtained through analysis of actual measurements or through pre-existing knowledge of the type of material being scanned. Using these parameters enables the generation of Perlin noise that replicates the noise introduced by speckle.

Surface roughness is a feature of the surface texture. It measures the variation between the actual surface and its ideal form in the direction of the normal vector. Roughness values can either be calculated on a profile (line) or on a surface (area). We consider the parameters profile by profile and generalize them to the surface by taking the average of the values of all the profiles. There are many different parameters to characterize roughness [18]. The most important parameters are the amplitude ones, which characterize the surface based on the vertical deviations of the roughness profile from the mean line.

The most common parameter to define roughness is the arithmetic average height (R_a). It is defined as the arithmetic average of the profile height deviation from the mean line. Other common parameters are the R_q (root mean square roughness), which represents the standard deviation of the distribution of surface heights; the R_p (maximum height of peaks); and the R_v (maximum depth of valleys), as well as the parameters R_{pm} and R_{vm}, which measure the mean of the height of the 10 highest peaks and the depth of the 10 lowest valleys, respectively. Figure 4 shows the most important parameters.

Another common parameter is the number of peaks in the profile P_c. This parameter calculates the number of peaks of the profile per unit length, given in peaks per millimeter. Peaks are only counted when the distance between the current peak and the previous one is greater than 10% of the maximum height of the profile.

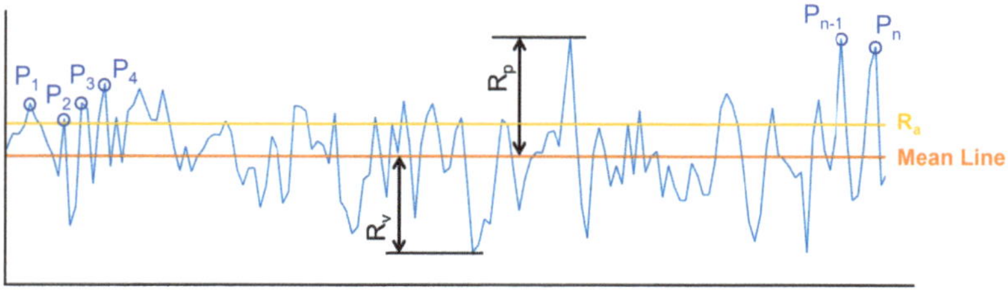

Figure 4. Common roughness parameters in a profile. P_i are all the peaks in the profile.

After analyzing the most common roughness parameters, those that align most effectively with the Perlin noise model can be determined. Equation (3), which illustrates the Perlin noise generation algorithm involving multiple scaled versions of a single noise function, reveals the critical parameters that require adjustment to correspond with the roughness parameters from the actual measurements. Subsequently, the chosen roughness parameters for modeling Perlin noise are outlined, along with the reason behind these decisions.

First, to generate the 2D Perlin noise function (N_m), as explained in the previous section, we need to adjust the size of the grid and the number of control points. The grid size is set to match the size of the resulting scan in pixels. This corresponds to the number of points per profile (ppf) and the number of scanned profiles ($n_{profiles}$). The number of control points per dimension is then determined based on P_c, which allows us to characterize different surface types. We consider that P_c provides a reliable approximation of the number of intersections with 0 that the first octave of Perlin noise will have, which corresponds to the number of control points. This parameter is selected because it is more robust than calculating the number of intersections with the mean line of the signal. The number of peaks per mm is estimated by knowing the pixel/mm ratio of a real scan in each direction.

In Perlin noise generation, the frequency represents the number of cycles between two adjacent control points. By aligning the number of control points in the grid with the number of peaks in the desired surface roughness, we can set the frequency to 1. This configuration ensures that there is precisely one complete cycle of the noise pattern between adjacent control points, simulating the spacing between two peaks of the roughness in the resulting noise.

The amplitude is set as the maximum value between R_{pm} and R_{vm}. So, the maximum Perlin amplitude corresponds to the highest value of the 10 sharpest peaks or valleys. In this way, the influence of possible outliers is minimized. The other parameters mentioned above will be used to give an estimate of the similarity between the real and simulated samples.

Experimentally, we have found that a number of octaves equal to 4 and a persistence of 0.5 provide satisfactory results, allowing us to add levels of detail to the noise. Increasing the number of octaves does not produce significant changes in the result. For octave five and above, their amplitude is no longer in the measurement range of the sensor and therefore it is no longer detectable.

Substituting the parameters of Equation (3) as explained in this section, Equation (9) is obtained. This equation presents our proposed model to simulate the speckle of the real measurements from Perlin noise. Remember that the amplitude of each octave is determined by the persistence factor as $A_m = A \cdot p^m$ and the frequency is typically set by multiplying the base frequency by a factor of 2 for each successive octave. N_m is the 2D Perlin noise function for each octave, with a grid size to match the size of the resulting scan in pixels and with a number of control points per dimension determined based on P_c and the dimensions of the scan.

$$PN(x,y) = \sum_{m=1}^{4} (0.5^{m-1} \cdot \max(R_{pm}, R_{vm}) \cdot N_m(2^{m-1}x, 2^{m-1}y)), \quad \forall x \in [0, ppf], \forall y \in [0, n_{profiles}] \tag{9}$$

2.3. Sensor Uncertainty Model: Gaussian Noise

To complete our simulation model, consideration is given to an additional set of sensor parameters: resolution Z (Δ_Z) and repeatability (σ_{rep}). The first one is related to the minimum change in depth that can be reliably detected by the sensor. The second one is the standard deviation of the readings provided by the sensor when repeating the measurement under exactly the same circumstances. The sensor resolution is modeled as a uniform distribution in the range $\left[-\frac{\Delta_Z}{2}, \frac{\Delta_Z}{2}\right]$. The model for the repeatability error is a zero-mean Gaussian distribution with a standard deviation equal to the repeatability parameter. In addition, we consider that both variables are independent, so we model their combined effect as a zero-mean Gaussian with a standard deviation:

$$\sigma = \sqrt{\left(\frac{\Delta_Z^2}{12} + \sigma_{rep}^2\right)} \tag{10}$$

To include the effect of this sensor uncertainty, we add to each measurement a value drawn from a zero-mean Gaussian with standard deviation σ. The final value is trimmed according to the Δ_Z parameter.

3. Results

This section presents the results of the laser triangulation profilometry simulator, which incorporates the geometrical and noise model. The algorithms have been implemented using the C++ running on the Ubuntu 20.04 LTS operating system. The open-source Visualization Toolkit software (VTK), version 9.0.0, [19] is used to manipulate the 3D model and simulate the inspection system, and MATLAB 2022b is used to process and analyze the results.

Different experiments are performed to test the reliability of the developed algorithms. Specifically, we replicated real measurements of three different steel products: bearing caps, car doors, and heavy steel plates. These products have different geometries, and, although they are made of the same material, they are manufactured using different processes, resulting in varying surface characteristics.

The results obtained are analyzed by comparing simulated scans with real ones. It is worth noting that quantifying the accuracy of the noise simulation presents significant challenges. The variability introduced by noise complicates the task of quantifying the reliability of the simulation. However, a quantitative analysis of the similarity between simulation and reality can be performed by comparing the distributions of the scans and their surface roughness parameters. Also, a visual comparison of scan appearances will be conducted, considering both individual profiles and the broader scan area.

3.1. Geometrical Model

Any commercial profilometric sensor can be simulated by modifying its parameters as needed. In these experiments, the parameters have been tuned to match those of the real inspection system. The actual profilometric sensor used in these experiments is the AT-C5-4090CS30-495 model from Automation Technology. The main parameters extracted from its datasheet [20] are listed in Table 1. Also, the travel speed of the sensor is set as 0.1 m/s.

Table 1. Parameters of the profilometric sensor from its datasheet and travel speed used in these experiments.

Working distance	400 mm
Z-Range	250 mm
FOV	63.5°
Points per profile	4096 pixels
Z resolution	3.8 µm
Profile speed	500 FPS
Travel speed	0.1 m/s

A 3D model of any part can be loaded as an STL file. The interface of our simulator, with the modeled profilometric sensor, can be seen in Figure 5, where a 3D model of a car door is loaded. Also, the inspection trajectories can be configured to simulate different scans. In this case, it is a straight line between the 2 points P_0 and P_1. The travel speed of the scan is set to match the real system. The result of this scan can be seen in Figure 6, as a point cloud and as a 2D image. The size of the image is the total number of profiles by the number of points per profile. The total number of profiles is determined by the travel speed sensor and the scan distance.

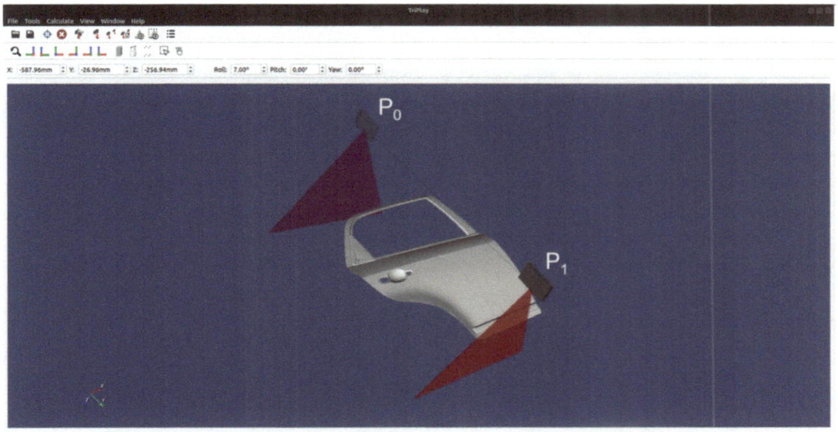

Figure 5. Interface of the developed simulator. The CAD model of a loaded car door and a scan path between the marked points P_0 and P_1 is shown.

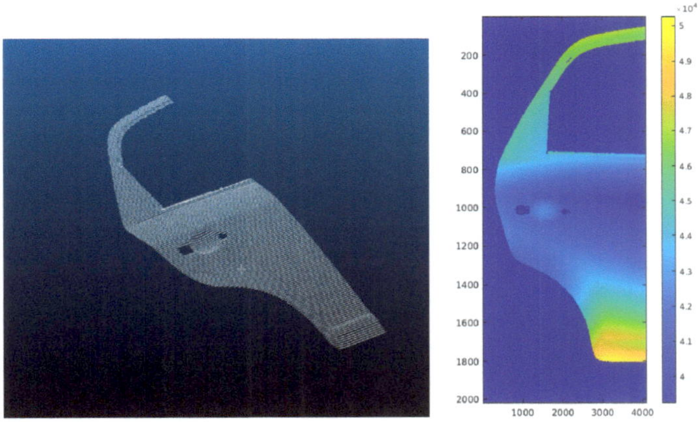

(**a**) (**b**)

Figure 6. Result obtained during the simulation of a scan of the section of the CAD model shown in Figure 5: (**a**) result in point cloud form; (**b**) result as 2D image. Each row corresponds to a scan profile, so the size of the image is determined by the total number of profiles scanned during the acquisition and the number of points per profile. (Size: 2000 × 4096.)

3.2. Noise Model

In this work, two different noises are presented: the uncertainty of the sensor and the speckle noise. The noise introduced by the uncertainty of the sensor is modeled as a Gaussian noise with an amplitude determined by the resolution of the measurement. The sensor resolution on the z-axis, according to the datasheet, is 3.8 µm. However, it is important to note that, based on prior experience and evidence, a more realistic value for the resolution is approximately 15 µm.

Noise introduced in the form of speckle is modeled as Perlin noise. The characteristics of this noise will depend on the material and the manufacturing process of the product. This model can be generalized from the R_{pm}, R_{vm}, and P_c values of each surface.

In Figure 7, a profile extracted from the scan shown in Figure 6 can be observed, specifically from the door handle area. The figures on the left provide a more comprehensive view of the profile, while those on the right are zoomed in on a specific part of it.

Figure 7a,b demonstrate that the scan without the addition of noise yields an ideal profile without any alterations. However, this is not desirable for replicating real scans. In Figure 7c,d, Gaussian noise is introduced to simulate the measurement noise of the sensor. The magnitude of this noise depends on the characteristics of the sensor being replicated, which in this case is 15 µm, as previously stated. Also, Perlin noise with an amplitude of 80 µm is included. It should be noted that this is a larger value than the actual roughness of the door material. However, it was calculated this way to see the differences between Gaussian noise and Perlin noise more clearly, only for this particular experiment. Figure 7e,f display the final simulated profile after incorporating Gaussian noise and Perlin noise. It is evident that the Perlin noise has a higher amplitude than the Gaussian noise, as mentioned above. This depends on the specific experiment, taking into account the sensor's resolution and the surface characteristics of the inspected part.

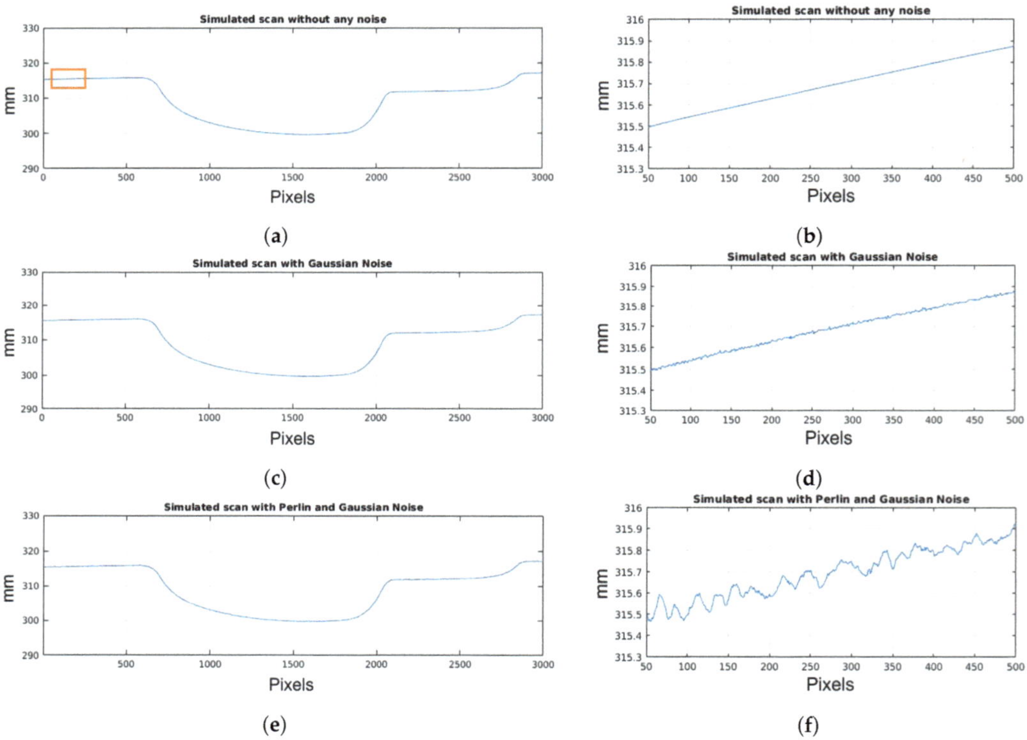

Figure 7. Noise simulation results. Images on the left show the complete profile and those on the right show a zoom of the section outlined in orange: (**a**,**b**) simulation profile before adding any noise; (**c**,**d**) simulation profile after Gaussian noise to simulate sensor uncertainty; (**e**,**f**) simulation profile after Perlin and Gaussian noise.

In the following, we will present the results obtained for the simulation of scans of three steel products: bearing caps, car doors and heavy steel plates. Each has a different manufacturing method: bearing caps are manufactured by casting, doors by stamping, and plates by rolling. Figure 8 shows an example of these types of products.

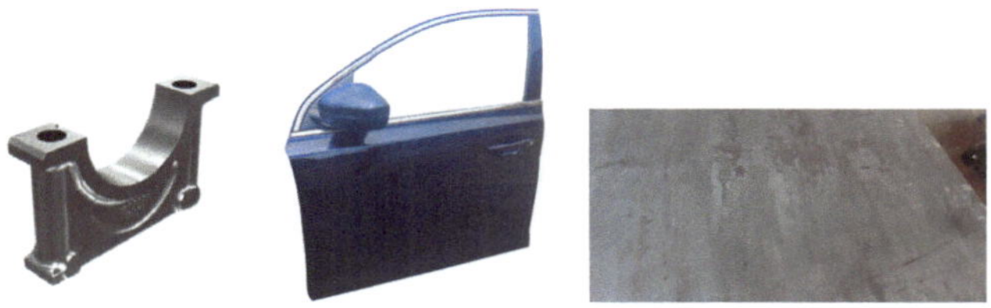

(a) (b) (c)

Figure 8. (**a**) Bearing cap (image obtained from [21]). (**b**) Car door. (**c**) Heavy steel plate.

Direct comparison between real and simulated scans is impossible, as noise patterns have a random component and vary with each measurement. Also, our objective is to model realistic measurements rather than replicate specific ones, simulating realistic scans of different products using any laser triangulation sensor. However, a comparison between the roughness parameters of real and simulated scans will be conducted in order to quantify the degree of similarity between scans. Also, their visual appearance will be evaluated, both at the profile and surface levels. In these experiments, scans of more or less flat areas are compared to facilitate the analysis of the results without affecting the shape of the part itself.

To emphasize the importance of simulating the noise introduced by the uncertainty and the noise introduced by the speckle, the same scan will be analyzed, on the one hand, with only the Perlin noise, and, on the other hand, with the full noise model. The roughness parameters of the simulations are estimated in order to compare them with the real sample, in particular, the parameters R_a, R_q, R_p, R_v, R_{pm}, and R_{vm}. These parameters are computed as the average of the values calculated for each profile.

Also, we compare the similarity between the real scan distribution and the simulated one using the two-sample Kolmogorov–Smirnov test [22]. It is a non-parametric hypothesis test that evaluates the difference between the CDFs (cumulative distribution functions) of the distributions of the two sample data vectors over the range of x in each dataset. The null hypothesis is that both samples come from the same distribution. A critical value for a 5% significance level is used, so the null hypothesis is rejected if the maximum absolute difference between the CDFs of the two samples is greater than 0.05.

3.2.1. Bearing Cap

This section presents the results obtained during the simulation of scans over a bearing cap. Figure 9 shows the comparison between the results of a simulation scan and a real one. It presents a profile and a 2D image of the obtained scan. It is possible to visually verify the similarity between simulated and real ones. The displayed areas correspond to a portion of the entire scan performed with a sensor of a 4096 pixels per profile. More precisely, it is an area of 100 profiles by 501 pixels, that represents an area of 22.5 × 22.5 mm. A section of the area is presented to offer a clearer view of the noise details and facilitate a more comprehensive analysis, also to preserve the confidentiality of the data.

Table 2 shows the estimated parameters obtained from the analysis of real scans, and compares them with the parameters obtained from the simulation results. Analyzing the table, we observe that the roughness parameters of real and simulated samples are very similar, moving in the same measurement range, with errors around 5%.

First row of Figure 10 shows the normalized histogram of real and simulated scan distributions for this experiments. It can be seen that they have similar distributions. After applying the Two-sample Kolmogorov-Smirnov Test, the result is that the null hypothesis is true and, therefore, real and simulated samples come from the same distribution. In Figure 10c the comparison between the cumulative distribution functions is shown.

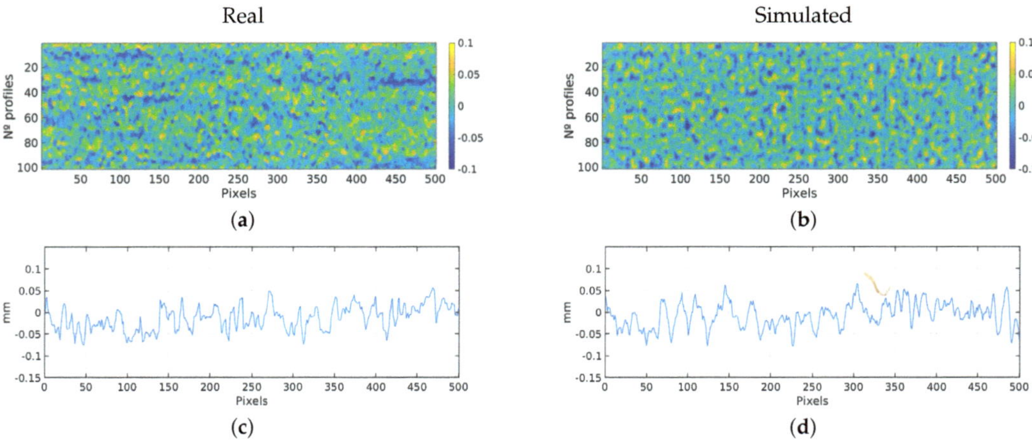

Figure 9. Bearing cap experiment: Comparison between real and simulated scan. (**a**) Real 2D scan image. (**b**) Simulated 2D scan image. (**c**) Real scan profile. (**d**) Simulated scan profile.

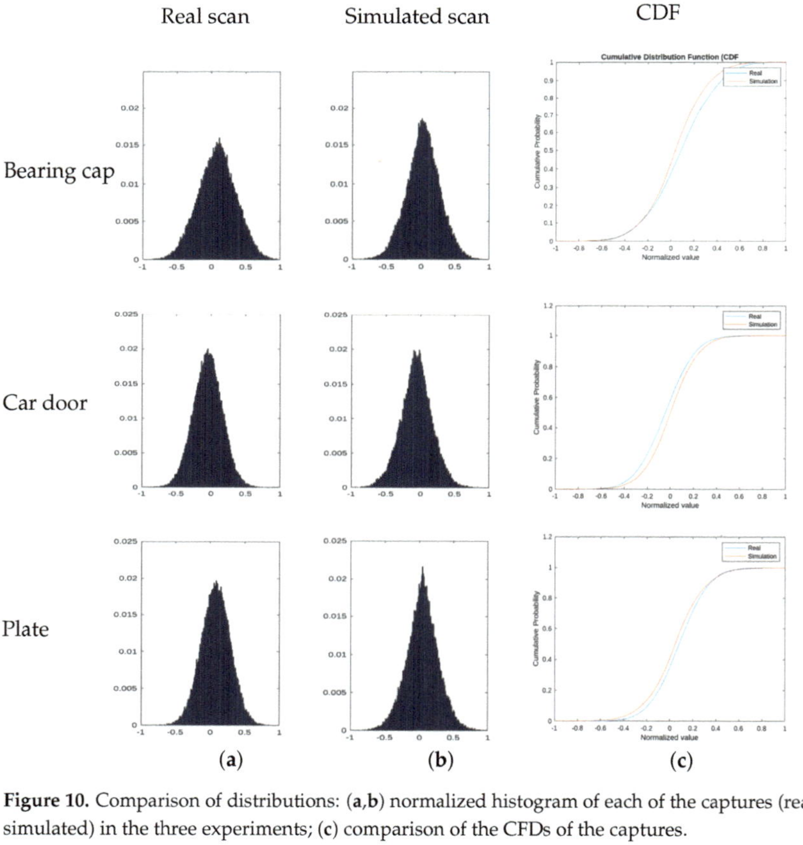

Figure 10. Comparison of distributions: (**a**,**b**) normalized histogram of each of the captures (real and simulated) in the three experiments; (**c**) comparison of the CFDs of the captures.

Table 2. Bearing Cap: Comparison of roughness parameters between real and simulated scans. Estimated as the average of the parameters calculated for each profile.

Bearing Cap			
	Real	**Simulation**	$\epsilon_r(\%)$
R_a (mm)	0.0211	0.0208	1.29
R_q (mm)	0.0263	0.0263	0.11
R_p (mm)	0.0731	0.0764	4.47
R_v (mm)	0.0742	0.0790	6.55
R_{pm} (mm)	0.0676	0.0702	3.76
R_{vm} (mm)	0.0700	0.0741	5.85
P_c (peaks/mm)	4.36	4.62	5.74

3.2.2. Car Door

This section presents the results of the simulation measurements on a car door. Analogous to the previous section, Table 3 presents a comparative between the roughness parameters of the simulated and real scan. Figure 11 shows the 2D image of the obtained scans. As the previous experiment, it is an area of 100 profiles by 501 pixels, also corresponding to 22.5 × 22.5 mm.

The differences in roughness parameters between the real and simulated scans of the car door surface are relatively small, with percentage errors ranging from 2.03% to 6.28%. This indicates a good agreement between the simulated results and the actual measurements. The simulation accurately captures the essential roughness characteristics of the car door surface, further validating the reliability of the simulation model.

In this case, it can be seen that the roughness values are lower than in the previous experiment. This is due to the characteristics of this particular surface. Both are steel materials, but the ingot is manufactured by casting and the car door by stamping. In general, stamping tends to result in lower surface roughness compared to steel casting. This is because stamping involves shaping the metal through controlled pressure and force, resulting in a smoother and more uniform surface. On the other hand, casting involves pouring molten metal into a mold, which can lead to surface irregularities and textures due to the solidification process.

Table 3. Car door: Comparison of roughness parameters between real and simulated scans. Estimated as the average of the parameters calculated for each profile.

Car Door			
	Real	**Simulation**	$\epsilon_r(\%)$
R_a (mm)	0.0134	0.0137	2.03
R_q (mm)	0.0169	0.0173	2.47
R_p (mm)	0.0523	0.0542	3.65
R_v (mm)	0.0522	0.0555	6.28
R_{pm} (mm)	0.0466	0.0482	3.50
R_{vm} (mm)	0.0477	0.0504	5.73
P_c (peaks/mm)	5.9446	6.2403	4.97

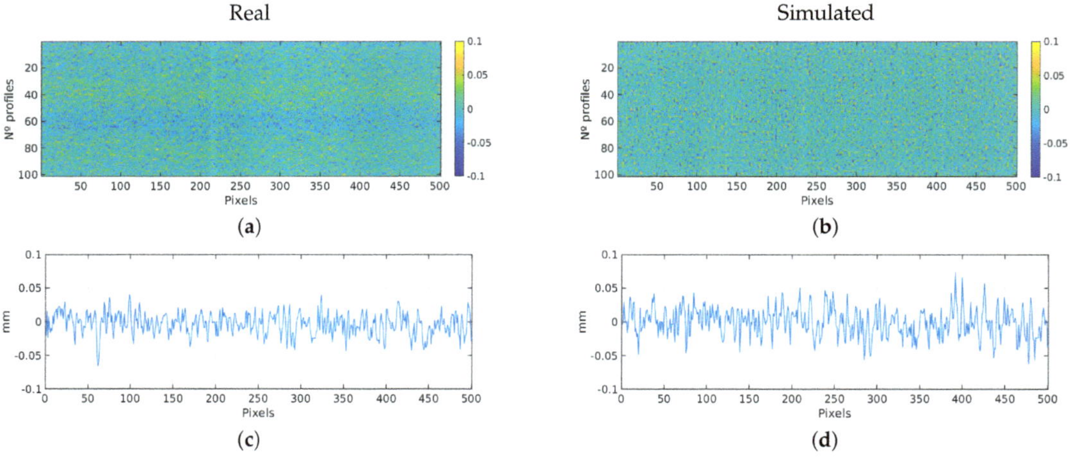

Figure 11. Car door experiment: Comparison between real and simulated scan. (**a**) Real 2D scan image. (**b**) Simulated 2D scan image. (**c**) Real scan profile. (**d**) Simulated scan profile.

The Two-sample Kolmogorov-Smirnov Test also confirms that the null hypothesis holds, indicating that both real and simulated samples originate from the same distribution. Figure 10 presents the histograms and the cumulative distribution functions.

3.2.3. Heavy Steel Plate

This section focuses on the simulation scans on a heavy steel plate. The roughness parameters of the simulated and real scans are compared in Table 4. Additionally, Figure 12 showcases a 2D image of the acquired scans, covering an area of 100 profiles with 501 pixels. In this case, the scanning parameters were different and cover an area of 20 × 40 mm.

As in previous experiments, the comparison between real and simulated scans shows that the roughness parameters generally align closely, with small percentage differences, around 4%. Overall, the similarity between the real and simulated roughness parameters indicates that the simulation accurately captures the characteristics of the heavy steel plate's surface.

Table 4. Heavy steel plate: Comparison of roughness parameters between real and simulated scans. Estimated as the average of the parameters calculated for each profile.

	Heavy Steel Plate		
	Real	**Simulation**	$\epsilon_r(\%)$
R_a (mm)	0.2484	0.2403	3.27
R_q (mm)	0.3118	0.3058	1.94
R_p (mm)	0.9397	0.9633	2.52
R_v (mm)	0.9834	0.9642	1.95
R_{pm} (mm)	0.8329	0.8550	2.65
R_{vm} (mm)	0.8925	0.8817	1.22
P_c (peaks/mm)	6.8758	6.2277	9.43

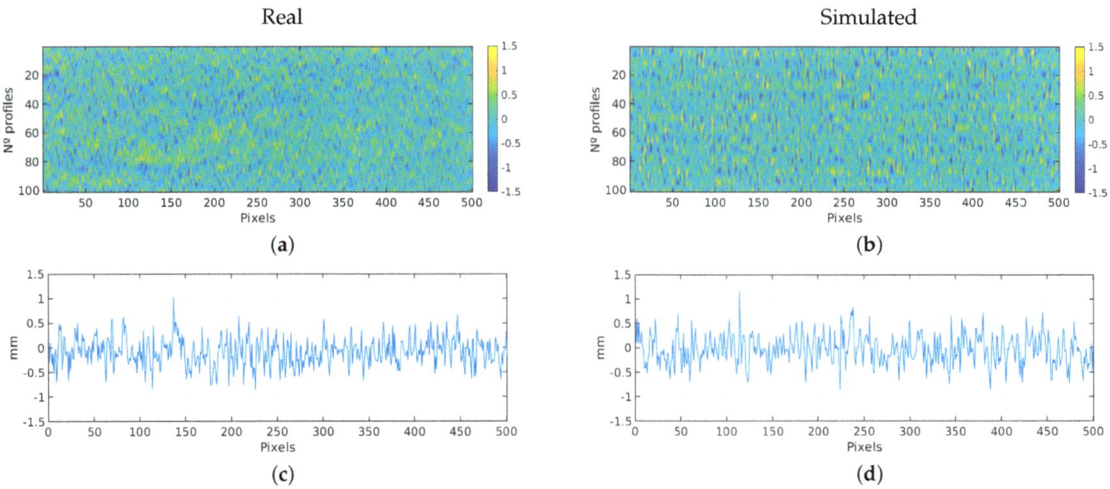

Figure 12. Plate experiment: comparison between real and simulated scans. (**a**) Real 2D scan image. (**b**) Simulated 2D scan image. (**c**) Real scan profile. (**d**) Simulated scan profile.

As in the previous experiments, we can confirm that both real and simulated samples have similar distributions using the two-sample Kolmogorov–Smirnov test. Figure 10 presents the histograms and the cumulative distribution functions.

Compared to the bearing cap and car door, heavy steel plates typically have higher surface roughness parameters. This is mainly due to the presence of scale or surface oxides, which can contribute to a rougher surface texture and irregularities.

4. Discussion

This work proposes a geometrical and noise model for simulating the readings of a commercial laser triangulation profilometric sensor, incorporating the sensor's physical properties and noise sources. The model accurately reproduces the geometrical characteristics of the scanned surface, as well as the noise patterns caused by speckle and sensor uncertainty. Validation experiments demonstrated a high degree of similarity between the generated scans and real sensor scans. Our research provides satisfactory results, enabling a realistic simulation of laser triangulation scans.

CAD models in STL format are used because is one of the most widely used files to represent the geometry of industrial parts. It stores data as a triangular mesh, which describes a surface by the unit normal and vertices of the triangles. One of the disadvantages is that the resolution of the sensor is higher than the triangular mesh. So, when we simulate the measurement of a profilometric sensor from a CAD model, a profile composed of small straight segments is obtained, corresponding to each of the faces of the triangles that make up the mesh of the model. This would not be a problem in the case of looking for coarse 3D reconstructions, but it is intended to simulate sensors working with resolutions of a few micrometers.

To avoid this problem and allow the simulation of realistic scans, we introduce the simulation of sensor measurement noise and speckle using a combination of Gaussian noise and Perlin noise. It is difficult to quantify the reliability of the noise simulation. In each real scan, the noise is different, both the noise introduced by the sensor and the speckle. However, we can verify the similarity between simulation and reality in our proposal.

This proposal allows the simulation of any commercial laser triangulation sensor from its datasheet, emulating its measurements, including both the measurement noise due to the limitations of the sensor itself and the noise generated by the surface of the scanned product. We allow a simulation of the noise in a relatively simple way, considering surface roughness parameters, without entering into complicated physical procedures.

We successfully simulated the speckle generated by the roughness of different types of materials, focusing specifically on three steel products manufactured using different fabrication methods. The results of our study highlight the accuracy of our modeling and simulation process for a wide range of surfaces. Comparison between the actual scans and our simulated results for the bearing cap, cab door, and plate surfaces consistently showed similar roughness characteristics. However, by generalizing the noise model using roughness parameters, we can simulate different surfaces, based on a priori knowledge of the type of material and surface finish or by a simple analysis of roughness parameters from a real scan.

The comparative analysis highlights small differences between the simulated and real scans for the roughness parameters in the three experiments. The small errors for the roughness parameters indicate that the simulation approach effectively captures the general roughness characteristics of the surfaces. Therefore, the results suggest that the current simulation method adequately reproduces the essential roughness features.

Overall, the results presented in this work validate the effectiveness of the simulation approach for reproducing the real scans. The similarity between the real and simulated scans, as observed in both the visual comparison and the quantitative analysis, indicates that the simulation accurately captures the essential roughness characteristics of each surface. The ability to accurately simulate surface roughness opens up possibilities for using simulated scans in different algorithms and applications, offering a cost-effective and efficient alternative to extensive experimental measurements.

The purpose of the proposed model is not to precisely replicate real measurements but rather to enable the simulation of real measurements obtained from commercial sensors. Speckle is an interference effect caused by the microtopology of the inspected surface, due to the spatial coherence of the illuminating source. This noise is simulated using a Perlin noise model, avoiding the complex real-world physical processes that produce it. This allows us to simulate measurements equivalent to actual measurements in a simpler and faster way. However, this limitation prevents us from using the simulator for the creation of new sensors or for other purposes, such as calculating the level of influence of error induced by speckle or other sources. However, this allows us to simulate realistic scans using 3D models that do not provide any roughness information.

Also, this work allows for a test bench to try different data acquisition methods. The most optimal trajectories of the sensors can be searched to scan different types of parts and see how this affects the 3D reconstruction and the identification and measurement of the defects. This is useful to allow the design of the acquisition system prior to its installation on product lines. To ensure measurement accuracy, companies need to build complex systems with expensive high-accuracy sensors on production lines. Also, the design of the acquisition system is performed before knowing the optimal configuration. This leads to increased costs and development time. Therefore, it is important to obtain a system that provides a realistic simulation of the acquisition system and the defects to be inspected.

As part of our future research directions, the focus will be directed towards surface defect detection in manufacturing using profilometric sensors. Many industries still rely on manual inspection due to the limited defect data for developing defect detection algorithms. Collecting authentic defect data is an arduous and costly task, especially during development. In addition, the sporadic nature of defects in industrial processes adds another layer of complexity. To address these problems, future work aims to enable the creation of a synthetic defect database including dimensional data. To this end, work is being conducted to simulate real defects in 3D models. From this, the scanning process of profilometric sensors will be reproduced using the simulator proposed in this paper. It is important to highlight that noise has an impact on measurements, so it is the key to accurately replicating it in the simulated measurements, as explained in this work. Having a bank of defects is crucial, especially when it comes to developing artifical intelligence algorithms effectively.

5. Conclusions

This work proposes a simulation model to replicate the readings of a real laser triangulation profilometric sensor, which provides high-accuracy depth measurements over a line. The model is composed of three different components: a geometrical model accounting for the sensor's physical attributes, a Gaussian noise model simulating sensor uncertainty, and a Perlin noise model mimicking speckle-induced noise.

The presented approach was validated through experiments that compare the 3D scans generated by the proposed model to those provided by a real laser triangulation profilometric sensor. Results found that the simulator is capable of accurately reproducing the geometrical characteristics of the scanned surface, as well as the noise patterns introduced by the material roughness and the sensor itself. The generated scans had a high degree of similarity to the real scans.

The results of the research have provided quite satisfactory results. Overall, the proposed geometrical and noise models allow for the creation of a simulation of an off-the-self sensor that accurately represents the behavior of the real sensor, including the effects of the different noise sources, especially speckle. This simulation model stands as a valuable asset for the design and evaluation of real-world applications, facilitating the assessment of data processing and analysis algorithms before physical system deployment. In this way, better solutions may be achieved while reducing the development time and costs.

Author Contributions: Conceptualization, I.Á. and S.R.-H.; methodology, S.R.-H. and I.Á.; software, S.R.-H.; validation, S.R.-H., M.R.-H. and R.C.G.; formal analysis, S.R.-H., R.C.G. and M.R.-H.; investigation, S.R.-H. and M.R.-H.; resources, M.R.-H. and S.R.-H.; data curation, S.R.-H. and M.R.-H.; writing—original draft preparation, S.R.-H. and R.C.G.; All authors have reviewed the manuscript. All authors have read and agreed to the published version of the manuscript.

Funding: This research was funded by the Severo Ochoa Program for predoctoral research and teaching with Ref: PA-20-PF-BP19-067, from the Asturias Regional Government.

Institutional Review Board Statement: Not applicable.

Informed Consent Statement: Not applicable.

Data Availability Statement: The data presented in this study are available from the corresponding authors on request.

Acknowledgments: The authors wish to thank Desarrollo de Soluciones Integrales Plus S.L. for providing the data used in these experiments.

Conflicts of Interest: The authors declare no conflict of interest.

Abbreviations

The following abbreviations are used in this manuscript:

A_m	Amplitude of Perlin noise function m-th octave
A_0	Perlin noise algorithm base amplitude
CAD	Computer-aided design
CDF	Cumulative distribution function
$C(x_i, y_i)$	Cell corners surrounding point (x, y)
$d(x, y)$	Depth from sensor to object at point (x, y)
Δ_z	Sensor depth resolution
f_m	Frequency of Perlin noise function m-th octave
f_0	Perlin noise algorithm base frequency
FOV	Field of view
$n(i, j)$	Influence value of corner $C(x_i, y_i)$ for point (x, y)
$N_m(f_m x, f_m y)$	m-th noise function component

$n_{octaves}$	Number of terms (octaves) used to build the Perlin noise
$n_{profiles}$	Number of scanned profiles
p	Perlin noise algorithm persistence factor
P_c	Number of peaks in the profile
$P_G(0, \sigma)$	Error due to sensor accuracy limits
$PN(x, y)$	Perlin noise value at point (x, y)
$\hat{r}(x, y)$	Simulated sensor reading at point (x, y)
ppf	Points per profile
R_a	Arithmetic average height
R_q	Root mean square roughness
R_p	Maximum height of peaks
R_v	Maximum depth of valleys
R_pm	Mean of the height of the 10 highest peaks
R_vm	Mean of the depth of the 10 lowest valleys
σ_{rep}	Sensor repeatability
STL	Standard Tessellation Language
W_d	Sensor working distance
$X - FOV$	Field of view along the horizontal direction
Z_R	Sensor Z-Range value

References

1. Mohammadikaji, M.; Bergmann, S.; Beyerer, J.; Burke, J.; Dachsbacher, C. Sensor-Realistic Simulations for Evaluation and Planning of Optical Measurement Systems with an Application to Laser Triangulation. *IEEE Sens. J.* **2020**, *20*, 5336–5349. [CrossRef]
2. Gamarra Acosta, M.R.; Vélez Díaz, J.C.; Castro, N.S. An Innovative Image-Processing Model for Rust Detection Using Perlin Noise to Simulate Oxide Textures. *Corros. Sci.* **2014**, *88*, 141–151. [CrossRef]
3. Häusler, G. Three-Dimensional Sensors - Potentials and Limitations. In *Handbook of Computer Vision and Applications, Volume 1: Sensors and Imaging*; Jähne, B., Haussecker, H., Geissler, P., Eds.; Academic Press: Cambridge, MA, USA, 1999; Chapter 19, pp. 485–506.
4. Álvarez, I.; Enguita, J.M.; Frade, M.; Marina, J.; Ojea, G. On-Line Metrology with Conoscopic Holography: Beyond Triangulation. *Sensors* **2009**, *9*, 7021–7037. [CrossRef] [PubMed]
5. Cajal, C.; Santolaria, J.; Samper, D.; Garrido, A. Simulation of laser triangulation sensors scanning for design and evaluation purposes. *Int. J. Simul. Model.* **2015**, *14*, 250–264. [CrossRef]
6. Abu-Nabah, B.A.; ElSoussi, A.O.; Al Alami, A.E.K. Virtual Laser Vision Sensor Environment Assessment for Surface Profiling Applications. *Measurement* **2018**, *113*, 148–160. [CrossRef]
7. Mohammadikaji, M.; Bergmann, S.; Irgenfried, S.; Beyerer, J.; Dachsbacher, C.; Wörn, H. A Framework for Uncertainty Propagation in 3D Shape Measurement Using Laser Triangulation. In Proceedings of the 2016 IEEE International Instrumentation and Measurement Technology Conference Proceedings, Taipei, Taiwan, 23–26 May 2016; pp. 1–6. [CrossRef]
8. Csencsics, E.; Schlarp, J.; Glaser, T.; Wolf, T.; Schitter, G. Simulation and Reduction of Speckle-induced Uncertainty in Laser Triangulation Sensors. In Proceedings of the 2022 IEEE International Instrumentation and Measurement Technology Conference (I2MTC), Ottawa, ON, Canada, 16–19 May 2022; pp. 1–6. [CrossRef]
9. Beermann, R.; Quentin, L.; Stein, G.; Reithmeier, E.; Kästner, M. Full Simulation Model for Laser Triangulation Measurement in an Inhomogeneous Refractive Index Field. *Opt. Eng.* **2018**, *57*, 114107. [CrossRef]
10. Perlin, K. An Image Synthesizer. In Proceedings of the 12th Annual Conference on Computer Graphics and Interactive Techniques, SIGGRAPH '85, Association for Computing Machinery: New York, NY, USA, 1985; pp. 287–296. [CrossRef]
11. Perlin, K. Improving Noise. *ACM Trans. Graph.* **2002**, *21*, 681–682. [CrossRef]
12. Lagae, A.; Lefebvre, S.; Cook, R.; DeRose, T.; Drettakis, G.; Ebert, D.; Lewis, J.; Perlin, K.; Zwicker, M. A Survey of Procedural Noise Functions. *Comput. Graph. Forum* **2010**, *29*, 2579–2600. [CrossRef]
13. Dong, J.; Liu, J.; Yao, K.; Chantler, M.; Qi, L.; Yu, H.; Jian, M. Survey of Procedural Methods for Two-Dimensional Texture Generation. *Sensors* **2020**, *20*, 1135. [CrossRef] [PubMed]
14. Li, H.; Yang, H.; Xu, C.; Cao, Y., Water Surface Simulation Based on Perlin Noise and Secondary Distorted Textures. In *Advances in Computer Science and Ubiquitous Computing*; Springer: Singapore, 2017; pp. 236–245. [CrossRef]
15. Conde-Rodríguez, F.; García-Fernández, L.; Torres, J.C. Modelling Material Microstructure Using the Perlin Noise Function. *Comput. Graph. Forum* **2021**, *40*, 195–208. [CrossRef]
16. Koutsoudis, A.; Ioannakis, G.; Vidmar, B.; Arnaoutoglou, F.; Chamzas, C. Using noise function-based patterns to enhance photogrammetric 3D reconstruction performance of featureless surfaces. *J. Cult. Herit.* **2015**, *16*, 664–670. [CrossRef]

17. Möller, T.; Trumbore, B. Fast, Minimum Storage Ray/Triangle Intersection. In Proceedings of the SIGGRAPH '05: ACM SIGGRAPH 2005 Courses, Los Angeles, CA, USA, 31 July–4 August 2005; Association for Computing Machinery: New York, NY, USA, 2005; p. 7-es. [CrossRef]
18. Gadelmawla, E.; Koura, M.; Maksoud, T.; Elewa, I.; Soliman, H. Roughness parameters. *J. Mater. Process. Technol.* **2002**, *123*, 133–145. [CrossRef]
19. Schroeder, W.; Martin, K.; Lorensen, B. *The Visualization Toolkit–An Object-Oriented Approach To 3D Graphics*, 4th ed.; Kitware, Inc.: New York, NY, USA, 2006.
20. Automation Technology. *High-Speed 3D Compact Sensors Fast, Accurate 3D Imaging for Industry.* 2019. Version 2.5. Available online: https://www.automationtechnology.de/cms/wp-content/uploads/2020/05/AT_BR_CS_Series_web.pdf (accessed on 28 August 202)
21. Crankshaft Bearing Cap [CAT Crankshaft Bearing Cap]. Available online: https://parts.cat.com/es/catcorp/167-8081 (accessed on 9 May 2023).
22. Massey, F.J. The Kolmogorov-Smirnov Test for Goodness of Fit. *J. Am. Stat. Assoc.* **1951**, *46*, 68–78. [CrossRef]

sensors

MDPI

Article

Sensor Fusion of GNSS and IMU Data for Robust Localization via Smoothed Error State Kalman Filter

Yuming Yin [1], Jinhong Zhang [1], Mengqi Guo [2], Xiaobin Ning [1,*], Yuan Wang [1] and Jianshan Lu [1]

[1] College of Mechanical Engineering, Zhejiang University of Technology, Hangzhou 310014, China
[2] College of Engineering, Beijing Forestry University, Beijing 100083, China
* Correspondence: nxb@zjut.edu.cn

Abstract: High−precision and robust localization is critical for intelligent vehicle and transportation systems, while the sensor signal loss or variance could dramatically affect the localization performance. The vehicle localization problem in an environment with Global Navigation Satellite System (GNSS) signal errors is investigated in this study. The error state Kalman filtering (ESKF) and Rauch–Tung–Striebel (RTS) smoother are integrated using the data from Inertial Measurement Unit (IMU) and GNSS sensors. A segmented RTS smoothing algorithm is proposed in order to estimate the error state, which is typically close to zero and mostly linear, which allows more accurate linearization and improved state estimation accuracy. The proposed algorithm is evaluated using simulated GNSS signals with and without signal errors. The simulation results demonstrate its superior accuracy and stability for state estimation. The designed ESKF algorithm yielded an approximate 3% improvement in long straight line and turning scenarios compared to classical EKF algorithm. Additionally, the ESKF−RTS algorithm exhibited a 10% increase in the localization accuracy compared to the ESKF algorithm. In the double turning scenarios, the ESKF algorithm resulted in an improvement of about 50% in comparison to the EKF algorithm, while the ESKF−RTS algorithm improved by about 50% compared to the ESKF algorithm. These results indicated that the proposed ESKF−RTS algorithm is more robust and provides more accurate localization.

Keywords: error state Kalman; RTS smoothing; robust and accurate localization

1. Introduction

The Global Navigation Satellite System (GNSS) has found widespread usage in vehicle measurement, intelligent driving, and robot navigation due to its stable long−term localization performance [1–3]. However, the susceptibility of satellite signals to external interference would lead to the GNSS having unreliable localization accuracy. Therefore, the combination of GNSS and Inertial Navigation System (INS) has been widely employed, in which both systems could compensate for each other's drawbacks and leverage their respective strengths to achieve continuous localization [4–6]. The reported studies have investigated the combined filtering algorithm of GNSS and INS, GNSS/IMU can provide position, velocity, and attitude information for vehicle control. Liu [7,8] presents a novel pitch and roll feedback mechanism that utilizes intrinsic information of the vehicle such as steering angle and wheel speed. To compensate for the cumulative velocity errors that occur during low sampling intervals of GNSS, the integration of reverse smoothing and grey prediction is employed. Shin and Naser [9] reported that the algorithm yielded better navigation and localization results, even in cases of short−term GNSS loss. Han [10] compared the combined satellite/inertial guidance parameter estimation results with those of a single satellite system, with the former yielding superior results. Li and Zhang [11] studied the combined GNSS/INS navigation algorithm and highlighted that the filtering algorithm still struggled to provide accurate parameter estimation in situations where few satellite data were available. Erfianti [12] researched a combined GNSS/INS navigation algorithm and discovered that the INS could only provide high−accuracy parameter estimation results for a short duration. During the satellite signal occlusion, the forward filtering algorithm alone thus cannot guarantee a robust localization.

Applying the extended Kalman filter (EKF) to estimate the motion of vehicle systems is well desirable due to the system nonlinearity [13–16]. The EKF linearizes the nonlinear model by approximating it with a first−order Taylor series around the state estimate and then estimates the state using the Kalman filter. M. M. Atia [17] proposed utilizing the extended Kalman filter to merge data from inertial sensors with GNSS data in a loosely coupled mode to improve the accuracy of road network maps in determining lanes. The resulting lane determination success rate was an impressive 97.14%. However, the local linearization operation of this approach can introduce significant estimation errors. Julier [18] introduced an Unscented Kalman filter algorithm that improves the estimation results, while its high complexity makes it unsuitable for most inertial guidance devices. Arasaratnam [19] et al. pointed out through theoretical analysis and simulation verification that the UKF has poor filtering performance or even divergence in solving the high−dimensional ($\geq$20) nonlinear state estimation problem. For this reason, they used the spherical radial rule to approximate the state posterior distribution in the optimal framework, and then proposed the Cubature Kalman Filter (CKF), but the CKF has higher computational complexity and requires more sampling and operations, resulting in poor real−time performance, which is not suitable for some applications with high real−time performance requirements. The extended Kalman filter thus remains the mainstream state estimation algorithm, and developing a low−complexity filter with high accuracy is still challenging [20,21].

To address this challenge, S. S Kourabbaslou [22] presents a flexible design framework utilizing symbolic engines to represent and linearize system and measurement models. A robust fixed−lag smoothing approach is proposed in case there is a mismatch between the nominal model and the actual model [23,24]. To improve the accuracy of vehicle stand−alone localization in highly dynamic driving conditions during GNSS outages, Gao [25] proposed a vehicle localization system based on vehicle chassis sensors considering vehicle lateral velocity. The Kalman filter combines vehicle states obtained from vehicle kinematics and dynamics to improve the reliability and accuracy of autonomous driving. A consensus−based and vehicle kinematics/dynamics integrated autonomous vehicle sideslip angle estimation algorithm based on GNSS/INS was proposed [26,27]. Madyastha [28] proposed a Kalman filtering method based on attitude error states. The error state Kalman filter (ESKF) is designed for covariance estimation, which serves as a weight for the inertial odometry optimization process [29–31]. These results validate the ESKF's effectiveness for error−state estimation, although a full−state estimation based on ESKF is not available yet. Further exploration of state estimation based on ESKF is thus favorable.

The major challenge in designing a sensor fusion algorithm for state estimation is to address GNSS signal errors and low−cost hardware limitations with a smoother approach. The errors of inertial navigation systems accumulate over time, while GNSS signals are heavily influenced by satellite signal quality and urban obstructions [32–34]. These would lead to significant localization errors. Using INS alone cannot provide accurate position estimation for a long term in signal−obscured environments [35–37]. In such cases, real−time high−accuracy localization is necessary, and data fusion from multiple sensors is often required [38,39]. In order to provide reliable localization information, simultaneous localization and map construction (SLAM) systems can be used, as well as multi−sensor integrated systems that combine LiDAR [40] and navigation systems to solve problems caused by GNSS antenna failures. To reduce GNSS and Inertial Measurement Unit (IMU) fusion costs and achieve desired accuracy, different navigation satellite [41–43] constellations need to be combined, especially in urban working environments.

In this paper, the low−cost sensor combination of only GNSS and IMU is considered, and effects of GNSS signal error and different fusion algorithm design are investigated. The extended Kalman filter and error state Kalman filter are designed in Section 2. The application of the error−extended Kalman filter and the RTS optimal smoothing algorithm is investigated in Section 3 for combined inertial guidance and GNSS localization in urban working environments. The insights and conclusions on the performance of the designed sensor fusion algorithms are further analyzed.

2. GNSS and IMU Integrated Filter Design

The overall sensor fusion framework integrating the GNSS and IMU sensor data with significant GNSS signal errors is illustrated in Figure 1. It mainly consists of four procedures, including data analysis, prediction process, update process and reverse smoothing, contributing to the developed ESKF−RTS smoothing localization algorithm. In particular, the algorithm defines a nominal state without considering the measurement noise of IMU and system disturbance, and an error state containing the noise and disturbance information is used for state estimation. The nominal state condition and error state prediction are updated simultaneously, and the error state is corrected using GNSS signal measurement and injected into the nominal state. The error state and its covariance matrix are subsequently reset. The acquisition frequency for GNSS data is 1 Hz, while the IMU data are acquired at a frequency of 100 Hz. The reverse time update and reverse segmentation smoothing are parallelly performed, which will be detailed below.

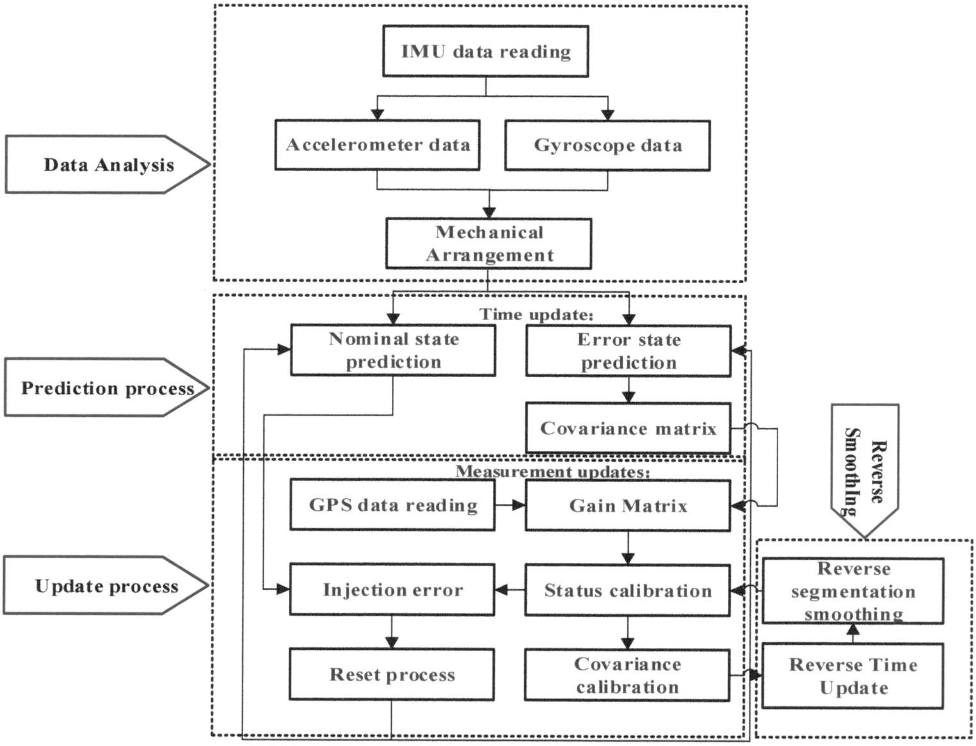

Figure 1. Flow chart of the designed ESKF−RTS algorithm.

2.1. Extended Kalman Filter

The extended Kalman filter (EKF) is also designed for comparison, using the sensor measurements directly. The EKF procedure is described briefly.

The developed discrete nonlinear vehicle system equations are:

$$\begin{cases} \boldsymbol{X}_k = f(\boldsymbol{X}_{k-1}) + \boldsymbol{W}_{k-1} \\ \boldsymbol{Z}_K = h(\boldsymbol{X}_k) + \boldsymbol{V}_k \end{cases}, \tag{1}$$

where $f(\cdot)$ and $h(\cdot)$ represent the state function and measurement function of nonlinear vehicle systems, $\boldsymbol{W}$ and $\boldsymbol{V}$ are Gaussian white noise of state $\boldsymbol{X}$ and measurement $\boldsymbol{Z}$, respectively, and k represents the discrete time step.

The state function of the nonlinear vehicle system in Equation (1) is approximated through local linearization. The nonlinear equation is expanded using a Taylor series, and only the first−order term is retained while ignoring the second and higher−order terms. This yields the following expression:

$$\boldsymbol{X}_k = f(\breve{\boldsymbol{X}}_{k-1}) + \boldsymbol{F}_{k-1}(\boldsymbol{X}_{k-1} - \breve{\boldsymbol{X}}_{k-1}) + \boldsymbol{W}_{k-1}, \tag{2}$$

where $\boldsymbol{F}_{k-1}$ is the Jacobian matrix of the function of the function $f(\boldsymbol{X}_{k-1})$.

The statistical characteristics of the predicted state can be obtained based on the results of the linearization in the form of

$$\widehat{\boldsymbol{X}}_{k,k-1} = E[f(\breve{\boldsymbol{X}}_{k-1}) + \boldsymbol{F}_{k-1}(\boldsymbol{X}_{k-1} - \breve{\boldsymbol{X}}_{k-1}) + \boldsymbol{W}_{k-1}] = f(\breve{\boldsymbol{X}}_{k-1}). \tag{3}$$

It is possible to further predict the variance matrix as

$$\begin{aligned} \boldsymbol{P}_{k,k-1} &= E[(\boldsymbol{X}_k - \breve{\boldsymbol{X}}_{k,k-1})(\boldsymbol{X}_k - \breve{\boldsymbol{X}}_{k,k-1})^T] \\ &= E\left\{[\boldsymbol{F}_{k-1}(\boldsymbol{X}_{k-1} - \breve{\boldsymbol{X}}_{k-1}) + \boldsymbol{W}_{k-1}][\boldsymbol{F}_{k-1}(\boldsymbol{X}_{k-1} - \breve{\boldsymbol{X}}_{k-1}) + \boldsymbol{W}_{k-1}]^T\right\}. \\ &= \boldsymbol{F}_{k-1}\boldsymbol{P}_{k-1}\boldsymbol{F}_{k-1}^T + \boldsymbol{Q}_{k-1} \end{aligned} \tag{4}$$

The same method is used for the measurement function $h(x_k)$ at the point $\hat{x}_k$, which can be obtained as

$$\boldsymbol{Z}_K = h(\breve{\boldsymbol{X}}_k) + \boldsymbol{H}_k(\boldsymbol{X}_k - \breve{\boldsymbol{X}}_k) + \boldsymbol{V}_k, \tag{5}$$

where $\boldsymbol{H}_k$ is the Jacobian matrix of the function of the function $h(\boldsymbol{X}_{k-1})$. The prediction of measurement is calculated as

$$\widehat{\boldsymbol{Z}}_{k,k-1} = E[h(\breve{\boldsymbol{X}}_{k-1}) + H_{k-1}(\boldsymbol{X}_{k-1} - \breve{\boldsymbol{X}}_{k-1}) + \boldsymbol{V}_{k-1}] = h(\breve{\boldsymbol{X}}_{k-1}). \tag{6}$$

Then, the collaborative variance can be further predicted as

$$\begin{aligned} \boldsymbol{P}_{k,k-1} &= E[(\boldsymbol{Z}_k - \breve{\boldsymbol{Z}}_{k,k-1})(\boldsymbol{Z}_k - \breve{\boldsymbol{Z}}_{k,k-1})^T] \\ &= E\left\{[H_{k-1}(\boldsymbol{X}_{k-1} - \breve{\boldsymbol{X}}_{k-1}) + \boldsymbol{V}_{k-1}][H_{k-1}(\boldsymbol{X}_{k-1} - \breve{\boldsymbol{X}}_{k-1}) + \boldsymbol{V}_{k-1}]^T\right\}. \\ &= \boldsymbol{H}_{k-1}\boldsymbol{P}_{k-1}\boldsymbol{H}_{k-1}^T + \boldsymbol{R}_{k-1} \end{aligned} \tag{7}$$

Using the information above, the iterative process of the EKF algorithm for the sensor fused localization can be summarized as

$$\begin{aligned} &\breve{\boldsymbol{X}}_k = f(\breve{\boldsymbol{X}}_{k-1}) \\ &\breve{\boldsymbol{P}}_k = \boldsymbol{F}_{k-1}\boldsymbol{P}_{k-1}\boldsymbol{F}_{k-1}^T + \boldsymbol{Q}_{k-1} \\ &\boldsymbol{K}_{k+1} = \breve{\bar{\boldsymbol{P}}}_{k+1}\boldsymbol{H}_{k+1}^T(\boldsymbol{H}_{k+1}\breve{\bar{\boldsymbol{P}}}_{k+1}\boldsymbol{H}_{k+1}^T + \boldsymbol{V}_{k+1})^{-1} \\ &\boldsymbol{X}_{k+1} = \boldsymbol{K}_{k+1}(\boldsymbol{Z}_{k+1} - \boldsymbol{h}(\breve{\boldsymbol{X}}_{k+1})) \\ &\boldsymbol{P}_{k+1} = (\boldsymbol{I} - \boldsymbol{K}_{k+1}\boldsymbol{H}_{k+1})\breve{\bar{\boldsymbol{P}}}_{k+1} \end{aligned}. \tag{8}$$

2.2. Error State Kalman filter

For the formulation of the error state filter, the system states are defined as the true, nominal, and error state values. The true state is expressed as a combination of the nominal and error states. The approach is to treat the nominal state as a dominant signal, which is highly nonlinear, and the error state as a small signal, which is linearly effective and suitable for linear Gaussian filtering. The nominal state vector of $[\boldsymbol{p}, \boldsymbol{v}, \boldsymbol{q}, \boldsymbol{a}_b, \boldsymbol{\omega}_b]^T$ is used in this study, where the additional dimension appears due to the quaternion representation used for rotation, the quaternion method is widely used in attitude update because it only requires the calculation of matrix differential equations, which is relatively small in computation and can avoid the singular value problem of Euler angles. In this paper, the quaternion method is used to solve the attitude of the carrier. The operators $\boldsymbol{q}$ and $\boldsymbol{R}$ represent, respectively, the quaternion corresponding to the axial angular vector $\boldsymbol{\theta}$ and its rotation matrix. In addition, the error state vector is $[\delta\boldsymbol{p}, \delta\boldsymbol{v}, \delta\boldsymbol{q}, \delta\boldsymbol{a}_b, \delta\boldsymbol{\omega}_b]^T$. The relevant symbol definitions of true, nominal, and error state are listed and compared in Table 1.

Table 1. The definition of variable symbols.

Name	True State	Nominal State	Error States
All states	x_t	x	δx
Location	p_t	p	δp
Speed	v_t	v	δv
Quaternion	q_t	q	δq
Rotation matrix	R_t	R	δR
Angular vectors	θ_t	θ	$\delta\theta$
Acceleration bias	a_{bt}	a_b	δa_b
Angular velocity bias	ω_{bt}	ω_b	$\delta\omega_b$

2.2.1. Continuous Time Kinetic Model

The nominal state kinematics refers to the system modelled with system noise or external perturbations in the form of

$$\begin{aligned} \dot{\boldsymbol{p}}_t &= \boldsymbol{v}_t \\ \dot{\boldsymbol{v}}_t &= \boldsymbol{R}_t(\boldsymbol{a}_m - \boldsymbol{a}_{bt} - \boldsymbol{a}_n) + \boldsymbol{g}_t \\ \dot{\boldsymbol{q}}_t &= \tfrac{1}{2}\boldsymbol{q}_t \otimes (\boldsymbol{\omega}_m - \boldsymbol{\omega}_{bt} - \boldsymbol{\omega}_n). \\ \dot{\boldsymbol{a}}_{bt} &= \boldsymbol{a}_\omega \\ \dot{\boldsymbol{\omega}}_{bt} &= \boldsymbol{\omega}_\omega \end{aligned} \tag{9}$$

The acceleration and angular velocity measurements are represented by $\boldsymbol{a}_m$ and $\boldsymbol{\omega}_m$, respectively, while $\boldsymbol{a}_n$ and $\boldsymbol{\omega}_n$ represent the corresponding noise vectors, and $\boldsymbol{a}_\omega$ and $\boldsymbol{\omega}_\omega$ represent the bias vectors for acceleration and angular velocity, respectively. The equations driving the system dynamics in the nominal state are as follows, where the nominal state refers to the modeled system without any noise or perturbations:

$$\begin{aligned} \dot{\boldsymbol{p}} &= \boldsymbol{v} \\ \dot{\boldsymbol{v}} &= \boldsymbol{R}(\boldsymbol{a}_m - \boldsymbol{a}_b) + \boldsymbol{g} \\ \dot{\boldsymbol{q}} &= \tfrac{1}{2}\boldsymbol{q} \otimes (\boldsymbol{\omega}_m - \boldsymbol{\omega}_b). \\ \dot{\boldsymbol{a}}_b &= 0 \\ \dot{\boldsymbol{\omega}}_b &= 0 \end{aligned} \tag{10}$$

Solving for the error state and simplifying all second−order infinitesimals. The kinetic equation for the error state can be obtained from Equations (9) and (10).

$$\begin{aligned} ffi\dot{p} &= ffiv \\ ffi\dot{v} &= \mathbf{R}(\boldsymbol{a}_m - \boldsymbol{a}_b)ffi\boldsymbol{\theta} - Rffi\boldsymbol{a}_b - R\boldsymbol{a}_n \\ ffi\dot{\boldsymbol{\theta}} &= -(\boldsymbol{\omega}_m - \boldsymbol{\omega}_b)ffi` - ffi\boldsymbol{\omega}_b - \boldsymbol{\omega}_n. \\ ffi\dot{\boldsymbol{a}}_b &= \boldsymbol{a}_\omega \\ ffi\dot{\boldsymbol{\omega}}_b &= \boldsymbol{\omega}_\omega \end{aligned} \tag{11}$$

2.2.2. Discrete Time Kinetic Model

In order to apply the derived state and measurement equations to the sensor fusion filter, they must be discretized based on the sampling time. The original continuous time equations are transformed into their discrete time form.

The recursive expression for the motion model of the nominal state is

$$\begin{aligned} \boldsymbol{p}_{(k+1)} &= \boldsymbol{p}_k + \boldsymbol{v}_k\Delta t + \tfrac{1}{2}[\mathbf{R}_k(\boldsymbol{a}_{mk} - \boldsymbol{a}_{bk}) + g]\Delta t^2 \\ \boldsymbol{v}_{(k+1)} &= \boldsymbol{v}_k + [\mathbf{R}_k(\boldsymbol{a}_{mk} - \boldsymbol{a}_{bk}) + g_k]\Delta t \\ \boldsymbol{q}_{(k+1)} &= \boldsymbol{q}_k \otimes \boldsymbol{q}_k((\boldsymbol{\omega}_{mk} - \boldsymbol{\omega}_{bk})\Delta t) \\ \boldsymbol{a}_b(k+1) &= \boldsymbol{a}_{bk} \\ \boldsymbol{\omega}_b(k+1) &= \boldsymbol{\omega}_{bk} \end{aligned} \tag{12}$$

The kinematic model of the error state is expressed recursively as a function of the non−negative integer k representing the kth time step:

$$\begin{aligned} ffi\boldsymbol{p}_{k+1} &= ffi\boldsymbol{p}_k + ffi\boldsymbol{v}_k\Delta t \\ ffi\boldsymbol{v}_{k+1} &= [-\mathbf{R}_k(a_{mk} - a_{bk})ffi\boldsymbol{\theta}_k - \mathbf{R}_k ffi\boldsymbol{a}_{bk}]\Delta t + ffi\boldsymbol{v}_k - \boldsymbol{w}_{vk} \\ ffi\boldsymbol{\theta}_{k+1} &= \mathbf{R}_k^T((\boldsymbol{\omega}_{mk} - \boldsymbol{\omega}_{bk})\Delta t)ffi\boldsymbol{\theta}_k - ffi\boldsymbol{\omega}_{bk}\Delta t + \boldsymbol{w}_{`k} \\ ffi\boldsymbol{a}_{b(k+1)} &= ffi\boldsymbol{a}_{bk} + \boldsymbol{w}_{ak} \\ ffi\boldsymbol{\omega}_{b(k+1)} &= ffi\boldsymbol{\omega}_{bk} + \boldsymbol{w}_{wk} \end{aligned} \tag{13}$$

The Gauss random noise for velocity, attitude, acceleration bias, and angular velocity bias are denoted as $\boldsymbol{w}_{vk}, \boldsymbol{w}_{`k}, \boldsymbol{w}_{ak}, , \boldsymbol{w}_{!k}$, their mean is zero, and their covariance matrices are obtained by integrating the covariances of $\sigma_{a_n}, \sigma_{\omega_n}, \sigma_{a_\omega}$ and σ_{ω_ω} over the step time Δt.

$$\begin{aligned} \boldsymbol{W}_v &= \sigma_{a_n}^2\Delta t^2\boldsymbol{I} \\ \boldsymbol{W}_` &= \sigma_{\omega_n}^2\Delta t^2\boldsymbol{I} \\ \boldsymbol{W}_A &= \sigma_{a_!}^2\Delta t\boldsymbol{I} \\ \boldsymbol{W}_\Omega &= \sigma_{\omega_!\Delta t}^2\boldsymbol{I} \end{aligned} \tag{14}$$

The standard deviation of Gaussian white noise for acceleration and angular velocity are denoted by σ_{a_n} and σ_{ω_n}, respectively. Similarly, σ_{a_ω} and σ_{ω_ω} are used to represent the standard deviation of Gaussian white noise for acceleration and angular velocity bias, respectively.

2.2.3. Development of the Error State Model

The discrete vector forms for all states, error states, IMU measurements, and noise are defined as follows:

$$\begin{aligned} \boldsymbol{x}_k &= [\boldsymbol{x}_k, \boldsymbol{v}_k, \boldsymbol{q}_k, \boldsymbol{a}_{bk}, \boldsymbol{\omega}_{bk}]^T, ffi\boldsymbol{x}_k = [ffi\boldsymbol{p}_k, ffi\boldsymbol{v}_k, ffi\boldsymbol{q}_k, ffi\boldsymbol{a}_{bk}, ffi\boldsymbol{\omega}_{bk}]^T, \\ \boldsymbol{u}_{mk} &= [\boldsymbol{a}_{mk}, \boldsymbol{\omega}_{mk}]^T, \boldsymbol{w}_k = [\boldsymbol{w}_{vk}, \boldsymbol{w}_{`k}, \boldsymbol{w}_{ak}, \boldsymbol{w}_{!k}]^T \end{aligned} \tag{15}$$

The recursive equation for all the error state is derived by combining Equations (5) and (7) in the form of

$$ffi\boldsymbol{x}_{k+1} = f_{ffi}(\boldsymbol{x}_k, ffi\boldsymbol{x}_k, \boldsymbol{u}_{mk}, \boldsymbol{w}_k) = \boldsymbol{F}_{xk}(\boldsymbol{x}_k, \boldsymbol{u}_{mk}) + \boldsymbol{F}_{wk}\boldsymbol{w}_k, \tag{16}$$

where $f_{\delta}(\cdot)$ represents the recursive function of the error state, the F_{xk} and F_{wk} are the Jacobian matrices corresponding to the error and noise states, respectively, which can be derived as

$$F_{xk} = \begin{bmatrix} I & I\Delta t & 0 & 0 & 0 \\ 0 & I & -R_k^T(\omega_{mk} - \omega_{bk})\Delta t & -R_k\Delta t & 0 \\ 0 & 0 & R_k^T(\omega_{mk} - \omega_{bk})\Delta t & 0 & -I\Delta t \\ 0 & 0 & 0 & I & 0 \\ 0 & 0 & 0 & 0 & I \end{bmatrix}, F_{wk} = \begin{bmatrix} 0 & 0 & 0 & 0 \\ I & 0 & 0 & 0 \\ 0 & I & 0 & 0 \\ 0 & 0 & I & 0 \\ 0 & 0 & 0 & I \end{bmatrix}. \tag{17}$$

2.2.4. The ESKF Prediction Process

The error states and the covariance prediction process can be obtained as

$$\begin{aligned} ffi\breve{\bar{x}}_{k+1} &= F_{xk}(\hat{x}_k, u_{mk}) ffix_k \\ \breve{\bar{P}}_{(k+1)} &= F_{xk}\breve{P}_k F_{xk}^T + F_{wk} Q_w F_{wk}^T \end{aligned} \tag{18}$$

where Q_w represents the noise covariance matrix of the form:

$$Q_w = \begin{bmatrix} W_v & 0 & 0 & 0 \\ 0 & W_\theta & 0 & 0 \\ 0 & 0 & W_A & 0 \\ 0 & 0 & 0 & W_\Omega \end{bmatrix}. \tag{19}$$

2.2.5. The ESKF Observation Process

Once the GNSS information becomes available, the observation is incorporated to continuously update the filter. This process also involves calibrating for the accelerometer and gyroscope biases.

The observation equation is typically expressed in a more uniform form within the filter in the form of

$$z_k = h(x_{tk}) + w_{mk}, \tag{20}$$

where z_k is the measurement signal vector, w_{mk} represents the Gaussian white noise of the measurement signal, and its covariance is V. The error calibration update equation is derived as

$$\begin{aligned} K_{k+1} &= \breve{\bar{P}}_{k+1} H_{k+1}^T (H_{k+1} \breve{\bar{P}}_{k+1} H_{k+1}^T + V_{k+1})^{-1} \\ ffi\hat{X}_{k+1} &= K_{k+1}(Z_{k+1} - H(\breve{X}_{k+1})) \\ \hat{P}_{k+1} &= (I - K_{k+1} H_{k+1}) \breve{\bar{P}}_{k+1} \end{aligned} \tag{21}$$

The Jacobian matrix H is required to be defined with respect to the error state $ffix$, and evaluated at the best true state estimate $\hat{x}_t = x \oplus ffi\hat{x}$. As the error state mean is zero at this stage (we have not observed it yet), we have $x_t = x$, and we can use the nominal error x as the evaluation point, leading to

$$H = \left.\frac{\partial h}{\partial x_t}\right|_x \left.\frac{\partial x_t}{\partial ffix}\right|_x = H_x X_{ffix}, \tag{22}$$

in which

$$X_{ffix} \triangleq \left.\frac{\partial x_t}{\partial ffix}\right|_x = \begin{bmatrix} I_6 & 0 & 0 \\ 0 & Q_{ffi\theta} & 0 \\ 0 & 0 & I_6 \end{bmatrix}, \tag{23}$$

$$Q_{ffi\theta} = \frac{1}{2}\begin{bmatrix} -q_x & -q_y & -q_z \\ q_w & -q_z & q_y \\ q_z & q_w & -q_x \\ -q_y & q_x & q_w \end{bmatrix}. \tag{24}$$

2.2.6. Combination of ESKF Error State and Nominal State Process

Based on the recurrence of the a priori system state and the calibration of the error state, the updated form is obtained as follows:

$$\widehat{x}_k = \widehat{x}_k^- \oplus ffi\widehat{x}_k. \tag{25}$$

Each of these system states corresponds to the following:

$$\begin{aligned} \widehat{p}_k &= \widehat{p}_k^- + ffi\widehat{p}_k \\ \widehat{v}_k &= \widehat{v}_k^- + ffi\widehat{v}_k \\ \widehat{q}_k &= \widehat{q}_k^- \otimes \widehat{q}ffi\widehat{\theta}_k. \\ \widehat{a}_{bk} &= \widehat{a}_{bk}^- + ffi\widehat{a}_{bk} \\ \widehat{\omega}_{bk} &= \widehat{\omega}_{bk}^- + ffi\widehat{\omega}_{bk} \end{aligned} \tag{26}$$

2.2.7. ESKF Error State Reset

After injecting the error state into the nominal state, the a priori error state and the corresponding covariance need to be reset, as

$$\begin{aligned} ffi\widehat{x} &= 0 \\ P &= G_k P_k G_k^T \end{aligned} \tag{27}$$

where G is the reset function $g(ffix) = ffix \ominus ffi\hat{x}$ of the Jacobi matrix, which is defined in the form of

$$G \triangleq \left.\frac{\partial g}{\partial ffix}\right|_{ffi\widehat{x}} = \begin{pmatrix} I_6 & 0 & 0 \\ 0 & I - (\frac{1}{2}ffi\widehat{\theta}) & 0 \\ 0 & 0 & I_9 \end{pmatrix}. \tag{28}$$

3. Robust Localization via RTS Smoothing

3.1. Reliability of Measurement Information

In the case of a GNSS signal loss, when there are no available GNSS measurements to update, the covariance matrix in Equation (21) tends to be infinite. The Kalman gain K tends to zero, so that when the measurement information is zero or differs too much from the predicted information, the update equation changes to the following form:

$$\begin{aligned} ffi\widehat{x}_{k+1} &= ffi\widehat{x}_k \\ \widehat{P}_{k+1} &= \breve{P}_k^- \end{aligned} \tag{29}$$

3.2. RTS Fundamental Design

The RTS smoothing can be regarded as a technique for obtaining an optimal state estimate when observations are available from moment 1 to moment N; it involves using previous estimates obtained through Kalman filtering in order to perform backward smoothing from moment k + 1 to moment k resulting in a more precise estimate. This method falls under the category of fixed interval smoothing and is considered as a fixed interval smoother, which is convenient for implementation.

The RTS smoother involves a two−step process: a forward filter followed by a backward smoothing. The forward filter is a standard Kalman filter described by Equation (18), which maintains all the predicted and updated estimates as well as their corresponding covariances for each epoch during the entire mission. The backward smoothing procedure begins at the end of the forward filter at time t_N, with an initial condition $ffix_{N,N}$ and $t_{N,N}$. Therefore, the backward smoothing can be seen as an update to the forward filter for obtaining an enhanced estimate. The equation for the RTS algorithm can be represented in the form of

$$\begin{aligned} ffi\widehat{\boldsymbol{x}}_{k,N} &= ffi\widehat{\boldsymbol{x}}_{k,k} + \boldsymbol{U}_k(\widehat{\boldsymbol{x}}_{k+1,N} - \widehat{\boldsymbol{x}}_{k+1,k}) \\ \widehat{\boldsymbol{P}}_{k,N} &= \widehat{\boldsymbol{P}}_{k,k} + \boldsymbol{U}_k(\widehat{\boldsymbol{P}}_{k+1,N} - \widehat{\boldsymbol{P}}_{k+1,k})\boldsymbol{U}_k^T \end{aligned} \tag{30}$$

where delta $ffix_{k,N}$ is the smoothed estimate of the state vector, and U_k is the smoothing gain matrix, which is calculated by the following equation:

$$\boldsymbol{U}_k = \widehat{\boldsymbol{P}}_{k,k}\boldsymbol{F}_{k+1,k}^T\widehat{\boldsymbol{P}}_{k+1,k}^{-1}, \tag{31}$$

where $k = N-1, N-2, \ldots\ldots, 0$. The RTS smoothing algorithm uses the results of the forward filter and backward smoothing to obtain an improved estimate. The inverse smoothing starts from the last filter result and proceeds forward one by one, so it must obtain N filter results to smooth the observations within the period [0, N]. However, if N is too large, the inverse smoothing process can be obviously lagged, which would limit the algorithm's application in real time. Therefore, a segmented RTS smoothing method is designed in this study.

3.3. Segmented RTS Smoothing Algorithm

Assuming the duration of a dynamic system is $k = 1, 2 \ldots \ldots, N$, the segmented RTS smoothing algorithm involves forward filtering and inverse smoothing of N observations in segments. The segment length is L, where $1 < L < N$. In other words, the RTS inverse smoothing process is performed immediately after obtaining the filtering results for the segmented period, without waiting for subsequent filtering results. This approach significantly reduces the lag time of the inverse smoothing process. Additionally, the segmentation length can be varied, avoiding the issue of poor real−time performance associated with RTS smoothing. In this study, the segmented RTS smoothing algorithm is applied to the potential probability assumption density filtering algorithm, effectively addressing the problem of poor real−time performance resulting from the use of RTS smoothing.

Step 1 Output variables are cleared: $ffi\hat{x}_k = [\,], \hat{p}_k = [\,]$.

Step 2 Local filtering result variable clearing $ffi\check{x}_k = [\,], \check{p}_k = [\,]$. The filtered result variable after local association is cleared $x_k = [\,]$, $Q_k = [\,]$, the segmented smoothing counter *count* is cleared to zero.

Step 3 If the tracking of the target is not finished, action continues down the track, otherwise the smoothing result $ffi\hat{x}_k$, $\hat{p}_k$ is output; then, the entire segmented RTS smoothing algorithm ends.

Step 4 Upon receiving the current filtering result, it is stored in the variables $ffi\check{x}_k$ and $\check{p}_k$. the Hungarian algorithm is used to correlate the trajectories and estimates, obtaining the filter value for each target. The correlated filter result is saved in variables x_k and Q_k, and they are counted using a counter called count.

Step 5 If $count == L$, after performing local correlation using x_k and Q_k for each target's filtering results, the inverse smoothing process uses the RTS smoothing algorithm. The results of the smoothing process are stored in variables delta $ffi\check{x}_k$ and $\check{p}_k$ before being transferred to Step 2. In order to significantly reduce the lag time of the inverse smoothing process, it is carried out immediately following the acquisition of the filtered data without waiting for subsequent data. Real−time performance of the fusion algorithm can be improved by decreasing the segment length L.

To correct the position, velocity, and attitude values that have been calculated in the forward filtering, it is recommended to use the state error smoothing values. These smoothing values can be utilized to derive the final optimal smoothed values for these states using the following equations.

4. Results and Discussions

The proposed sensor fusion algorithm is demonstrated in a relatively open environment, which allows for uninterrupted satellite signal and individualized GNSS localization. The acquisition frequency for GNSS data is 1 Hz, while the IMU data are acquired at a frequency of 100 Hz; the smooth dimension L is selected as 10. These parameters provide acceptable conditions for analyzing the accuracy of combined GNSS/IMU localization, even in the event of GNSS signal loss, as demonstrated by the trajectory shown in Figure 2. In order to evaluate the localization accuracy of the GNSS/IMU combined localization process, the simulation involves artificially inducing the loss of lock by introducing exaggerated errors to the GNSS satellite observation information based on raw GNSS data.

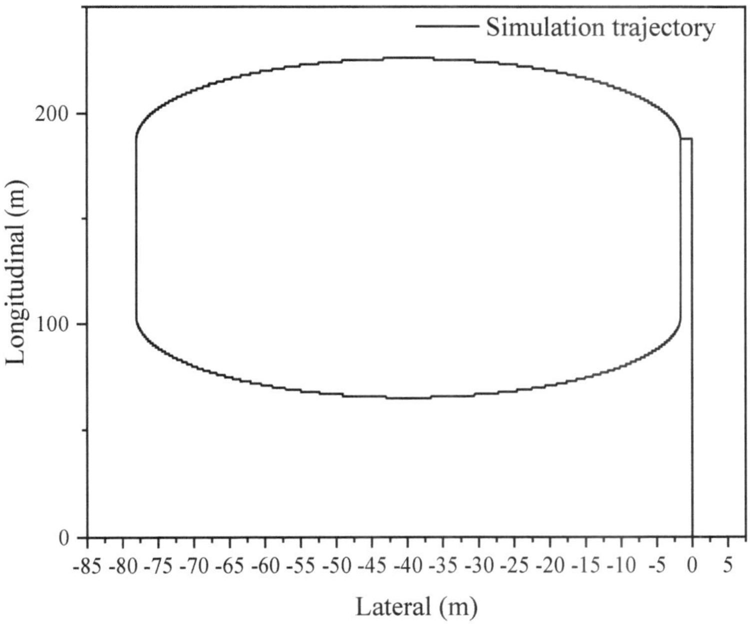

Figure 2. Oval shape motion trajectory.

4.1. Oval Track Simulation Analysis

Figure 2 illustrates the simulated oval shape trajectory, which comprises a long straight line with small angle turns in two dimensions.

The localization results for the various filtering algorithms are shown in Figures 3–5 and Table 2. It can be seen that the ESKF algorithm exhibits better target tracking accuracy compared to the EKF algorithm, as evidenced by its RMS values of 1.633 m, 1.782 m, and 1.476 m for Lateral, Longitudinal, and Vertical directions, respectively. By reducing the error state parameter, the accuracy of the three poses is improved by 2.8%, 2.1%, and 52.0%, respectively. After performing the backward smoothing filtering process, the RMS values of the three attitudes are further optimized to 1.463 m, 1.588 m, and 1.393 m, respectively. Moreover, the target tracking accuracy of ESKF–RTS is superior to ESKF, with the attitude accuracy in the three directions improving by 10.4%, 10.9%, and 5.6%, respectively. The trajectory also reveals that the curve of ESKF–RTS is smoother, further highlighting the advantages of ESKF–RTS over the other two algorithms in addressing the problem of smooth estimation of nonlinear states.

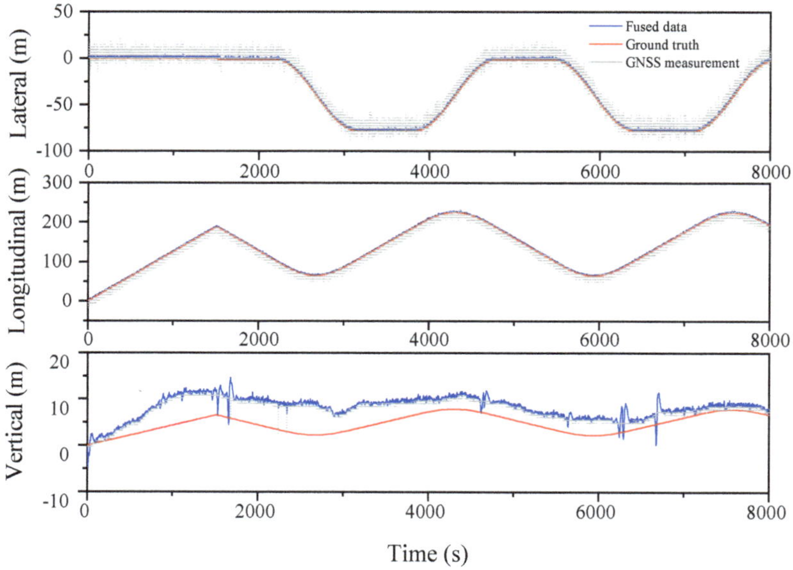

Figure 3. Lateral, Longitudinal and Vertical positions of EKF (Oval).

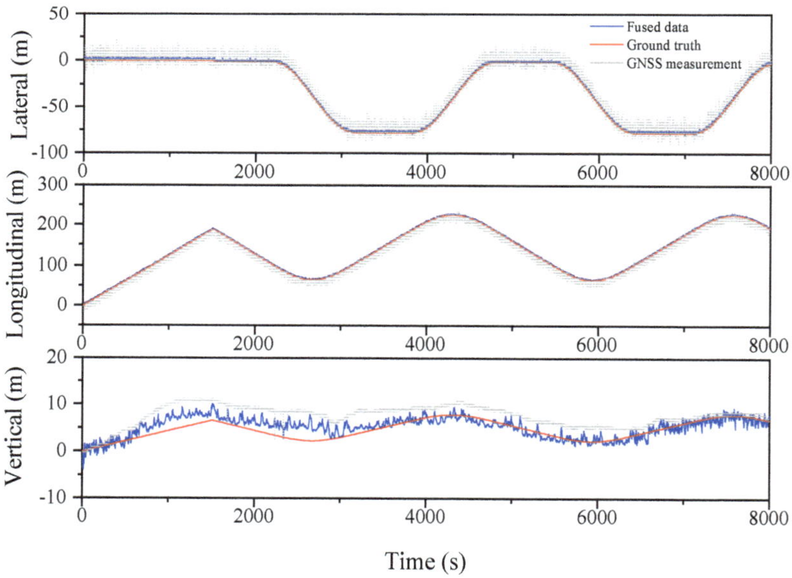

Figure 4. Lateral, Longitudinal and Vertical positions of ESKF (Oval).

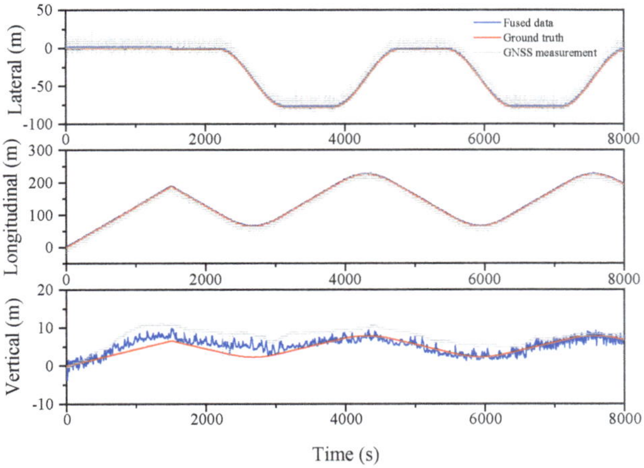

Figure 5. Lateral, Longitudinal and Vertical positions of ESKF–RTS (Oval).

Table 2. RMS values of localization errors for different algorithms (Oval).

Position Error RMS (m)	Lateral	Longitudinal	Vertical
GNSS	2.049	2.598	2.619
EKF	1.680	1.820	3.075
ESKF	1.633	1.782	1.476
ESKF–RTS	1.463	1.588	1.393

4.2. Serpentine Track Simulation Analysis

Further verification of the designed algorithms is performed in a serpentine trajectory, as shown in Figure 6, and the localization results are compared as shown in Figures 7–9 and Table 3.

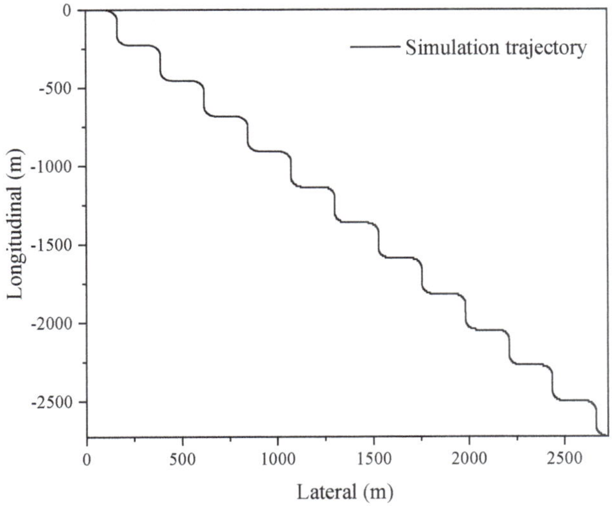

Figure 6. Serpentine shape motion trajectory.

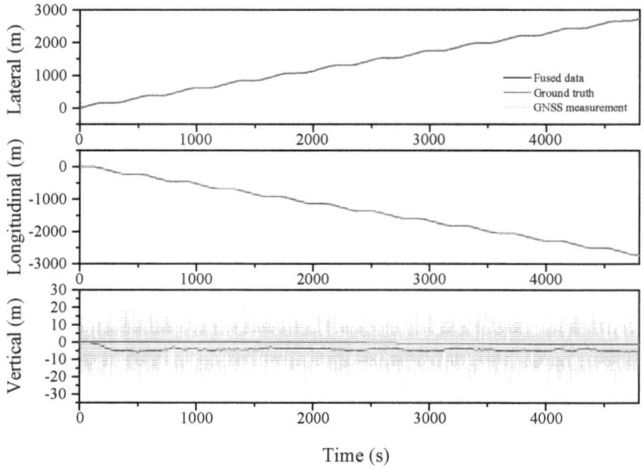

Figure 7. Lateral, Longitudinal and Vertical positions of EKF (Serpentine 1).

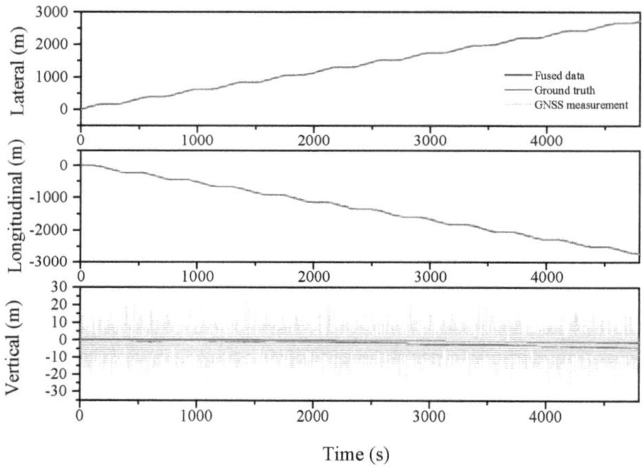

Figure 8. Lateral, Longitudinal and Vertical positions of ESKF (Serpentine 1).

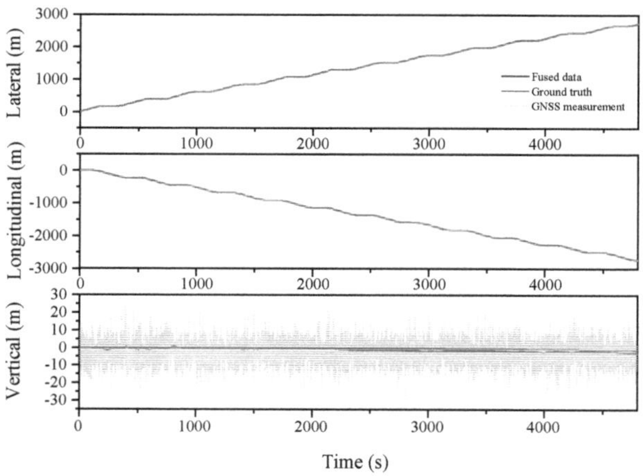

Figure 9. Lateral, Longitudinal and Vertical positions of ESKF−RTS (Serpentine 1).

Table 3. RMS values of localization errors for different algorithms (Serpentine 1).

Position Error RMS (m)	Lateral	Longitudinal	Vertical
GNSS	1.912	1.793	5.680
EKF	0.955	1.585	5.823
ESKF	0.464	0.641	1.700
ESKF−RTS	0.206	0.243	0.912

Table 3 illustrates the improved performance of the ESKF and ESKF−RTS algorithms compared to the EKF algorithm in the serpentine condition. The RMS of Lateral, Longitudinal, and Vertical in EKF are 0.955 m, 1.585 m, and 5.823 m, respectively. After reducing the state value, the RMS of the three positions for the ESKF algorithm are 0.464 m, 0.641 m, and 1.700 m, which are 51.4%, 59.6%, and 70.8% better than the EKF algorithm. The ESKF−RTS algorithm further improves the position accuracy, with RMS values of 0.206 m, 0.243 m, and 0.912 m for the three directions, respectively, which are 55.6%, 62.1%, and 46.4% higher than those of the ESKF algorithm. Figures 7–9 show the comparison of the root mean square errors, which further demonstrates the superior performance and stability of the ESKF−RTS algorithm. However, since the simulation was conducted with a relatively good GNSS signal, the robustness of the algorithm could not be well evaluated. Subsequently, the GNSS data for the serpentine condition are partitioned into eight segments, and the error magnitude is amplified by a factor of two at intervals of 600 s in order to evaluate the algorithm's robustness.

The serpentine working condition with further exaggerated GNSS signal error is analyzed, as shown in Figures 10–12 and Table 4. The RMS values for the Lateral, Longitudinal, and Vertical directions are 1.231 m, 1.735 m, and 1.453 m, respectively, for the EKF algorithm. Meanwhile, the ESKF algorithm improves these values by 48.4%, 48.7%, and 34.1%, respectively, resulting in the RMS values of 0.635 m, 0.890 m, and 0.957 m. Moreover, the ESKF−RTS algorithm improves the position accuracy in three directions by 42.1%, 52.6%, and 52.1%, respectively, compared to the ESKF algorithm, with RMS values of 0.368 m, 0.422 m, and 1.456 m. Even though the accuracy of ESKF−RTS decreases on the Vertical axis, the higher degree of smoothing in Figure 12 suggests that the ESKF−RTS smoothing algorithm can significantly enhance robustness and achieve precise localization, even with lower accuracy of GNSS sensors.

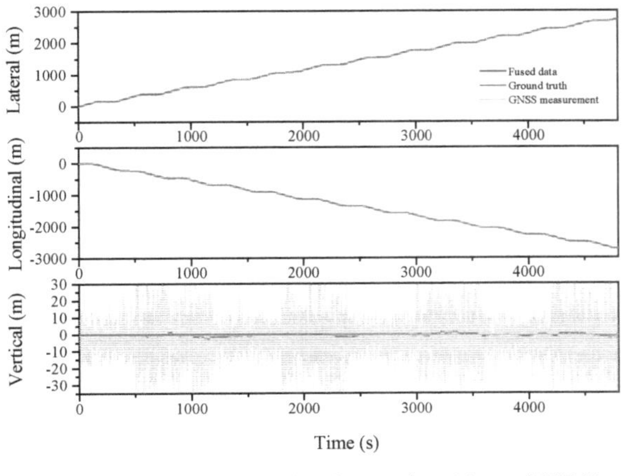

Figure 10. Lateral, Longitudinal and Vertical positions of EKF (Serpentine 2).

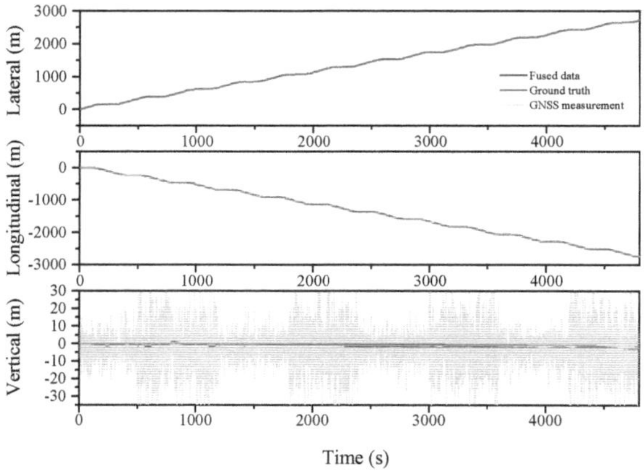

Figure 11. Lateral, Longitudinal and Vertical positions of ESKF (Serpentine 2).

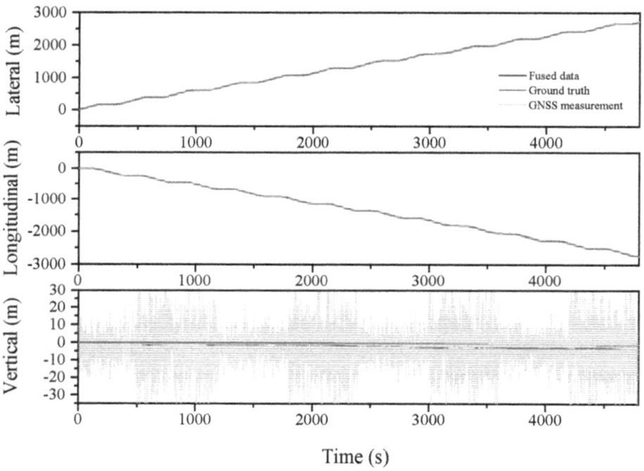

Figure 12. Lateral, Longitudinal and Vertical positions of ESKF−RTS (Serpentine 2).

Table 4. RMS values of localization errors for different algorithms (Serpentine 2).

Position Error RMS (m)	Lateral	Longitudinal	Vertical
GNSS	3.370	3.308	8.988
EKF	1.231	1.735	1.453
ESKF	0.635	0.890	0.957
ESKF−RTS	0.368	0.422	1.456

Next, we continued the simulation of circular operating conditions and divided the entire simulation process into four sections to validate the feasibility of the algorithm by doubling the GNSS signal error, as shown in Figures 13–16. The experimental results shown in Table 5 once again demonstrated the feasibility and robustness of the error state Kalman filtering algorithm, indicating that the algorithm can achieve stable and accurate integrated navigation under various operating conditions.

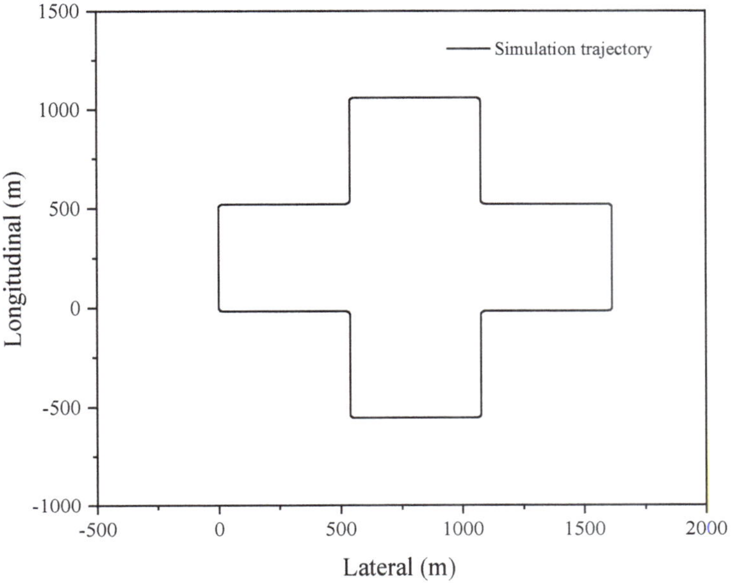

Figure 13. Motion trajectory of polygon.

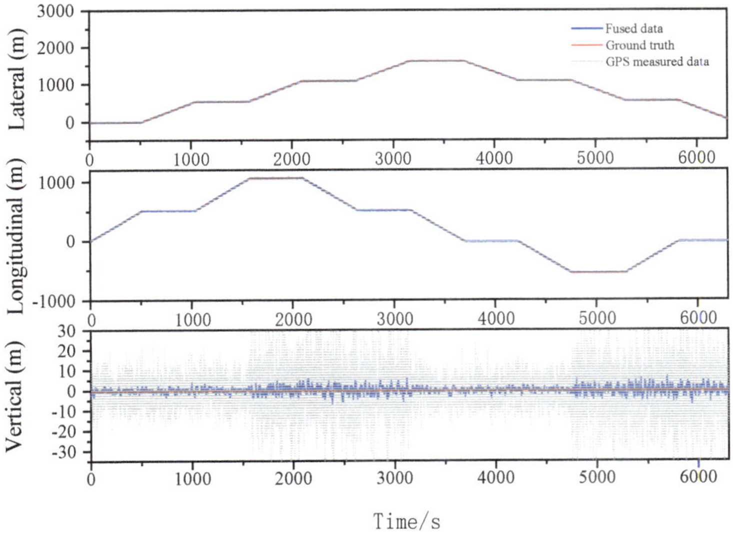

Figure 14. Lateral, Longitudinal and Vertical positions of EKF (Polygonal).

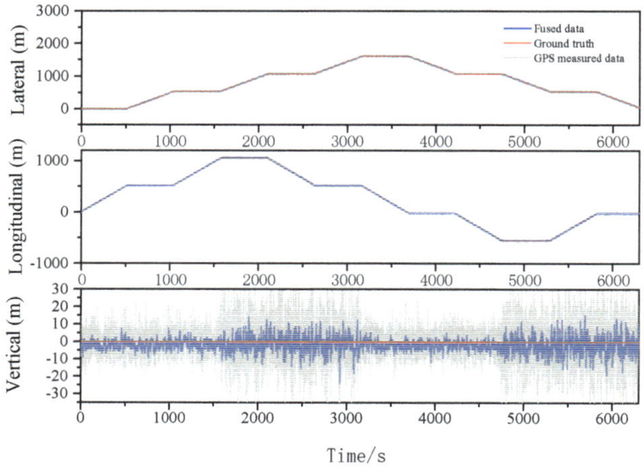

Figure 15. Lateral, Longitudinal and Vertical positions of ESKF (Polygonal).

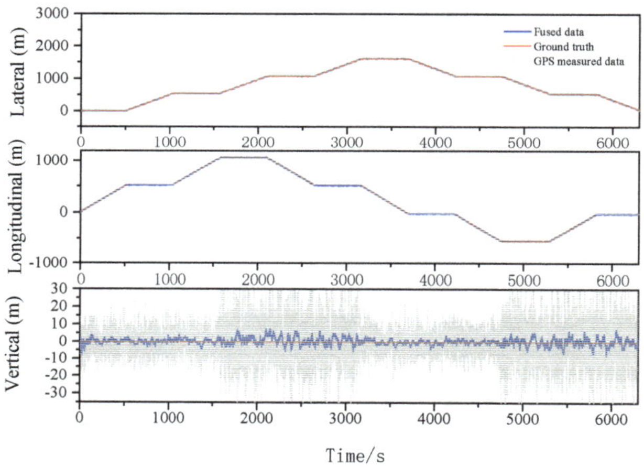

Figure 16. Lateral, Longitudinal and Vertical positions of ESKF−RTS (Polygonal).

Table 5. RMS values of localization errors for different algorithms (Polygonal).

Position Error RMS (m)	Lateral	Longitudinal	Vertical
GNSS	2.601	2.677	8.516
EKF	1.243	1.289	4.328
ESKF	1.253	1.320	1.237
ESKF−RTS	0.900	0.896	1.999

The above data show the results of testing the navigation system using EKF, ESKF, and ESKF−RTS algorithms. By analyzing these results, we can draw the following conclusions. The above data demonstrate the performance of three different filtering algorithms in GNSS navigation. First, we can see that the performance of the EKF and ESKF algorithms is relatively similar, with position errors around 1.3 m. This is because both algorithms use Kalman filtering, but the ESKF algorithm introduces error states while considering errors, which can more accurately estimate errors and improve accuracy. At the same time, the ESKF algorithm can also reduce the impact of errors in future time by predicting error states, thus improving the stability of the algorithm.

On the other hand, the ESKF−RTS algorithm performs even better, with position errors even lower than 1 m. This is because the ESKF−RTS algorithm uses segmental smoothing to optimize the filtering results, which can more accurately estimate and correct errors. In the ESKF−RTS algorithm, RTS stands for "Recursive Least Squares Smoothing" which can combine the prediction results of the ESKF algorithm with the observation results to obtain more accurate state estimation results.

Overall, the above data indicate that the ESKF−RTS algorithm performs well in GNSS navigation. The advantage of this algorithm is that it not only considers error states, but also further optimizes the filtering results through smoothing algorithms. Therefore, in practical applications, selecting the ESKF−RTS algorithm for navigation filtering can achieve more accurate and stable results, improving the reliability and accuracy of the navigation system.

5. Conclusions and Future Work

This paper introduces an error state extended Kalman filter algorithm and segmented Rauch–Tung–Striebel (RTS) smoothing algorithm to enhance the localization accuracy and robustness of GNSS and IMU sensors. The cumulative error of INS over time are overcome when GNSS signal is disturbed. The simulation results show that the proposed method is more linear and has higher localization accuracy than the traditional EKF algorithm. It also demonstrates good robustness in achieving better accuracy with low−quality GNSS signals. The algorithm can serve as a foundation for low−cost sensor fusion processing and is a valuable reference for further research.

Over the next few years, combined navigation systems will see increased usage and development across a range of applications. As sensor and communication technologies progress, and the demands for navigation safety and reliability continue to rise, there will be a greater need for more accurate and dependable navigation systems.

Furthermore, the RTS smoothing algorithm−assisted combined navigation algorithm presented in this paper was simulated for offline computation. For practical applications in the future, real−world testing will be necessary to verify the hardware feasibility of the algorithm, and to consider real−time data transmission and computation to achieve unmanned driving. Overall, positioning is a crucial component of intelligent driving, and its potential impact on society and the economy is vast. Combined navigation will be more and more widely used in production and practical life.

Author Contributions: Conceptualization, Y.Y. and J.Z.; methodology, J.Z. and Y.Y.; software, M.G. and X.N.; validation, Y.W. and J.L.; formal analysis, X.N. and Y.W.; investigation, M.G.; resources, J.L.; data curation, M.G. and J.L.; writing—original draft preparation, J.Z.; writing—review and editing, X.N., Y.Y. and M.G.; visualization, Y.W. and J.L.; supervision, X.N.; project administration, J.L.; funding acquisition, X.N., Y.Y. and J.L. All the authors have read and agreed to the published version of the enclosed manuscript. All authors have read and agreed to the published version of the manuscript.

Funding: This research was funded by the Zhejiang Provincial Natural Science Foundation of China, grant number LY22E050018; This research was funded by the Basic Public Welfare Research Program of Zhejiang Province, grant number LGG22E050019; the National Natural Science Foundation of China (NSFC), grant number 51905483.

Institutional Review Board Statement: Not applicable.

Informed Consent Statement: Not applicable.

Data Availability Statement: The data that support the findings of this study are available from the corresponding author, Xiaobin Ning, upon reasonable request.

Conflicts of Interest: The authors declare no conflict of interest.

Abbreviations

GNSS	Global Navigation Satellite System
IMU	Inertial Measurement Unit
INS	Inertial Navigation System
RTS	Rauch–Tung–Striebel
EKF	Extended Kalman filter
CKF	Cubature Kalman Filter
ESKF	Error state Kalman filter
RMS	Root mean square
Symbols	
X	System state
$\breve{X}$	Prediction of the system state
$\widehat{\boldsymbol{X}}$	Update of the system state
$\delta \widehat{x}_{k,N}$	The smoothed estimate of the state vector
U_k	The smoothing gain matrix
$\widehat{\boldsymbol{P}}_{k,N}$	The covariance of smoothing estimates
W	Gaussian white noise
Z	System measurement
V_k	Measurement of Gaussian white noise
$f(\cdot)$	State functions of nonlinear systems
$h(\cdot)$	Measurement functions for nonlinear systems
F_k	The Jacobi matrix of $f(\cdot)$ at x_k
H_k	The Jacobi matrix of $h(\cdot)$ at x_k
p_k	Covariance matrix of states
q_k	Covariance matrix of noise
δ	Error character
x_t	True state
x	Nominal state
p_t	Position at time t
v_t	Velocity at time t
q_t	Quaternion at time t
R_t	Rotation matrix at time t
θ_t	The angular vector at time t
a_{bt}	Acceleration bias at time t
ω_{bt}	Angular velocity bias at time t
a_m	Acceleration measurement
ω_m	Angular velocity measurement
a_n	Acceleration noise vector
ω_n	Angular velocity noise vector
a_ω	Acceleration bias vector
ω_ω	Angular velocity bias vector
Δt	Time interval from k to k+1
w_{vk}	Velocity Gaussian random noise
$w_{\theta k}$	Angular Gaussian random noise
w_{ak}	Acceleration bias Gaussian random noise
$w_{\omega k}$	Velocity biased Gaussian random noise
W_v	Corresponding covariance matrix of $\boldsymbol{w}_{vk}$
v	Corresponding covariance matrix of $\boldsymbol{w}_{mk}$
σ_{a_n}	Acceleration Gaussian white noise
F_{xk}	Error state Jacobi matrix
F_{wk}	Noise state Jacobi matrix
q_w	Noise covariance matrix
U_k	Smoothing gain matrix
$count$	Smooth counter

References

1. Jiang, H.; Shi, C.; Li, T.; Dong, Y.; Li, Y.; Jing, G. Low-cost GNSS/INS integration with accurate measurement modeling using an extended state observer. *GNSS Solut.* **2020**, *25*, 17.
2. Jiang, W.; Liu, D.; Cai, B.; Rizos, C.; Wang, J.; Shangguan, W. A Fault-Tolerant Tightly Coupled GNSS/INS/OVS Integration Vehicle Navigation System Based on an FDP Algorithm. *IEEE Trans. Veh. Technol.* **2019**, *68*, 6365–6378. [CrossRef]
3. Shu, Y.; Xu, P.; Niu, X.; Chen, Q.; Qiao, L.; Liu, J. High-Rate Attitude Determination of Moving Vehicles With GNSS GPS, BDS, GLONASS, and Galileo. *IEEE Trans. Instrum. Meas.* **2022**, *71*, 5501813. [CrossRef]
4. Li, T.; Zhang, H.; Niu, X.; Gao, Z. Tightly-Coupled Integration of Multi-GNSS Single-Frequency RTK and MEMS-IMU for Enhanced Positioning Performance. *Sensors* **2017**, *17*, 2462. [CrossRef]
5. Miller, I.; Campbell, M. Sensitivity Analysis of a Tightly-Coupled GNSS/INS System for Autonomous Navigation. *IEEE Trans. Aerosp. Electron. Syst.* **2012**, *48*, 1115–1135. [CrossRef]
6. Shi, B.; Wang, M.; Wang, Y.; Bai, Y.; Lin, K.; Yang, F. Effect Analysis of GNSS/INS Processing Strategy for Sufficient Utilization of Urban Environment Observations. *Sensors* **2021**, *21*, 620. [CrossRef] [PubMed]
7. Liu, W.; Xiong, L.; Xia, X.; Lu, Y.; Gao, L.; Song, S. Vision-aided intelligent vehicle sideslip angle estimation based on a dynamic model. *IET Intell. Transp. Syst.* **2020**, *14*, 1183–1189. [CrossRef]
8. Liu, W.; Xia, X.; Xiong, L.; Lu, Y.; Gao, L.; Yu, Z. Automated Vehicle Sideslip Angle Estimation Considering Signal Measurement Characteristic. *IEEE Sens. J.* **2021**, *21*, 21675–21687. [CrossRef]
9. Shin, E.H.; El-Sheimy, N. Accuracy Improvement of Low Cost INS/GNSS for Land Applications. In Proceedings of the 2002 National Technical Meeting of the Institute of Navigation, San Diego, CA, USA, 28–30 January 2002.
10. Han, H.; Wang, J.; Wang, J.; Tan, X. Performance Analysis on Carrier Phase-Based Tightly-Coupled GNSS/BDS/INS Integration in GNSS Degraded and Denied Environments. *Sensors* **2015**, *15*, 8685–8711. [CrossRef]
11. Li, T.; Zhang, H.; Niu, X.; Zhang, Q. Performance Analysis of Tightly Coupled RTK/INS Algorithm in Case of Insufficient Number of Satellites. *Geomat. Inf. Sci. Wuhan Univ.* **2018**, *43*, 478–484.
12. Erfianti, R.; Asfihani, T.; Suhandri, H.F. GNSS/IMU Sensor Fusion Performance Comparison of a Car Localization in Urban Environment Using Extended Kalman Filter. *IOP Conf. Ser. Earth Environ. Sci.* **2023**, *1127*, 012006. [CrossRef]
13. Sabatelli, S.; Galgani, M.; Fanucci, L.; Rocchi, A. A Double-Stage Kalman Filter for Orientation Tracking With an Integrated Processor in 9-D IMU. *IEEE Trans. Instrum. Meas.* **2012**, *62*, 590–598. [CrossRef]
14. Shao, Y.; Xu, F.; Chen, J.; Lu, J.; Du, S. Engineering surface topography analysis using an extended discrete modal decomposition. *J. Manuf. Process.* **2023**, *90*, 367–390. [CrossRef]
15. Zhao, C.; Lui, C.F.; Du, S.; Wang, D.; Shao, Y. An Earth Mover's Distance based Multivariate Generalized Likelihood Ratio Control Chart for Effective Monitoring of 3D Point Cloud Surface. *Comput. Ind. Eng. Comput. Ind. Eng.* **2023**, *175*, 108911. [CrossRef]
16. Zhao, C.; Du, S.; Lv, J.; Deng, Y.; Li, G. A novel parallel classification network for classifying three-dimensional surface with point cloud data. *J. Intell. Manuf.* **2023**, *34*, 515–527. [CrossRef]
17. Atia, M.M.; Hilal, A.R.; Stellings, C.; Hartwell, E.; Toonstra, J.; Miners, W.B.; Basir, O.A. A Low-Cost Lane-Determination System Using GNSS/IMU Fusion and HMM-Based Multistage Map Matching. *IEEE Trans. Intell. Transp. Syst.* **2017**, *18*. 3027–3037. [CrossRef]
18. Julier, S.J.; Uhlmann, J.K. New Extension of the Kalman Filter to Nonlinear Systems. *Signal Processing Sens. Fusion Target Recognit. VI. SPIE* **1997**, *3068*, 182–193.
19. Arasaratnam, I.; Haykin, S. A numerical-integration perspective on Gaussian filters. *IEEE Trans Autom. Control.* **2009**, *54*, 1254–1269. [CrossRef]
20. Jeon, H.; Min, J.; Bang, H.; Youn, W. Quaternion-Based Iterative Extended Kalman Filter for Sensor Fusion of Vision Sensor and IMU in 6-DOF Displacement Monitoring. *IEEE Sens. J.* **2022**, *22*, 23188–23199. [CrossRef]
21. Lee, J.-C.; Chen, C.-C.; Shen, C.-T.; Lai, Y.-C. Landmark-Based Scale Estimation and Correction of Visual Inertial Odometry for VTOL UAVs in a GPS-Denied Environment. *Sensors* **2022**, *22*, 9654. [CrossRef]
22. Kourabbaslou, S.S.; Zhang, A.; Atia, M.M. A Novel Design Framework for Tightly Coupled IMU/GNSS Sensor Fusion Using Inverse-Kinematics, Symbolic Engines, and Genetic Algorithms. *IEEE Sens. J.* **2019**, *19*, 11424–11436. [CrossRef]
23. Yi, S.; Zorzi, M. Robust Kalman Filtering Under Model Uncertainty: The Case of Degenerate Densities. *IEEE Trans. Autom. Control* **2021**, *67*, 3458–3471. [CrossRef]
24. Yi, S.; Zorzi, M. Robust fixed-lag smoothing under model perturbations. *J. Frankl. Inst.* **2023**, *360*, 458–483. [CrossRef]
25. Gao, L.; Xiong, L.; Xia, X.; Lu, Y.; Yu, Z.; Khajepour, A. Improved Vehicle Localization Using On-Board Sensors and Vehicle Lateral Velocity. *IEEE Sens. J.* **2022**, *22*, 6818–6831. [CrossRef]
26. Xia, X.; Hashemi, E.; Xiong, L.; Khajepour, A. Autonomous Vehicle Kinematics and Dynamics Synthesis for Sideslip Angle Estimation Based on Consensus Kalman Filter. *IEEE Trans. Control. Syst. Technol.* **2022**, *31*, 179–192. [CrossRef]
27. Xia, X.; Xiong, L.; Huang, Y.; Lu, Y.; Gao, L.; Xu, N.; Yu, Z. Estimation on IMU yaw misalignment by fusing information of automotive onboard sensors. *Mech. Syst. Signal Process.* **2022**, *162*, 107993. [CrossRef]
28. Madyastha, V.K.; Ravindra, V.C.; Mallikarjunan, S.; Goyal, A. Extended Kalman filter vs. error state Kalman filter for aircraft attitude estimation. In Proceedings of the AIAA Guidance, Navigation, and Control Conference and Exhibit, Portland, OR, USA, 8–11 August 2011.

29. He, J.; Sun, C.; Zhang, B.; Wang, P. Adaptive Error-State Kalman Filter for Attitude Determination on a Moving Platform. *IEEE Trans. Instrum. Meas.* **2021**, *70*, 9513110. [CrossRef]
30. Critchley-Marrows, J.J.R.; Wu, X.; Cairns, I.H. Treatment of Extended Kalman Filter Implementations for the Gyroless Star Tracker. *Sensors* **2022**, *22*, 9002. [CrossRef]
31. Xia, M.; Sun, P.; Guan, L.; Zhang, Z. Research on Algorithm of Airborne Dual-Antenna GNSS/MINS Integrated Navigation System. *Sensors* **2023**, *23*, 1691. [CrossRef]
32. Ke, Y.; Lv, Z.; Zhang, C.; Deng, X.; Zhou, W.; Song, D. Tightly Coupled GNSS/INS Integration Spoofing Detection Algorithm Based on Innovation Rate Optimization and Robust Estimation. *IEEE Access* **2022**, *10*, 72444–72457. [CrossRef]
33. Chen, Q.; Zhang, Q.; Niu, X. Estimate the Pitch and Heading Mounting Angles of the IMU for Land Vehicular GNSS/INS Integrated System. *IEEE Trans. Intell. Transp. Syst.* **2020**, *22*, 6503–6515. [CrossRef]
34. Gao, B.; Hu, G.; Zhong, Y.; Zhu, X. Cubature Kalman Filter With Both Adaptability and Robustness for Tightly-Coupled GNSS/INS Integration. *IEEE Sens. J.* **2021**, *21*, 14997–15011. [CrossRef]
35. Qin, T.; Li, P.; Shen, S. VINS-Mono: A Robust and Versatile Monocular Visual-Inertial State Estimator. *IEEE Trans. Robot.* **2018**, *34*, 1004–1020. [CrossRef]
36. Yang, C.; Cheng, Z.; Jia, X.; Zhang, L.; Li, L.; Zhao, D. A Novel Deep Learning Approach to 5G CSI/Geomagnetism/VIO Fused Indoor Localization. *Sensors* **2023**, *23*, 1311. [CrossRef]
37. Akeila, E.; Salcic, Z.; Swain, A. Reducing Low-Cost INS Error Accumulation in Distance Estimation Using Self-Resetting. *IEEE Trans. Instrum. Meas.* **2013**, *63*, 177–184. [CrossRef]
38. Chen, C.-L.; He, R.; Peng, C.-C. Development of an Online Adaptive Parameter Tuning vSLAM Algorithm for UAVs in GPS-Denied Environments. *Sensors* **2022**, *22*, 8067. [CrossRef]
39. López, E.; García, S.; Barea, R.; Bergasa, L.M.; Molinos, E.J.; Arroyo, R.; Romera, E.; Pardo, S. A Multi-Sensorial Simultaneous Localization and Mapping (SLAM) System for Low-Cost Micro Aerial Vehicles in GPS-Denied Environments. *Sensors* **2017**, *17*, 802. [CrossRef]
40. Elamin, A.; Abdelaziz, N.; El-Rabbany, A. A GNSS/INS/LiDAR Integration Scheme for UAV-Based Navigation in GNSS-Challenging Environments. *Sensors* **2022**, *22*, 9908. [CrossRef]
41. Revert Calabuig, N.; Laarossi, I.; Álvarez González, A.; Pérez Nuñez, A.; González Pérez, L.; García-Minguillán, A.C. Development of a Low-Cost Smart Sensor GNSS System for Real-Time Localization and Orientation for Floating Offshore Wind Platform. *Sensors* **2023**, *23*, 925. [CrossRef]
42. Liu, Y.; Liu, F.; Gao, Y.; Zhao, L. Implementation and Analysis of Tightly Coupled Global Navigation Satellite System Precise Point Localization/Inertial Navigation System (GNSS PPP/INS) with Insufficient Satellites for Land Vehicle Navigation. *Sensors* **2018**, *18*, 4305. [CrossRef]
43. Li, Z.; Zhang, Y.; Shi, Y.; Yuan, S.; Zhu, S. Performance Enhancement of INS and UWB Fusion Localization Method Based on Two-Level Error Model. *Sensors* **2023**, *23*, 557. [CrossRef] [PubMed]

sensors

MDPI

Article

Sensitivity Analysis of RV Reducer Rotation Error Based on Deep Gaussian Processes

Shousong Jin, Shulong Shang, Suqi Jiang, Mengyi Cao and Yaliang Wang *

College of Mechanical Engineering, Zhejiang University of Technology, Hangzhou 310032, China
* Correspondence: wangyaliang@zjut.edu.cn

Abstract: The rotation error is the most important quality characteristic index of a rotate vector (RV) reducer, and it is difficult to accurately optimize the design of a RV reducer, such as the Taguchi type, due to the many factors affecting the rotation error and the serious coupling effect among the factors. This paper analyzes the RV reducer rotation error and each factor based on the deep Gaussian processes (DeepGP) model and Sobol sensitivity analysis(SA) method. Firstly, using the optimal Latin hypercube sampling (OLHS) approach and the DeepGP model, a high-precision regression prediction model of the rotation error and each affecting factor was created. On the basis of the prediction model, the Sobol method was used to conduct a global SA of the factors influencing the rotation error and to compare the coupling relationship between the factors. The results show that the OLHS method and the DeepGP model are suitable for predicting the rotation error in this paper, and the accuracy of the prediction model constructed based on both of them is as high as 95%. The rotation error mainly depends on the influencing factors in the second stage cycloidal pinwheel drive part. The primary involute planetary part and planetary output carrier's rotation error factors have little effect. The coupling effects between the matching clearance between the pin gear and needle gear hole (δJ) and the circular position error of the needle gear hole (δt) is noticeably stronger.

Keywords: RV reducer; rotation error; deep Gaussian processes; sensitivity analysis

1. Introduction

The term "rotate vector (RV) reducer" typically refers to a two-stage reduction mechanism consisting of the first-stage involute planetary gear transmission and the second-stage cycloidal pin-wheel transmission. This mechanism has the advantages of a compact structure, a high transmission accuracy and a large reduction ratio, and it is frequently used in high precision equipment in Computerized Numerical Control (CNC) machine tools, aerospace and industrial robotics industries [1,2]. The most crucial performance metric for an RV reducer is the positioning transmission precision, which is often quantified by the output rotation error. The rotation error is determined by the precision of the parts processing and the assembly of the transmission chain from the input to the output of the RV reducer, as well as the state of motion and force; even under the same operating conditions, the quality characteristics of each link in the transmission chain do not have the same degree of influence on the RV reducer's output rotation error. It is challenging to accurately optimize the design of RV reducers, such as the Taguchi type, that will handle the problem satisfactorily because of the transmission chain's influence from a variety of parameters and the presence of random coupling effects. A sensitivity analysis (SA) of a precise RV reducer's rotation error is urgently required.

Scholars both at home and abroad are currently conducting extensive research on rotation error analysis. YANG et al. provided an analytical error modeling approach for the RV reducer's over-constrained structure and demonstrated the link between the original error and the transmission precision [3]. Hidaka et al. used the equivalent spring approach to model the exact system, and they qualitatively examined the transmission accuracy of the entire RV reducer [4–6]. Using the dynamic sub-structure method, HE and SHAN developed the transmission error dynamics model and mathematical equations for the RV-40E reducer for robots [7]. CAO et al. created an equivalent model of the transmission error of the RV reducer and solved the model's parameters using the empirical stiffness formula to determine the theoretical transmission error of the reducer [8]. By analyzing the structural characteristics of the RV reducer, the parts processing and assembly process, and the influence of the manufacturing errors on the clearance, CHU et al. proposed a selective assembly method to make the clearance of the RV reducer meet the requirements [9]. LIU analyzed the transmission accuracy of the reducer from multiple angles, such as manufacturing and assembly errors and external working conditions, using theoretical analysis, numerical simulation, dynamics, and transmission accuracy tests [10]. Meng et al. provided transmission error models for positive and negative transmission by analyzing the static analysis of the actual meshing process between the gear pairs while taking friction into account [11]. The accuracy of the RV reducer's equivalency error model significantly increased when Liu et al. introduced an RV reducer transmission error modeling and optimization approach and used the particle swarm algorithm to refine the empirical parameters of the identification model [12]. Based on the Adams software and virtual prototype technology, Tong created many virtual prototypes of the RV reducer and investigated the effects of the load, part elastic deformation, and component manufacturing error on the transmission error [13]. JIN et al. proposed a new method for predicting the rotation error based on improved grey wolf optimal support vector regression. This method overcomes the shortcomings of the previous rotation error research methods, such as their time-consuming nature and low computational precision [14]. On the basis of error sensitivity analysis, Hu et al. suggested an elastic transmission error compensation approach for a rotating vector to boost its transmission accuracy [15]. Wu et al. used the dynamics analysis program ADAMS to create a virtual prototype of the RV reducer employed in the robot and tested the transmission accuracy of the Japan RV-40E reduction using the Grating method [16].

The studies mentioned above provide many valuable insights for the analysis and evaluation of the RV reducer rotation error, but the majority of them are based on qualitative analysis results, which have some limitations and do not provide an accurate analysis of the RV reducer rotation error. For example: the calculation process of the pure geometric method is relatively complicated and does not take into account the actual manufacturing error of the parts. Due to a large amount of simplification in the modelling process, the equivalent spring method has a large gap between the final modelling result and the actual situation, and the model construction of the virtual prototype analysis method is complicated and time-consuming. In order to address the limitations of the traditional modelling and analysis methods of the rotation error, this paper proposes a novel modelling method for predicting the rotation error of a RV reducer based on deep Gaussian processes (DeepGP) and conducts sensitivity analysis on each factor influencing the rotation error.

To begin with, the optimal Latin hypercube sampling (OLHS) method is combined with a DeepGP model to develop a high-precision rotation error prediction model. The model's accuracy and efficiency are validated by comparing it to the spring equivalence method and the virtual prototype. Then, using the variance-based Sobol method, a global sensitivity analysis of the factors influencing the rotation error was performed based on the prediction model to identify a number of factors that had a substantial impact on the rotation error. Lastly, quantitative analysis of the influence of the coupling effect between the factors on the rotation error will provide reference significance for RV reducer quality improvement, cost reduction, and other optimization designs.

2. Analysis of the Factors Influencing the Rotation Error of RV-40E Reducer

2.1. RV Reducer Transmission Principle and Structural Composition Analysis

An RV reducer is a two-stage transmission mechanism compounded by the first-stage involute planetary gear transmission and the second-stage cycloidal pin-wheel transmission, which has many different composition structures and transmission forms and can be applied to different occasions. A typical RV-40E reducer structure is shown in Figure 1a below, including the input shaft with a sun gear, planetary gear, crankshaft, cycloidal pinwheel, needle tooth shell, planet carrier, needle gear, and other parts. The RV reducer transmission principle as shown in Figure 1b. Through the involute transmission, the input shaft with a sun gear transfers the power to the planetary wheel to complete the first stage of deceleration; the crankshaft and the planetary wheel are fixedly connected at the same speed, the crankshaft drives the cycloid wheel to turn around the needle tooth shell through a slewing bearing, and the rotation of the cycloid gear transmits the planet carrier and flange plate. The planet carrier and flange plate are bolted together at the same speed and act as the output mechanism to complete the second stage deceleration.

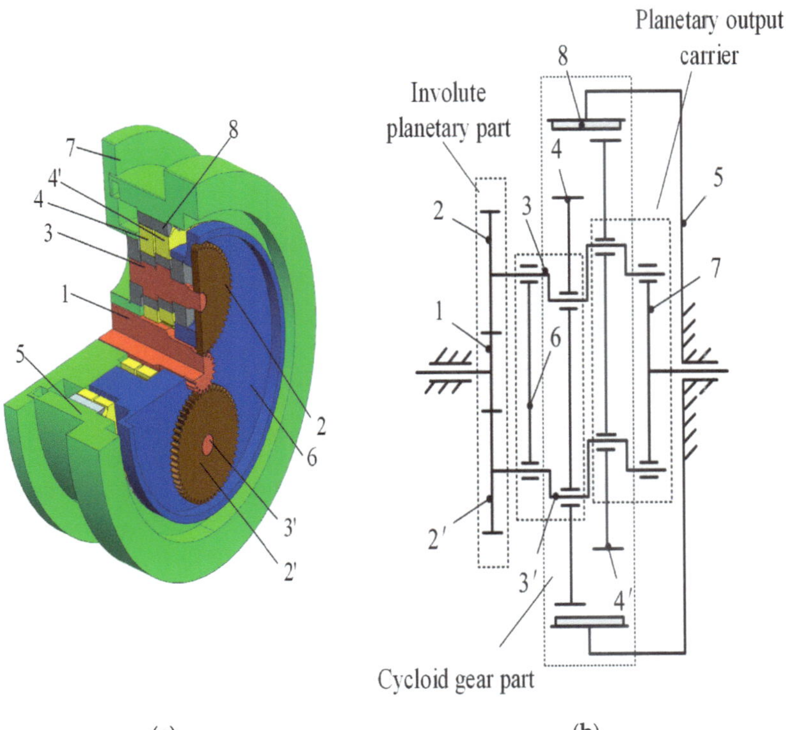

Figure 1. Schematic diagram of RV-40E reducer structure principle. (**a**) The structure of RV reducer; (**b**) Description of what is contained in the second panel. 1. Input shaft with sun gear; 2 (2′). Planetary gear; 3 (3′). Crankshaft; 4 (4′). Cycloid gear; 5. Needle tooth shell; 6. Planet carrier; 7. Flange plate; 8. Needle gear.

The rotation error is the difference between the theoretical output rotation angle and the actual rotation angle; these are calculated by the following formula:

$$\varphi_{er} = \varphi_{in} \div i - \varphi_{out} \quad (1)$$

where φ_{er} is the rotation error; φ_{in} is the input angle of input shaft; φ_{out} is the actual angle of the output shaft; i is the transmission ratio. The main structural composition parameters of the RV-40E reducer are shown in Table 1.

Table 1. The main parameter of RV-40E reducer.

Parameters	Value
Number of teeth of sun gear and planetary gear	18/54
Number of teeth of cycloid gear (z_c) and needle gear	39/40
Eccentricity of crankshaft (a) (mm)	1.5
Circle center radius of needle tooth shell (δr_p) (mm)	82
Radius of needle gear (δr_{rp}) (mm)	4
Total transmission ratio (i)	121
Short amplitude factor (k_1)	0.73
Input power	1.05 (kW)
Nominal torque	377 (N·m)

2.2. Analysis of Factors Influencing the Rotation Error of RV Reducer

The rotation error is a major quality performance metric used to evaluate the performance of the speed reducers. The influencing factors might be classified as static or dynamic. Assembly errors, fit clearance between pieces, and machining faults are examples of static variables. Despite the fact that there are many static elements that affect the rotation error, just one region of the needle tooth shell has more than ten different types of mistakes. If every static piece is evaluated, the work would be vast and unrealistic. Based on a study of the reducer's transmission chain and actual production research, 15 frequently static influencing elements were selected from the transmission chain of the RV reduction at all levels, and refer to [17] to create the parameter ranges for each factor, as shown in Table 2. Factors 1–3 are generated by the primary involute planetary gear drive, factors 4–12 are generated by the second-stage cycloidal pin-wheel drive, and factors 13–15 are generated by the planetary output carrier. The dynamic factor is primarily the influence of the elastic deformation of each part on the rotation error of the reducer when the reducer is loaded, and when the reducer is running, the dynamic factor consists primarily of the load and the speed. As the purpose of this research paper is to determine the factors influencing the high sensitivity of the RV reducer to the rotation error in order to facilitate the Taguchi-type optimized design of the RV reducer, the factors influencing this sensitivity will be identified. The dynamic influence factor is an external influence factor of the speed reducer rotation error that is not currently considered in this paper.

Table 2. RV reducer rotation error static influence factors and parameters to take the value range.

Sources of Influencing Factors	No.	Influencing Factor of Rotation Error	Code Name	The Range/mm
The primary involute planetary part	1	Center distance error of input shaft and planetary gear	Δf_a	0.15~0.50
	2	Radial error between gear and ring	ΔF_r	0.20~0.30
	3	Fit clearance of gear and shaft	ΔC_k	0.001~0.003
The second stage cycloidal pinwheel part	4	Amount of equidistant modification of cycloid gear	Δr_{rp}	0.016~0.048
	5	The amount of radial-moving modification of cycloid gear	Δr_p	0.026~0.078
	6	Clearance of inside hole of cycloid gear and slewing bearing	σ	0~0.01
	7	Radius error of needle tooth center circle	δr_p	0.005~0.015
	8	Needle gear radius error	δr_{rp}	0.01~0.03
	9	Matching clearance between the pin gear and needle gear hole	δJ	0.001~0.003
	10	Circular position error of the needle gear hole	δt	0~0.01
	11	Eccentricity error of crankshaft	δa	0.01~0.03
	12	Clearance of crankshaft and slewing bearing	Δr	0~0.002
Planetary output carrier	13	Eccentricity error of planet carrier	Eh	0.005~0.015
	14	Carrier wheel crankshaft center distance error	ΔFa	0~0.01
	15	Bearing hole phase error of planet carrier	Δ_2	0.03~0.05

The aforementioned static influencing factors combine to generate the RV reducer rotation error, which can be represented as follows using the fuzzy comprehensive evaluation mathematical method:

$$\begin{aligned}\varphi_{er} = {} & \omega_{\Delta f_a}\delta_{\Delta f_a} + \omega_{\Delta F_r}\delta_{\Delta F_r} + \omega_{\Delta C_k}\delta_{\Delta C_k} + \omega_{\Delta r_{rp}}\delta_{\Delta r_{rp}} + \omega_{\delta\Delta r_p}\delta_{\delta\Delta r_p} + \omega_{\sigma}\delta_{\sigma} \\ & + \omega_{\delta r_p}\delta_{\delta r_p} + \omega_{\delta r_{rp}}\delta_{\delta r_{rp}} + \omega_{\delta J}\delta_{\delta J} + \omega_{t}\delta_{t} + \omega_{\delta a}\delta_{\delta a} + \omega_{\Delta r}\delta_{\Delta r} \\ & + \omega_{Eh}\delta_{Eh} + \omega_{\Delta Fa}\delta_{\Delta Fa} + \omega_{\Delta_2}\delta_{\Delta_2}\end{aligned} \tag{2}$$

where φ_{er} is the rotation error of the RV reducer, δ_i is the rotation error of each influencing factor, and ω_i is the weight of each factor. The weights represent the overall degree of influence of each influencing factor on the rotation error of the reducer [18].

2.3. RV Reducer Rotation Error Sample Data Acquisition

In this study, the factor-error data were obtained directly from the RV reducer production and processing facility to ensure the practicality of the rotation error prediction model and sensitivity analysis results. Each reducer's rotation error is measured using a dedicated RV reducer rotation error test bench, and the deviation of each influencing factor is measured using the double cylindric rod measuring method, the common normal line method, and other technologies. Figure 2 depicts the schematic diagram of the rotation error test bench. The servo motor can be set to rotate at different speeds, the magnetic powder brake can be used to apply different loads. Two circular encoders record the angular signals at the input and output of the reducer; using the phase difference information of the angular signals, calculate the rotation error of the reducer.

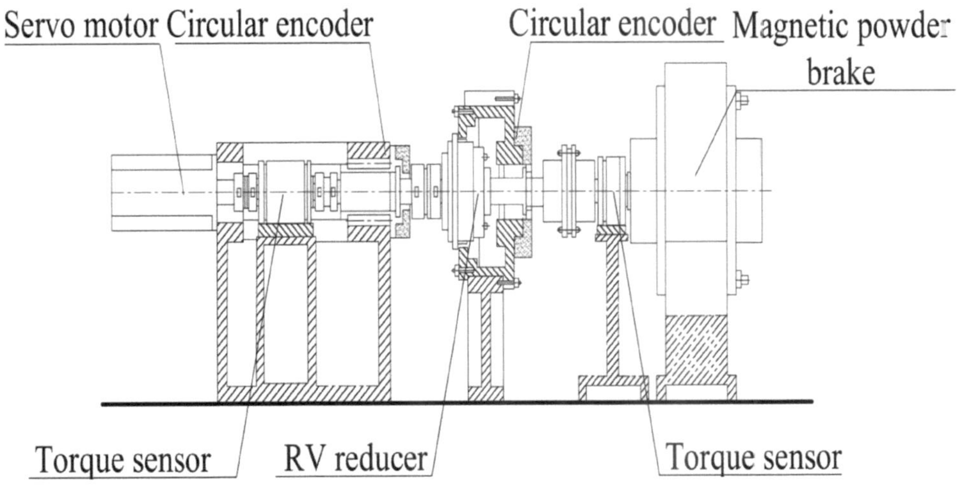

Figure 2. Schematic diagram of RV reducer rotation error test bench.

When testing each reducer, the test bench is configured with the same speed and load, the rotation error data of each RV reducer in the stable operation is recorded, and the average value is considered the rotation error value. Adjust the servo motor speed to 1200 rpm, set the load bit to 200 N·m, record the rotation error after the reducer has stabilized, and select at random the test curve chart for three reducers, as depicted in Figure 3. The sample point of the factor-rotation error is the sum of the value of the rotation error received from the test and the deviation of each influencing factor.

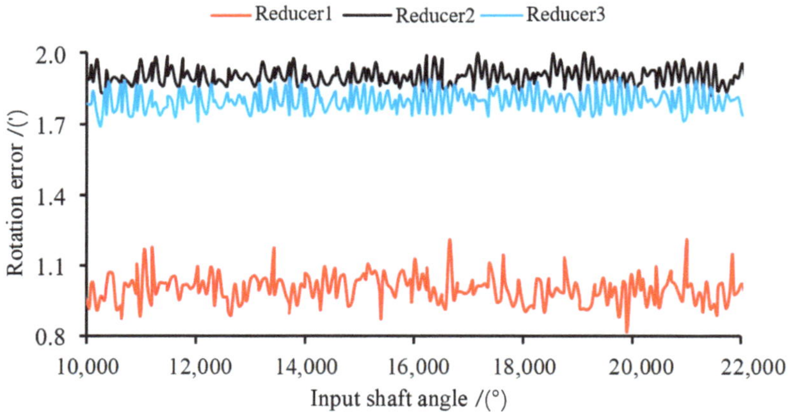

Figure 3. Test curve chart for reducers.

3. Construction of Prediction Model of RV Reducer Rotation Error Based on the DeepGP Model

Due to the multiplicity of the components and their interactions, the regression of the reducer's rotation error exhibits high-dimensional nonlinearity. Surrogate models, which have the advantages of simple model development, quick processing, and high prediction accuracy, are now frequently used to solve nonlinear problems. Traditional surrogate models, such as Kriging [19] and the Gaussian process (GP) [20], are frequently constrained by the network structure and kernel functions, resulting in poor fitting accuracy. The DeepGP model stacks multiple Gaussian process regression models, and by combining probabilistic models and deep learning, it retains the excellent performance of Gaussian process regression while having more powerful mapping capabilities in fitting high-dimensional problems [21], and we decided to build a rotation error prediction model based on the DeepGP model in this paper.

There are three essential steps in building the RV reduction rotation error prediction model:

1. Determine the range of values for each rotation error-influencing factor based on the actual production research, and then use the OLHS technique to collect the sample points for the prediction model's training set.
2. Construct a rotation error prediction model using the DeepGP model and determining the structure and parameters of the model.
3. Use a validation set to evaluate the prediction model's accuracy in order to make it easier to conduct a sensitivity analysis of each individual influencing element in the following section.

3.1. Sample Point Extraction Based on the OLHS Method

During the construction phase of the prediction model, the distribution of the sample points in the modeling space and the total number of sample points have a significant impact on the model's performance. An adequate number of uniformly distributed sample points can significantly reduce the computation costs while accurately reflecting the change trend and information of the real model in space. In all other respects, the proxy model created using these sample points can accurately and credibly reflect the sample space [22].

The Latin hypercube sampling (LHS) method, which is based on the concept of probabilistic stratification, divides the probability distribution space of the test factors into N nonoverlapping subregions based on the range of values for each influencing factor, and then conducts independent equal-probability sampling in each subregion. This method is more capable of filling empty spaces than the orthogonal test method. Due to the numerous factors affecting the rotation error in this paper, if 100 sample points are to be collected, the sample space must be divided into 100^{15} sub-regions, which significantly increases the computer's workload by excessively subdividing the spatial sample area. Moreover, there may be a part of the design space missing due to too many areas, so the LHS method is obviously not suitable for the task of sample point sampling in this paper. Based on the LHS approach, the OLHS method can ensure that all design points are uniformly distributed across the design space by evenly and orthogonally distributing them in the design space region of the test factors [23].

The parameter range of each rotation error influence factor was determined in Table 2, and 200 sample points of the influencing factor data were extracted from each parameter range using the OLHS method. The following are the specific sampling steps:

1. Each dimensional axis is subdivided into 10 equal intervals, thereby dividing the sample space range into 10^{15} sub-regions with dimensions based on the 15 variables that affect the rotation error of the reducer, such as the Δr_{rp}.
2. Obtaining 200 sub-regions of an optimal Latin hypercube sampling matrix with uniform distribution based on the space-filling optimality criterion, space-filling optimality criterion is shown in the following equation:

$$\sum_{i=1}^{n}\sum_{j=1}^{n} d_{ij}^2 = min \sum_{i=1}^{n}\sum_{j=1}^{n}\sum_{l=1}^{n}\left(x_i^{(l)} - x_j^{(l)}\right)^2 \tag{3}$$

In this equation, x_i and x_j indicate any two points, d_{ij} indicates the distance between any two points, and l indicates the dimension of the sample points, and this equation is used for the OLHS matrix calculation.

3. Following that, a sample point is selected within each of the extracted sub-regions and each parameter sampling, point is then used as the coordinate component of each dimension of the sampling point.

Using the sample point's coordinate set, each sample point is located in the sample database of the rotation error. Due to the discrete nature of the database, some sample points cannot be located. Instead, sample points near the spatial locations of these coordinate points are chosen. The final extracted partial data of the factor-rotation error sample points are shown in Table 3, which uses millimeters(mm) as the unit of measurement for each influencing factor. These data comprise the training set for the prediction model. Concurrently, 200 sample points were selected at random from the database to serve as a validation set for testing the accuracy of the prediction model developed in the subsequent paper.

Table 3. The set for training rotation error prediction model.

Sample Point	$(\Delta f_a, \Delta F_r, \Delta C_k, \Delta r_{rp}, \Delta r_p, \sigma, r_p, r_{rp}, \delta J, \delta t, a, \Delta r, Eh, \Delta Fa, \Delta_2)$	Rotation Error/″
1	(0.037, 0.200, 0.001, 0.047, 0.044, 0.009, 82.015, 4.027, 0.001, 0.001, 1.519, 0.001, 0.011, 0.006, 0.037)	0.059
2	(0.164, 0.297, 0.002, 0.038, 0.072, 0.003, 82.011, 4.016, 0.002, 0.005, 1.525, 0.001, 0.011, 0.001, 0.041)	1.238
3	(0.325, 0.277, 0.001, 0.022, 0.047, 0.001, 82.013, 4.030, 0.001, 0.004, 1.528, 0.002, 0.010, 0.008, 0.034)	1.538
...	...	...
100	(0.164, 0.224, 0.003, 0.030, 0.076, 0.002, 82.009, 4.026, 0.002, 0.003, 1.516, 0.002, 0.008, 0.005, 0.041)	1.483
101	(0.420, 0.258, 0.002, 0.028, 0.063, 0.005, 82.012, 4.024, 0.002, 0.002, 1.529, 0.001, 0.008, 0.002, 0.044)	2.161
...	...	...
199	(0.436, 0.242, 0.003, 0.036, 0.042, 0.004, 82.015, 4.018, 0.003, 0.001, 1.515, 0.001, 0 011, 0.03, 0.047)	1.217
200	(0.286, 0.233, 0.002, 0.044, 0.068, 0.002, 82.008, 4.026, 0.001, 0.003, 1.510, 0, 0.007, 0.008, 0.050)	1.040

3.2. Process of Building Rotation Error Prediction Model

3.2.1. Deep Gaussian Processes Model

The DeepGP employs the Gaussian process and deep learning as a hierarchical network to address the nonlinear relationships between the nodes in the hierarchy [24]. Layering numerous single-layer Gaussian processes on top of one another results in a deep network structure. These layers are then linked together. The DeepGP model's organizational structure is depicted in Figure 4.

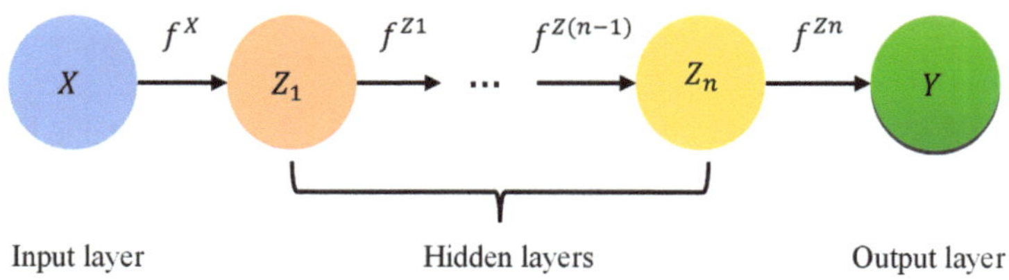

Figure 4. The structure of the DeepGP model.

The DeepGP model can be expressed as the following function:

$$y = f^{Zn}\left(f^{Z(n-1)}(\ldots\ldots f^{x}(x))\right) + \varepsilon \tag{4}$$

where x is the input layer, y is the output layer, Z_i is the hidden layer, and ε is the noise. A Gaussian process with only input and output layers and no hidden layers is standard.

The information transfer rules for the deep Gaussian process model are as follows:

$$Z_1 = f^{x}(x) + \varepsilon \tag{5}$$

$$Z_2 = f^{Z_1}(Z_1) + \varepsilon_1^{Z_1} \tag{6}$$

$$\ldots$$

$$Z_n = f^{Z_{(n-1)}}\left(Z_{(n-1)}\right) + \varepsilon_{(n-1)}^{Z_{(n-1)}} \tag{7}$$

$$y = f^{Z_n}(Z_n) + \varepsilon_n^{Z_n} \tag{8}$$

The layer's nodes represent the input values of the network's lower layer and the output values of the network's higher layer. A Gaussian process is utilized to map the two layers of neurons. The layers are connected by a kernel function that automatically modifies the connections between a large number of nodes and determines the ideal model parameters while assuming different weights for each conceivable dimension. According to Table 4, the most popular varieties of kernels are as follows.

Table 4. The commonly used kernel functions.

Kernel Function	Expression
The Linear kernel	$k(x, x'; \theta) = xx'$
The absolute-exponential kernel (RBF)	$k(x, x'; \theta) = \sigma^2\left(-\frac{1}{2}\left[\frac{(x-x')}{l}\right]^2\right)$
The absolute-exponential kernel	$k(x, x'; \theta) = \sigma^2 exp\left(\frac{\lvert x-x'\rvert}{l}\right)$
The Periodic covariance kernel function (PER)	$k(r) = \sigma^2 exp\left[-\frac{2sin^2(\pi\lvert x-x'\rvert/p)}{l^2}\right]$

Where, l denotes the characteristic length scale. p denotes the period.

DeepGP is no longer a straightforward Gaussian process due to its multi-layered network structure and complex mapping relationships. Bayesian inference methods are no longer applicable to DeepGP model parameter estimation. Due to the complexity and high dimensionality of the parameter space and the space where the hidden variables are located, variational inference is frequently employed in deterministic approximation inference to compute the model's hidden variables [25]. Figure 5 depicts the training optimization flowchart of the DeepGP model. The training of the model consists primarily of variational inference within the DeepGP layers, variational inference between the DeepGP layers, and DeepGP parameter optimization, and the specific steps are as follows:

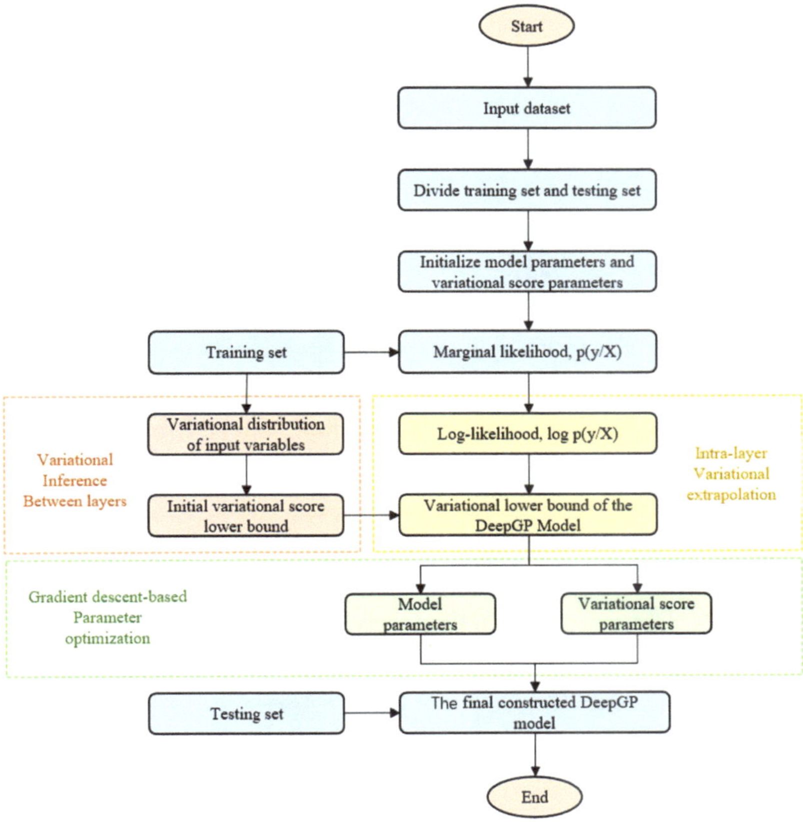

Figure 5. DeepGP model training flow chart.

Step1: The sample data are separated into training and test sets, and the number of input variables is determined.

Step2: Initialize the model parameters and variational parameters of the DeepGP model.

Step3: Input sample training data to train the model.

Step4: Variational inference within the DeepGP model layer: using a sparse variational inference method to derive an initial variational lower bound on the original probability space.

Step5: Variational inference between DeepGP model layers: maximizing the log-edge likelihood of the target data and obtaining the best approximation of $\log(p(y/X))$.

Step6: DeepGP model parameter optimization: after determining the induced variables, the model structure and model parameters are determined using the conjugate gradient algorithm with the lower bound of the variance as the objective function.

Step7: Lastly, using the validation data to assess the constructed model's accuracy.

The mean absolute percentage error (MAPE), the coefficient of determination (R^2), the mean absolute error (MAE), and the root mean squared error (RMSE) are also used to evaluate the local and global accuracy of the developed model. The formulas for calculating the four performance indicators are provided below:

$$\begin{cases} MAPE = \frac{\sum_{i=1}^{n}|\hat{y}_i - y_i|}{y_i} \\ R^2 = 1 - \frac{\sum_{i=1}^{n}(\hat{y}_i - y_i)^2}{\sum_{i=1}^{n}(\hat{y}_i - \overline{y_i})^2} \\ MAE = \frac{\sum_{i=1}^{n}|\hat{y}_i - y_i|}{n} \\ RMSE = \sqrt{\frac{\sum_{i=1}^{n}(y_i - \hat{y}_i)^2}{n}} \end{cases} \tag{9}$$

where n denotes the number of sample points, y_i denotes the measured value of sample points, $\hat{y}_i$ denotes the predicted value of sample points, and $\overline{y_i}$ denotes the average value of sample points.

3.2.2. Construction of RV Reducer Rotation Error Prediction Model

The hyperparameters of the DeepGP model consist of the number of hidden layers, the number of nodes within each layer, and the type of kernel between each layer. These hyperparameters are critical for the model's accuracy and effectiveness. Theoretically, the more hidden layers there are in a deep network, the smaller the model's regression error will be. However, if there are too many layers, the model's training time will be excessively long, it will be overly intricate, and it will have overfitting issues. The literature [26] suggests that DeepGP models with two or three hidden layers perform well in terms of prediction. In this work, the hidden layers of the model are chosen as two layers to maintain the prediction speed while reducing the model complexity.

Erickson et al. [27] concluded that GPy: a Gaussian Process framework written in Python outperforms other packages. As a result, this study uses GPy packages to build a DeepGP model. The number of nodes in the input layer corresponds to the number of variables that influence the rotation error of the reducer. The output layer variable is the rotation error, and the output layer contains a total of one layer. Therefore, the model-based DeepGP model and the input X and output layers Y can be expressed as follows:

$$X = [x_1, x_2, \ldots, x_{15}] = [\Delta f_a, \Delta F_r, \ldots, \Delta F_a, \Delta_2] \tag{10}$$

$$Y = \varphi_{er} \tag{11}$$

$$y = f^{Z2}\left(f^{Z1}(f^{x}(x))\right) + \varepsilon \tag{12}$$

As the rotation error increases gradually as the error value of each influencing factor increases, the rotation error is linear with each influencing factor as a whole. As a result, the linear kernel was selected as the kernel function between the input layer and the initial hidden layer; The RBF kernel is selected as the kernel function between the first and second hidden layers and between the second hidden layer and the output layer due to the severe coupling effect between the influencing factors.

In this paper, the ideal number of nodes in the two hidden layers is determined by the grey wolf optimization (GWO) algorithm [28]. The training set is utilized to construct the prediction model, and the fitness function is the R^2 of the model's ability to predict the validation set collected above. The wolf count (P) of the population is set to 30, and the number of iterations (N) is set to 100. It is determined through an iterative search that the prediction model with the highest R^2 value is constructed when the number of nodes in the first hidden layer is six and the number of nodes in the second hidden layer is three for the validation set, as well as the prediction accuracy of this model. As a result, Figure 6 depicts the structure of the RV reducer rotation error prediction model developed using the DeepGP model.

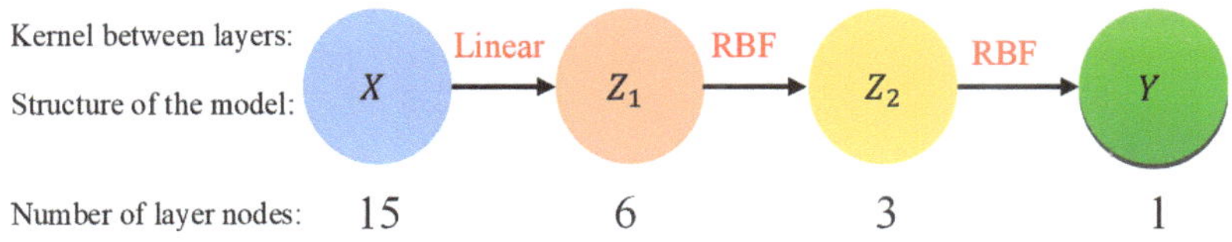

Figure 6. Structure diagram of the rotation error prediction model.

3.2.3. Accuracy Check of Rotation Error Prediction Model

Using the 200 validation sample points collected above, the accuracy of the prediction model developed for this study is evaluated. MAPE, R^2, MAE, and RMSE are chosen as the performance metrics to evaluate the model's regional and overall accuracy. In order to reduce the chance factor in the test data, the validation set was evenly divided into two for the model regression testing. Table 5 shows the test results for the two validation sets. In this study, the prediction model has a prediction accuracy of more than 95% for both validation sets. The prediction model in this paper has a low prediction error and good fitting accuracy.

Table 5. Results of model regression accuracy test.

Performance Metrics	Validation Set 1	Validation Set 2
MAPE	0.05533	0.07584
R^2	0.97871	0.95913
MAE	0.06059	0.07085
RMSE	0.07053	0.08143

The rotation error dynamics model proposed in the literature [7] based on the equivalent spring error model and the virtual prototype constructed in the literature [13] were selected for comparison in order to validate the applicability and superiority of the rotation error prediction model constructed based on the DeepGP model proposed in this paper. The equivalent spring schematic model of the RV reducer constructed using the dynamic substructure method is depicted in Figure 7a, below, where *Ki* represents the stiffness of each part. The virtual prototype of the RV reducer constructed in the Adams dynamics software is depicted in Figure 7b below.

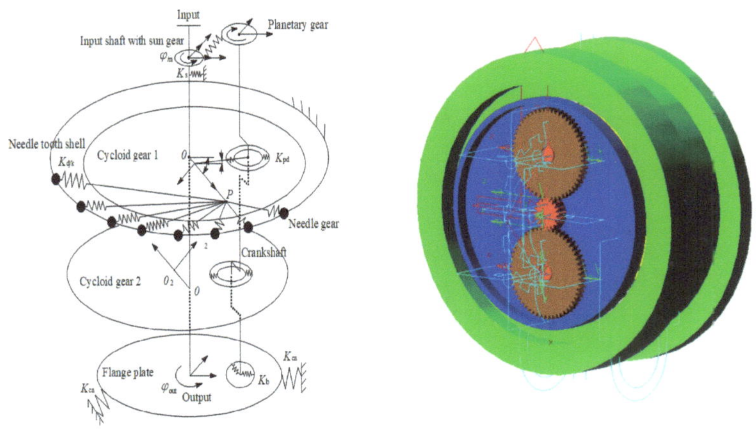

(a) (b)

Figure 7. Two model diagrams of RV-40E Reducer. (**a**) Equivalent spring error model of RV-40E reducer. (**b**) Virtual prototype of RV-40E reducer.

As the virtual prototype simulation is time-consuming and inefficient, this paper arbitrarily selects 40 points from the validation set to compare the three models' accuracy verification. The test results of the three models are depicted in Table 6 and Figure 8 below.

Table 6. Comparison of regression accuracy of three models.

Performance Metrics	The Proposed Method	Equivalent Spring Method	Virtual Prototype
MAPE	0.06533	0.17326	0.20461
R^2	0.96846	0.78542	0.76941
MAE	0.08059	0.13212	0.12941
RMSE	0.07053	0.16819	0.20851

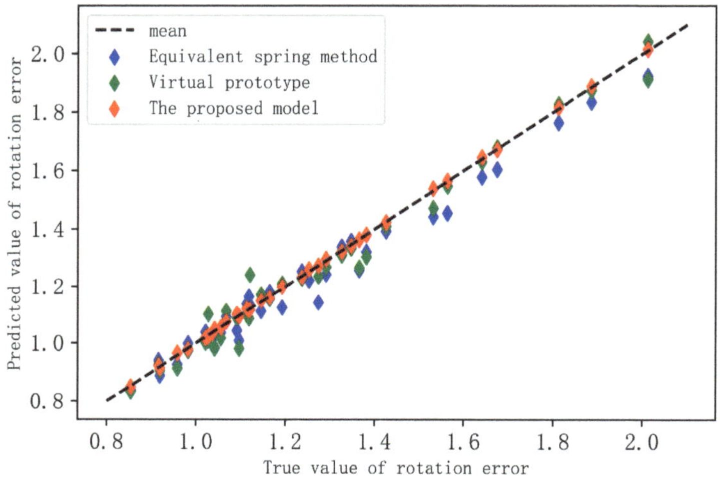

Figure 8. Test results of this model and two other models.

It can be seen from the Table 6 that the R^2 value of the prediction model presented in this article reaches 96%. The model presented in this paper has stable mapping and high prediction accuracy, whereas the R^2 value of the other two models is just over 75%. Figure 8 shows that the values predicted in this article are more in line with the true value, while the rotation error predicted by the other two models deviates from the measured values and are, in most cases, smaller than the measured values. Due to the fact that this paper is based on the actual production and processing plant data, the other two models may not account for the influence of the friction force and disregard the influence of certain components, such as the bearings, in the modeling process.

According to the model evaluation results, the samples collected using the OLHS approach were uniform and representative, which matched the criteria for the sampling points required by the prediction model and could guarantee the model's construction accuracy. This paper extracts the data directly from the actual production and processing plant to assure the applicability and precision of its analysis results. The RV reducer rotation error modelling method presented in this paper, which is based on the DeepGP model, is faster and more precise than the traditional modelling techniques. The mapping stability of the prediction model constructed in this paper paves the way for a later sensitivity analysis of the factors affecting rotation error.

4. Sensitivity Analysis of Factors Influencing the Rotation Error of RV Reducers Based on Sobol Method

The examination of the rotation error prediction model reveals that it is stable in mapping and has a high level of prediction accuracy, meeting the requirements of sensitivity analysis (SA). As a result, the SA can use the newly developed rotation error prediction model to reduce the computation costs and improve the analysis efficiency.

4.1. Sobol Sensitivity Analysis Method

SA focuses on the relative magnitude of the effect of the model input on the output uncertainty, which identifies the important design variables and is crucial for the optimal design of engineering products. SA is divided into local SA and global SA. In the current process, global sensitivity is frequently used to analyze the issue. The common sensitivity analysis techniques used today include the Fourier amplitude sensitivity test (FAST) [29], Morris [30], one-factor-at-a-time methods with Latin-hypercube sampling (LH-OAT) [31], Generalized Likelihood Uncertainty Estimation (GLUE) [32], and the Sobol method [33]. According to the analysis principles, the sensitivity analysis methods can be divided into qualitative and quantitative methods. Table 7 lists the characteristics of the different sensitivity analysis methods.

Table 7. Characteristics of sensitivity methods.

Sensitivity Analysis Methods		Characteristics of Sensitivity Methods
Local sensitivity analysis	FAST	Simple operation, ignores parameter interactions, low model applicability.
Global sensitivity analysis	Morris	Compared the output results of adjacent parameters in the parameter space, inefficient.
	LH-OAT	Although the benefits of both the random one-factor-at-a-time and LHS sampling techniques are taken into account, the computer program is complex.
	GLUE	Combining the benefits of Rivest-Shamir-Adleman techniques and fuzzy mathematics to rank sensitivities as scatter plots.
	Sobol	The sensitivity indices are solved using the Monte Carlo sampling technique, which can distinguish independently between parameter-independent and parameter-interacting sensitivities.

Comparing the characteristics of the aforementioned sensitivity analysis methods, the variance-based Sobol method considers the influence of the parameter interactions on the results and can not only solve the local and global sensitivity indexes of each uncertainty variable, but can also quantify the influence of the parameter interactions on the results. In this paper, the Sobol method is employed to quantify the sensitivity indexes of each factor influencing the rotation error and to examine the influence of the interaction and coupling between the factors on the rotation error.

For the multivariate function $y = g(x)$, it can be decomposed into the following summation form:

$$g(x) = g_0 + \sum_{i=1}^{k} g_i(x_i) + \sum_{1 \le i \le i \le k} g_{ij}(x_i, x_j) + \ldots + g_{1,2,\ldots,k}(x_1, x_2, \ldots, x_k) \quad (13)$$

The right term of the above equation is expressed in the form of a multiple integral as follows:

$$g_0 = \int \ldots \int g(x)dx = \int \ldots \int g(x_1, \ldots, x_k)dx_1 \ldots dx_k \quad (14)$$

$$g_i(x_i) = \int \ldots \int g(x)dx_{-\mathrm{i}} - g_0 \tag{15}$$

$$g_{ij}(x_i, x_j) = \int \ldots \int g(x)dx_{-(ij)} - g_0 - g_i(x_i) - g_j(x_j) \tag{16}$$

where $dx_{-\mathrm{i}}$ denotes the integral of variables other than x_i, $dx_{-(ij)}$ denotes the integral of variables other than x_i and x_j, the remaining terms of Equation (13) can be derived to find.

The variance V and bias a of y can be expressed as:

$$V = \int \ldots \int g^2(x_1, \ldots, x_k)dx_1 \ldots dx_k - g_0^2 \tag{17}$$

$$V_{i_1 \ldots i_s} = V[g_{i_1 \ldots i_s}] = \int \ldots \int g_{i_1 \ldots i_s}^2(x_{i_1}, \ldots, x_{i_s})d_{x_{i_1}} \ldots d_{x_{i_s}} \tag{18}$$

where $1 \leq i_1 < \ldots < i_s \leq k$, $V_{i_1 \ldots i_s}/V$ denotes the sensitivity of $\{i_1, \ldots, i_s\}$. Sensitivity indicators for different order variables can be obtained by adjusting the number of $\{i_1, \ldots, i_s\}$. Among them, the first-order Sobol sensitivity indicator S_i and the global sensitivity indicator S_{T_i} are commonly used.

The first-order indices are also known as the main effects, and the corresponding expression is shown in the equation below:

$$S_i = V_i/V \tag{19}$$

Other higher-order Sobol sensitivity indices can also be obtained based on the bias variance, and higher-order Sobol indices indicate the cross-correlation between two or more factors. First-order and higher-order indices have the following properties:

$$\sum_{i=1}^{k} S_i + \sum_{i<j} S_{ij} + \ldots + S_{ij \ldots k} = 1 \tag{20}$$

The total-order indices are also known as the total effects, and can be expressed as:

$$S_{T_i} = 1 - S_{-_i} \tag{21}$$

where S_{-_i} denotes the sum of all $S_{ij \ldots k}$ excluding variable x_i.

S_i denotes the first-order effect of the output parameter x_i alone on the output result, and the S_i does not account for covariate cross-talk. S_{T_i} denotes the sum of the first-order and higher-order effects of the input parameter x_i on the output results, and it evaluates the parameter x_i's overall effect on the model output. When $S_{T_i} = 0$, the variable x_i has no effect on the model output. In this paper, we will calculate each sensitivity index using a Monte Carlo sampling method based on variance [34].

4.2. Sensitivity Analysis Process for Rotation Error Influencing Factors Based Sobol Method

The Python package SALib is used as the primary research tool in this study, and the Sobol method is employed to conduct a sensitivity analysis of the variables influencing the RV reducer's rotation error. As a result of the small sample size used in this study, the results of the SA using these sample points alone may not be accurate enough. The inhomogeneity of the sample location space will lead to numerous inaccuracies if a large number of samples are collected at random. As a result, the previously mentioned rotation error prediction model is extracted first, followed by the use of SALib's proprietary Saltelli sampler to create uniform samples, and finally the Sobol method to calculate the SA for each influencing component. Each test was repeated three times under the same conditions to ensure that the sampling had no effect on the SA results, and the average of all the values was used to determine each SA's final result.

4.2.1. Results of Total Effects of Factors Influencing the Rotation Error of RV Reducer

The SA method was built in Python software. Firstly, the prediction model of the reducer rotation error constructed above was imported; then, each influencing variable and its value range were determined, and the SA problem model of this paper was defined; finally, 2400 uniform samples were generated by the Saltelli sampler for the quantitative analysis of 15 influencing factors of the reducer rotation error.

To begin, the total effects of each rotation error influence factor are analyzed, and the total effects indicators of the reducer rotation error influencing factors are displayed in Table 8, where a and the other identifiers represent each component influencing the rotation error. As shown in the table, the total effect of δt(j) is 0.568193, making it the factor with the highest index. δt is the most sensitive to the rotation error's effects. The effects of Δr_{rp}, Δr_p, σ, δr_p, δr_{rp}, δJ, δt, Δr are more than ten times as much as the remaining influencing factors' effects. Analyzing the structure and transmission principle of the reducer reveals that its first-stage planetary gear reduction mechanism is located far from the output end, and its influence on the rotation error has been greatly reduced due to the error it produces after the second stage reduction transmission. The combined force of two 180° out of phase cycloid wheels is applied to the planetary frame. The impact of the planetary output carrier on the RV reducer's rotation error is significantly mitigated [35]. The rotation error of the RV reducer can be attributed to the error affecting the factor of the secondary cycloidal pin-wheel transmission mechanism.

Table 8. The total effects indicators of the reducer rotation error influencing factors.

Label	Influencing Factor of Rotation Error	Total Effects
a	Δf_a	0.012414
b	ΔF_r	0.025427
c	ΔC_k	0.007129
d	Δr_{rp}	0.269588
e	Δr_p	0.144105
f	σ	0.387645
g	δr_p	0.112615
h	δr_{rp}	0.531029
i	δJ	0.111341
j	δt	0.568193
k	δa	0.021295
l	Δr	0.228339
m	Eh	0.014437
n	ΔFa	0.012762
o	Δ_2	0.009721

4.2.2. Analysis of the Coupling Effect of Factors Influencing the Rotation Error of RV Reducer

The involute planetary transmission component and the planetary output carrier have a very small and negligible impact on the rotation error, as demonstrated by the above study; thus, we will ignore these aspects for the time being. Figure 9 depicts the primary influences and overall effects of each contributing element on the rotation error of the RV reducer. Where each influence component, denoted by a data label, is the same as in Table 8. As depicted in the figure, δr_{rp}(h) and δt(j) have higher sensitivity indices and their impact on the rotation error is marginally greater. Their overall and primary effects are comparable. The remaining factors, such as Δr_{rp} (d) and Δr (l), vary substantially in terms of their main effects and total effects, despite having relatively smaller sensitivity indicators. As these factors interact significantly with other factors, it is vital to study each factor's coupling relationship. The second-order sensitivity index of the factor influencing the rotation error is then analyzed.

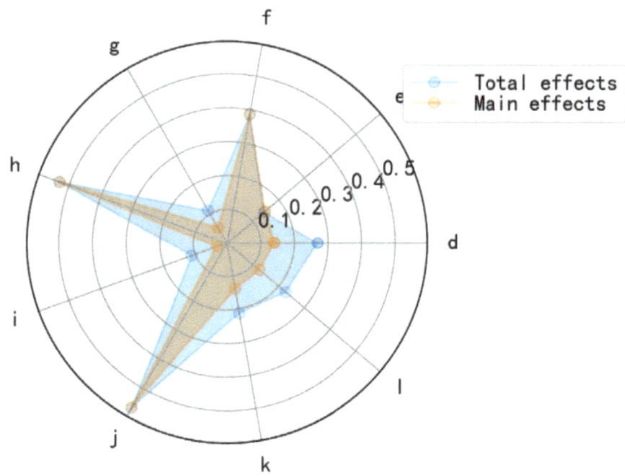

Figure 9. The main effects and total effects of each influencing factor of RV reducer rotation error.

Figure 10 shows a heat map of the second-order sensitivity index between the variables affecting the rotation error of the RV reducer, with the labels of the rows and columns representing the same error factors as in Table 8. An increase in the thermogram readings or a darkening of the color of the two elements indicates that the coupling between the two components has a greater influence on the reducer's rotation error. As shown in the picture, the second-order sensitivity index between δJ(i) and δt(j) is 0.029, and the second-order sensitivity index between δJ(i) and δr_{rp}(h) is 0.025. These two coupling effects are noticeably stronger. The second-order sensitivity index between Δr_{rp}(d) and Δr_p(e) is 0.0077, and the coupling effects between Δr_{rp}(d) and Δr_p(e) have less of an impact on the rotation error. As the second-order sensitivity index between the other two components was either zero or negative, none of the coupling effects were significant. By comparing the first-order, second-order, and total effects of the factors influencing the rotation error, we can conclude that the total effects of each factor is produced by the combination of the main effects and the coupling between itself and the other influences.

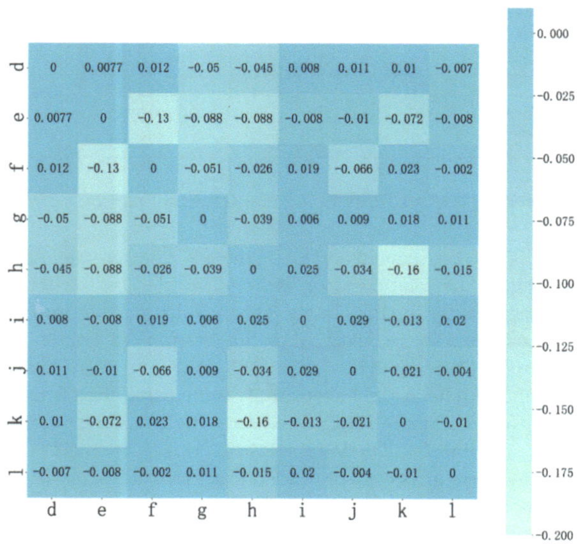

Figure 10. Heat map of second-order sensitivity of RV reducer rotation error.

4.2.3. Comparison of Sensitivity Analysis Results

We compare the Sobol method with the established Taylor midvalue theorem sensitivity analysis method [13] to more thoroughly assess the Sobol method's accuracy. In the literature [13], the equivalence error method based on the principle of the equivalent spring method was first used to derive the value of the rotation error caused by each factor of the rotation error, followed by the Taylor midvalue theorem to obtain the sensitivity coefficient for each factor, and finally the sensitivity index for each factor. The results of the sensitivity comparison of some of the factors influencing the rotation errors calculated by the Taylor and Sobol methodologies are shown in Table 9.

Table 9. Comparison of the sensitivity indexes of the factors influencing the rotation error obtained by the two methods.

Infuencing Factor of Ratation Error	Taylor Sensitivity Method			Sobol
	Rotation Error Caused by Each Factor	Sensitivity Coefficient	Sensitivity Index	Total Effects
Δf_a	$\frac{\Delta f_a \times 180 \times 60}{\pi \times i \times r_1}$	-	0.025	0.0124
ΔF_r	$\frac{\Delta F_r \times 180 \times 60}{\pi \times i \times r_1}$	-	0.025	0.0254
Δr_{rp}	$\frac{2 \times \Delta r_{rp}}{a \times Z_c}$	1	1	0.269588
Δr_p	$\frac{2 \times \Delta r_p \times \sqrt{1-k_1^2}}{a \times z_c}$	$\sqrt{1-k_1^2}$	0.68	0.144105
δr_p	$\frac{2 \times \delta r_p \times \sqrt{1-k_1^2}}{a \times z_c}$	$\sqrt{1-k_1^2}$	0.68	0.112615
δr_{rp}	$\frac{-2 \times \delta r_{rp}}{a \times z_c}$	1	1	0.5312
δJ	$\frac{\delta J}{a \times z_c}$	0.5	0.5	0.1111
δt	$\frac{k_1 \times \delta J}{a \times z_c}$	k_1	0.73	0.56819
Δr	$\frac{180 \times 60 \times \Delta r}{\pi \times a}$	$\frac{ez_c}{2a}$	0.679487	0.228339

Where r_1 denotes the radius of the input shaft with sun gear.

According to Table 9, when comparing the normalized sensitivity index of each factor deviation derived by the equivalent error method with the Taylor midvalue theorem and the sensitivity index calculated in this paper, the calculation results of the two methods have little difference, the sensitivity index of each factor deviation is in the same order, and the sensitivity qualitative analysis of each factor deviation is the same. The approach proposed in the literature [13] does not consider the effect of the coupling effects between factors and is more of a qualitative analysis. In contrast, this paper analyzes the combined effect of each factor on the rotation error, solves the sensitivity index of each factor quantitatively, and examines how the coupling between the factors affects the rotation error.

As a result, the rotation error prediction model and the sensitivity analysis method presented in this paper are correct.

5. Conclusions

Using the DeepGP model and the Sobol sensitivity approach, this paper investigates the sensitivity of each contributing component of the RV reducer rotation error. The preceding analysis leads to the following conclusions:

- Using the OLHS method, actual production facilities were sampled to ensure the accuracy of the created prediction models, as well as the dependability and applicability of the SA results. Using the Deep GP model, a high-precision prediction model for the rotation error of each RV reducer was developed, and the accuracy was compared with the equivalent method and virtual prototypes to demonstrate the validity and accuracy of the prediction model, thereby establishing the conditions for the sensitivity analysis of the RV reducer's rotation error.

- On the basis of the prediction model, a global sensitivity analysis of the variables affecting the rotation error of the RV reducer was conducted using the Sobol method. The primary cause of the reducer's rotation error is the second-stage cycloidal pin-wheel transmission mechanism, with the influence of the planetary gear transmission and the planetary output carrier being small and negligible.
- The second-order sensitivity index is used to evaluate how multiple influencing elements interact to affect the rotation error of the reducer. There is significantly more coupling between δJ and δt, as well as between δr_{rp} and δJ. The coupling effect between other variables has little effect on the rotation error. The total order sensitivity index of each factor is calculated by adding its main effects and the coupling of those effects to other influences. The results of this paper's sensitivity analysis were contrasted with the results of Taylor's midvalue theorem to confirm the originality and precision of the analysis results.

Author Contributions: Methodology and project administration, S.J. (Shousong Jin); Writing original draft and developing software, S.S., S.J. (Suqi Jiang) and M.C.; Supervision and data curation, Y.W. All authors have read and agreed to the published version of the manuscript.

Funding: This research was supported by National High-tech R&D Program of China (Grant. No. 2015AA043002), the Natural Science Foundation of Zhejiang Province (Grant No. LQ22E050017).

Institutional Review Board Statement: Not applicable.

Informed Consent Statement: Not applicable.

Data Availability Statement: The data presented in this study are available on request from the corresponding author. The data are not publicly available, because they involve corporate data privacy.

Acknowledgments: The authors thank the Zhejiang Shuanghuan Transmission Machinery Co., Ltd. for providing the RV reducer sample and some of the data.

Conflicts of Interest: The authors declare no conflict of interest in the publication of this article.

References

1. He, W.D.; Shan, L.J. Status and Development of RV Reduce. *J. Dalian Jiaotong Univ.* **2016**, *37*, 13–18.
2. Pham, A.D.; Ahn, H.J. High Precision Reducers for Industrial Robots Driving 4th Industrial Revolution: State of Arts, Analysis, Design, Performance Evaluation and Perspective. *Int. J. Precis. Eng. Manuf. Green Technol.* **2018**, *5*, 519–533. [CrossRef]
3. Yang, Y.H.; Zhou, G.C.; Chang, L.; Chen, G. A Modelling Approach for Kinematic Equivalent Mechanism and Rotational Transmission Error of RV Reducer. *Mech. Mach. Theory* **2021**, *163*, 104384. [CrossRef]
4. Hidaka, T.; Wang, H.Y.; Ishida, T.; Matsumoto, K.; Hashimoto, M. Rotational Transmission Error of K-H-V Planetary Gears with Cycloid Gear (1st Report, Analytical Method of the Rotational Transmission Error). *JSME Int. J. C-Mech. Syst.* **1994**, *60*, 645–653.
5. Ishida, T.; Wang, H.Y.; Hidaka, T.; Matsumoto, K.; Hashimoto, M. Rotational Transmission Error of K-H-V Planetary Gears with Cycloid Gear (2nd Report, Effects of Manufacturing and Assembly Errors on Rotational Transmission Error). *JSME Int. J. C-Mech. Syst.* **1994**, *60*, 278–285. [CrossRef]
6. Wang, H.Y.; Ishida, T.; Hidaka, T.; Matsumoto, K.; Hashimoto, M. Rotational Transmission Error of K-H-V Planetary Gears with Cycloid Gear (3rd report, Mutual Effects of Errors of the Elements on the Rotational Transmission Error). *JSME Int. J. C-Mech. Syst.* **1994**, *60*, 286–293. [CrossRef]
7. He, W.D.; Shan, L.J. Research and Analysis on Transmission Error of RV Reducer Used in Robot. In *Recent Advances in Mechanism Design for Robotics*; Springer: Berlin/Heidelberg, Germany, 2015; pp. 231–238.
8. Cao, K.; Shao, G.; Liu, T. Equivalent Error Modeling and Parameter Optimization of RV Reducer. *J. Xiamen Univ. Nat. Sci.* **2022**, *61*, 246–252.
9. Chu, X.Y.; Xu, H.H.; Wu, X.M.; Tao, J.P.; Shao, G.F. The Method of Selective Assembly for the RV Reducer Based on Genetic Algorithm. *Proc. Inst. Mech. Eng. Part C-J. Mech. Eng. Sci.* **2018**, *232*, 921–929. [CrossRef]
10. Liu, K. Analysis and Evaluation of Transmission Accuracy of RV Reducer. Master's Thesis, Xiamen University of Technology, Xiamen, China, 2022.
11. Meng, Y.H.; Wu, C.L.; Ling, L.P. Mathematical Modeling of the Transmission Performance of 2k-H Pin Cycloid Planetary Mechanism. *Mech. Mach. Theory* **2007**, *42*, 776–790. [CrossRef]
12. Liu, T.D.; Lu, M.; Shao, G.F.; Wang, R.Y. Modeling and Optimization of Rotate Vector Reducer Transmission Error. *Control Theory Appl.* **2020**, *37*, 215–221.

13. Tong, X.T. Research on Dynamic Transmission Error of RV Reducer Based on Virtual Prototype Technology. Master's Thesis, Zhejiang University of Technology, Hangzhou, China, 2019.
14. Jin, S.S.; Cao, M.Y.; Qian, Q.C.; Zhang, G.; Wang, Y.L. Study on an Assembly Prediction Method of RV Reducer Based on Igwo Algorithm and Svr Model. *Sensors* **2023**, *23*, 366. [CrossRef]
15. Hu, Y.H.; Li, G.; Zhu, W.D.; Cui, J.K. An Elastic Transmission Error Compensation Method for Rotary Vector Speed Reducers Based on Error Sensitivity Analysis. *Appl. Sci.* **2020**, *10*, 481. [CrossRef]
16. Wu, X.; He, W. Transmission Accuracy Test and Virtual Prototype Simulation of the RV Reducer Used in the Robot. *J. Mach. Des.* **2017**, *34*, 73–77.
17. Wen, B.C. *Handbook of Mechanical Design*; China Machine Press: Beijing, China, 2017; Volume 2, pp. 114–137.
18. Yang, D.C.H.; Blanche, J.G. Design and Application Guidelines for Cycloid Drives with Machining Tolerances. *Mech. Mach. Theory* **1990**, *25*, 487–501. [CrossRef]
19. Martin, J.D.; Simpson, T.W. Use of Kriging Models to Approximate Deterministic Computer Models. *AIAA J.* **2005**, *43*, 853–863. [CrossRef]
20. Marrel, A.; Iooss, B.; Van Dorpe, F.; Volkova, E. An Efficient Methodology for Modeling Complex Computer Codes with Gaussian Processes. *Comput. Stat. Data Anal.* **2008**, *52*, 4731–4744. [CrossRef]
21. Damianou, A.C.; Lawrence, N.D. Deep Gaussian Processes. *J. Mach. Learn. Res.* **2013**, *31*, 207–215.
22. McKay, M.D.; Beckman, R.J.; Conover, W.J. A Comparison of Three Methods for Selecting Values of Input Variables in the Analysis of Output from a Computer Code. *Technometrics* **2000**, *42*, 55–61. [CrossRef]
23. Lai, Y.Y.; Jiang, X.; Fang, L.Q.; Li, M.; Li, G.Q. *Parametric Optimization Theory and Examples in Detail of Isight*; Beihang University Press: Beijing, China, 2012; pp. 90–95.
24. Hebbal, A.; Brevault, L.; Balesdent, M.; Talbi, E.; Melab, N.; IEEE. Efficient Global Optimization Using Deep Gaussian Processes. In Proceedings of the 2018 IEEE Congress on Evolutionary Computation (CEC), Rio de Janeiro, Brazil, 8–13 July 2018; pp. 2625–2632.
25. Damianou, A. Deep Gaussian Processes and Variational Propagation of Uncertainty. Ph.D. Thesis, University of Sheffield, Sheffield, UK, 2015.
26. Calandra, R.; Peters, J.; Rasmussen, C.E.; Deisenroth, M.P.; Ieee. Manifold Gaussian Processes for Regression. In Proceedings of the 2016 International Joint Conference on Neural Networks (IJCNN), Vancouver, BC, Canada, 24–29 July 2016; pp. 3338–3345.
27. Erickson, C.; Ankenman, B.E.; Sanchez, S.M.; Ieee. Comparison of Gaussian Process Modeling Software. In Proceedings of the 2016 Winter Simulation Conference (WSC), Washington, DC, USA, 11–14 December 2016; pp. 3692–3693.
28. Mirjalili, S.; Mirjalili, S.M.; Lewis, A. Grey Wolf Optimizer. *Adv. Eng. Softw.* **2014**, *69*, 46–61. [CrossRef]
29. Schaibly, J.H.; Shuler, K.E. Study of the Sensitivity of Coupled Reaction Systems to Uncertainties in Rate Coefficients. II Applications. *J. Chem. Phys.* **1973**, *59*, 3879–3887. [CrossRef]
30. Campolongo, F.; Cariboni, J.; Saltelli, A. An effective screening design for sensitivity analysis of large models. *Environ. Model. Softw.* **2007**, *22*, 1509–1518. [CrossRef]
31. Griensven, A.G.; Meixner, T.; Grunwald, S.; Bishop, T.; Diluzio, M.; Srinivasan, R. A global sensitivity analysis tool for the parameters of multi-variable catchment models. *J. Hydrol.* **2006**, *324*, 10–23. [CrossRef]
32. Beven, K.; Binley, A. The future of distributed models: Model calibration and uncertainty prediction. *Hydrol. Process.* **1992**, *6*, 279–298. [CrossRef]
33. Zhang, M.X.F.; Pandey, M.D. An Effective Approximation for Variance-Based Global Sensitivity Analysis. *Reliab. Eng. Syst. Saf.* **2014**, *121*, 164–174. [CrossRef]
34. Zhai, Q.Q.; Yang, J.; Zhao, Y. Space-Partition Method for the Variance-Based Sensitivity Analysis: Optimal Partition Scheme and Comparative Study. *Reliab. Eng. Syst. Saf.* **2014**, *131*, 66–82. [CrossRef]
35. Wang, X.L. Development of RV Reducer Performance Tester. Master's Thesis, Beijing University of Technology, Beijing, China, 2015.

sensors

MDPI

Article

A Point Cloud Data-Driven Pallet Pose Estimation Method Using an Active Binocular Vision Sensor

Yiping Shao [1], Zhengshuai Fan [1], Baochang Zhu [2], Jiansha Lu [1,*] and Yiding Lang [3]

[1] College of Mechanical Engineering, Zhejiang University of Technology, Hangzhou 310023, China
[2] Noblelift Intelligent Equipment Co., Ltd., Huzhou 313100, China
[3] Ningbo Fujia Industrial Co., Ltd., Ningbo 330200, China
* Correspondence: ljs@zjut.edu.cn

Abstract: Pallet pose estimation is one of the key technologies for automated fork pickup of driverless industrial trucks. Due to the complex working environment and the enormous amount of data, the existing pose estimation approaches cannot meet the working requirements of intelligent logistics equipment in terms of high accuracy and real time. A point cloud data-driven pallet pose estimation method using an active binocular vision sensor is proposed, which consists of point cloud preprocessing, Adaptive Gaussian Weight-based Fast Point Feature Histogram extraction and point cloud registration. The proposed method overcomes the shortcomings of traditional pose estimation methods, such as poor robustness, time consumption and low accuracy, and realizes the efficient and accurate estimation of pallet pose for driverless industrial trucks. Compared with traditional Fast Point Feature Histogram and Signature of Histogram of Orientation, the experimental results show that the proposed approach is superior to the above two methods, improving the accuracy by over 35% and reducing the feature extraction time by over 30%, thereby verifying the effectiveness and superiority of the proposed method.

Keywords: driverless industrial trucks; pose estimation; adaptive Gaussian weight-based fast point feature histogram; point cloud registration

1. Introduction

The application of intelligent logistics equipment has played a key role in the transformation and upgrading of the manufacturing industry in recent years. Driverless industrial trucks are common intelligent logistics equipment [1] and widely used in warehousing, production, medical treatment, the service industry and other fields. It can realize automatic material handling and improve production efficiency and lower production costs for intelligent logistics systems [2]. However, as shown in Figure 1, due to the influence of obstacles, uneven lighting, human intervention and other factors, a certain deviation is caused between the actual pose and the correct pose of the pallet. As a result, collision and incomplete forks will occur when driverless industrial trucks forklift pallets, which will lead to dumping, damage and safety problems. Therefore, it is necessary to upgrade the technology of traditional driverless industrial trucks and introduce some external industrial vision sensors [3] to achieve the purpose of adaptive and automatic production.

In order to solve the problem shown in Figure 1, based on the data collected by the vision sensor, it is necessary to estimate the pose of the pallet when the driverless industrial truck is driven to a certain distance in front of the pallet, so as to correct the pose deviation of the pallet and ensure the efficiency and safety of the logistics process. There are two main kinds of target pose estimation methods: LIDAR based and vision based. The LIDAR-based methods are mature and have high estimation accuracy. Baglivo et al. [4] proposed an efficient scheme, which combined a laser range-based object localization approach with PC-Sliding. Mohamed et al. [5] presented a novel architecture allowing a robot to detect, localize and track pallets using machine learning techniques based on an onboard 2D laser rangefinder. Zhang et al. [6] improved the matching degree of multi-modal features of the target and achieved accurate target pose estimation by fusing the range of view, aerial view and RGB view of the LIDAR.

(a) (b)

Figure 1. Diagram of pallet position deviation. (**a**) Correct pose. (**b**) Deviation pose.

However, due to the problems of high cost, limited range and difficulty in eliminating cumulative errors of LIDAR-based methods, vision-based methods are often adopted for pose estimation in indoor environments with sufficient illumination. The vision-based pose estimation methods mainly include 2D vision and 3D vision. In certain work environments, the 2D vision-based methods have been widely used in the pose estimation of the target object [7–10]. Varga et al. [11] obtained the position and pose information of pallets by intensity image and a stereo camera, and the LBP featured descriptor was introduced to realize automatic fork picking of pallets. Monocular images and matching of 2D deformation patterns were used to estimate the pose of known objects in dynamic environments by Casado et al. [12]. With the rapid development of low-cost depth sensors, object pose estimation has been transformed from traditional single-point and segmented measurements to dense point clouds and full-profile measurements [13–16]. Compared with 2D vision, 3D vision can obtain one more dimension of target information, which solves the problem of information loss in the process of mapping from 3D space to 2D space and has gradually become a hot topic in current research [17,18]. The current popular 3D vision solution is to estimate the pose of the target by point cloud registration. Common point cloud registration algorithms include the Normal Distributions Transform (NDT) [19] algorithm, Principal Component Analysis (PCA) algorithm [20], Iterative Closest Point (ICP) algorithm [21], and many other improved algorithms. The principle of the traditional ICP algorithm is popular and easy to understand, and the registration effect is remarkable; thus, it is widely used in point cloud registration. However, the ICP algorithm also has drawbacks, such as long calculation time and inability to solve local optimal problems [22]. Wu et al. [23] proposed a novel nearest neighbor search algorithm to improve the iteration speed of point cloud registration. Fotsing et al. proposed a novel region growing-based approach [24] for plane detection in unorganized point clouds extract reliable seeds by the Iterative Closest Point algorithm to enhance the performance and the quality of the results.

In addition, the initial pose of the point cloud to be registered has an important influence on the accuracy of point cloud registration. The coarse registration based on feature matching is used to obtain a better initial pose, which is beneficial to improve the pose estimation accuracy [25]. Rusu [26] proposed the Fast Point Feature Histogram (FPFH) descriptor, which can be used to describe the neighborhood geometry information of the query point and is often used to estimate the target pose. SaltiS et al. [27] proposed a Signature of Histograms of Orientation (SHOT), which can represent topological features, has rotation invariance and is robust to noise. Li et al. [28] proposed real-time path planning based on a VFH feature descriptor, which was robust against a large degree of surface noise and missing depth information; however, pose estimation could fail when the object was placed symmetrically with the viewpoint. Compared with other feature descriptors, the FPFH has the characteristics of fast computation and high accuracy, and it is used to describe the geometric features of the pallet point cloud in this study. However, the current FPFH descriptor also has some drawbacks. Firstly, when selecting the weight coefficient of the FPFH, only the Euclidean distance between the query point and the neighborhood point is considered, which makes the weight order difference too large and reduces the robustness of the FPFH feature descriptor. In addition, the current calculation approach of the FPFH feature descriptor does not consider the selection criteria of neighborhood radius and is usually debugged based on experience to determine neighborhood radius, which reduces the efficiency and accuracy of pose estimation.

In view of the above problems, a pallet pose estimation approach based on an Adaptive Gaussian Weight-based Fast Point Feature Histogram is proposed. On one hand, when determining the neighborhood radius of the descriptor, the optimal neighborhood radius of each point is obtained based on the minimum rule of neighborhood feature entropy function, which overcomes the randomness of neighborhood radius parameters debugging manually. On the other hand, when determining the weight of neighborhood points, the weight calculation formula is redefined according to the average distance and standard deviation between key points and their neighborhood points, which makes the weight of each neighborhood point controlled within a certain range.

The remainder of the paper is organized as follows: In Section 2, an overview of the proposed approach and the specific steps of the method are described. In Section 3, two cases are presented to verify the effectiveness of the proposed approach in engineering application, and the experimental results are analyzed and discussed. Finally, a conclusion is drawn in Section 4.

2. The Proposed Pallet Pose Estimation Method

2.1. Overview of the Proposed Approach

In order to realize the accurate position of pallets for driverless industrial trucks in the storage environment, a pallet pose estimation method based on an Adaptive Gaussian Weight-based Fast Point Feature Histogram is proposed; the procedure involves the following steps, and the flowchart is shown in Figure 2.

Step 1: Point cloud preprocessing. The source point cloud of the pallet and the scene point cloud containing the pallet are collected using an active binocular vision sensor. The redundant scene information in the scene cloud is removed through the pass-through filtering and voxel mesh downsampling method [29], and the plane segmentation algorithm is used to obtain the target pallet point cloud. The key points in the target point cloud are extracted by the Intrinsic Shape Signatures (ISS) algorithm [30].

Step 2: Adaptive Gaussian Weight-based Fast Point Feature Histogram definition. Adaptive optimal neighborhood radius is used to determine the neighborhood range of each point in the target point cloud. By calculating the mean value and variance of the distance between each key point and its neighborhood, the Adaptive Gaussian weight-based Fast Point Feature Histogram (AGWF) of each key point is obtained.

Step 3: Point cloud registration. According to the AGWF feature descriptor, the SAC-IA (sample consensus initial alignment) algorithm is used to coarsely register the source point cloud with the target point cloud. Then, the ICP algorithm is used to transform the point cloud iteratively and obtain the optimal rigid transformation matrix. The matrix parameters are converted into horizontal deviation and angle to realize pallet pose estimation.

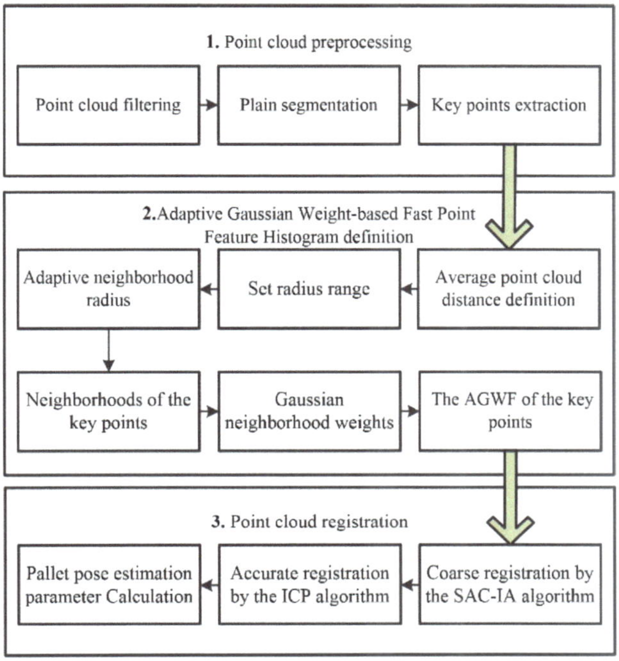

Figure 2. Flowchart of the pallet pose estimation method.

2.2. Point Cloud Preprocessing

2.2.1. Point Cloud Filtering

The original source point cloud P containing the pallet and the original scene point cloud Q_{so} (about millions of points) are collected by an active binocular vision sensor. The efficiency and accuracy of point cloud processing will be reduced due to the large number of acquired scene point clouds and a large amount of redundant scene information. The filtering interval is determined according to the spatial position relationship between the pallet and the driverless industrial trucks in the standard state, and the invalid point clouds and background information in the original scene point cloud Q_{so} are removed by the classical pass-through filter. A large number of redundant points are removed by voxel grid downsampling, and the complete geometric features of the point cloud are retained to obtain the filtered scene point cloud Q_s.

The specific steps of the specific steps of voxel grid downsampling are as follows: (1) In point cloud Q_{so}, the maximum and minimum values in X, Y and Z directions are x_{max}, x_{min}, y_{max}, y_{min}, z_{max}, z_{min}, respectively. Set the dimensions of the voxel grid d_0, where $m = \lceil (x_{max} - x_{min})/d_0 \rceil$,$n = \lceil (y_{max} - y_{min})/d_0 \rceil$, $l = \lceil (z_{max} - z_{min})/d_0 \rceil$, and $\lceil\rceil$ represents round up. (2) Encode each point $Q_v(x_v, y_v, z_v)$ in Q_{so} as (m_v, l_v, n_v) to determine which grid each point belongs to, where $m_v = \lfloor (x_v - x_{min})/d_0 \rfloor$,$n_v = \lfloor (y_v - y_{min})/d_0 \rfloor$, $l_v = \lfloor (z_v - z_{min})/d_0 \rfloor$,and $\lfloor\rfloor$ represents round down. If there are some points in a voxel grid, calculate its center of gravity $C_0 = (x_0, y_0, z_0)$, and replace the points in each grid by the point nearest to the center of gravity to obtain the filtered scene point cloud Q_s, where $x_0 = \frac{1}{k}\sum_{i=1}^{k} x_i$,$y_0 = \frac{1}{k}\sum_{i=1}^{k} y_i$,$z_0 = \frac{1}{k}\sum_{i=1}^{k} y_i$, and k represents the number of points in the grid.

2.2.2. Plain Segmentation

Because the filtered scene point cloud Q_s still contains pallet, wall, ground and other information, and the pallet needs to be separated from the wall and ground, the plane segmentation method based on Random Sampling Consensus (RANSAC) is used to find the points belonging to the plane iteratively according to the set plane model. Meanwhile, the ground with different degrees of fluctuation can be detected by setting the model distance threshold. The specific steps are as follows: (1) The initial plane model $A_x + B_y + C_z + D = 0$ is constructed by selecting any three points from the filtered scene point cloud. (2) The distance d_i from point q_i (points in the point cloud Q_s) to the initial plane and the angle α_i between the coordinates of point q_i and the normal vector of the initial plane are calculated. A distance threshold d_e and an angle threshold α_e are set. If $d_i < d_e$ and $\alpha_i < \alpha_e$, then point q_i is considered an in-plane point. (3) Iterations are carried out continuously until the number of in-plain points reaches the threshold t, and the final fitting plane model representing the wall and ground is removed, obtaining the target pallet point cloud Q.

2.2.3. Key Points Extraction

After filtering and plane segmentation, there are still tens of thousands of point clouds in the target pallet point cloud Q, which will reduce the pose estimation efficiency and fail to meet the requirements of the operation of driverless industrial trucks. Therefore, the points with obvious geometric features are selected from the target point cloud Q to form the key point set Q_t, and only the features of the key points are extracted, which can significantly improve the efficiency of feature extraction of the point cloud. Due to the advantages of high speed, accuracy and robustness, an intrinsic Shape Signature (ISS) algorithm is developed to extract key points, and it is suitable for various applications [31]. The main steps are as follows: (1) Establish a local coordinate at each point q_v in point cloud Q and set a neighborhood search radius r_f. (2) Search for the neighborhoods of the query point q_v with r_f and obtain their neighborhood points q_j; the weight $w_{vj} = \left(\|q_j - q_v\|\right)^{-1}$ is calculated according to the Euclidean distance between q_v and q_j. (3) Calculate the weighted neighborhood covariance matrix $C_i = \sum_{|q_i - q_j| < r_{frame}} w_{ij}(q_i - q_j)(q_i - q_j)^T / \sum_{|q_i - q_j| < r_{frame}} w_{ij}$ of q_v. (4) The eigenvalues λ_v^1, λ_v^2 and λ_v^3 are obtained by eigenvalue decomposition of the covariance matrix, and they are arranged in descending order. (5) Set the thresholds ε_1 and $\varepsilon_2(\varepsilon_1, \varepsilon_2 \leq 1)$. If $\frac{\lambda_1^2}{\lambda_v^1} \leq \varepsilon_1, \frac{\lambda_1^3}{\lambda_v^2} \leq \varepsilon_2$, and the points are supposed to be the key points, otherwise, iterate over the next point. Repeat the process until all the points have been traversed, and finally obtain the key point set Q_k of the target point.

2.3. *Adaptive Gaussian Weight-based Fast Point Feature Histogram Definition*

2.3.1. Adaptive Neighborhood Radius

The premise of accurate pose estimation is to construct a feature descriptor of the pallet point cloud with high efficiency, strong robustness and high accuracy; the neighborhood radius is an important factor affecting the performance of feature descriptors. The neighborhood radius of traditional FPFH is usually set to a fixed value according to experience, which reduces the speed of feature extraction. A neighborhood radius selection criterion based on adaptive neighborhood feature entropy is proposed to obtain the neighborhood radius of each key point q_k (points in the key point set Q_k) adaptively. The detailed steps are as follows:

1. Set the range of point cloud neighborhood search radius r_j from lower limit r_{min} to upper limit r_{max} with radius interval r_d. The upper and lower limits of the radius range are determined by the average point cloud distance d_p, which is defined as follows [32]:

$$d_p = \frac{1}{N}\sum d_m \tag{1}$$

   where N represents the total number of points in the pallet point cloud, and d_m represents the distance of each key point q_k from its nearest point. Set $r_{min} = d_p$, $r_{max} = 2d_p$, $r_{j+1} = r_j + \Delta r$, where $j = 1, 2, \ldots$.

2. Calculate the covariance matrix and eigenvalues corresponding to different neighborhood radius r_j. The neighborhood covariance matrix M is defined as follows:

$$M = \begin{bmatrix} e_1 & e_2 & e_3 \end{bmatrix} \begin{bmatrix} \lambda_1 & 0 & 0 \\ 0 & \lambda_2 & 0 \\ 0 & 0 & \lambda_3 \end{bmatrix} \begin{bmatrix} e_1^T \\ e_2^T \\ e_3^T \end{bmatrix} \quad (2)$$

   where $\lambda_1, \lambda_2, \lambda_3$ are the eigenvalues of the neighborhood covariance matrix, and e_1, e_2, e_3 are the corresponding eigenvectors.
3. According to the eigenvalues, the neighborhood feature entropy function E_ξ is constructed:

$$E_\xi = -\xi_1 \ln(\xi_1) - \xi_2 \ln(\xi_2) - \xi_3 \ln(\xi_3) \quad (3)$$

   where $\xi_i = \frac{\lambda_i}{\sum \lambda_i}, i = 1, 2, 3$.
4. The adaptive optimal neighborhood radius r_{opt} of point cloud is determined based on the minimum criterion of neighborhood feature entropy function, that is, when E_ξ reaches the minimum value, the corresponding neighborhood radius r is the optimal neighborhood radius r_{opt}:

$$ropt = \text{argmin}(E_\xi) \quad (4)$$

2.3.2. Gaussian Weight-Based Fast Point Feature Histogram

The features of traditional FPFH are determined by the neighborhood in the radius r of the query point itself and the neighborhood of its neighborhood points, whose maximum range is 2r. The FPFH neighborhood of a query point is shown in Figure 3, where p_{a1}–p_{a5} are the neighborhood points of the query point P_a within the neighborhood radius r, and p_b–p_i are the neighborhood points of the points p_{a1}–p_{a5}.

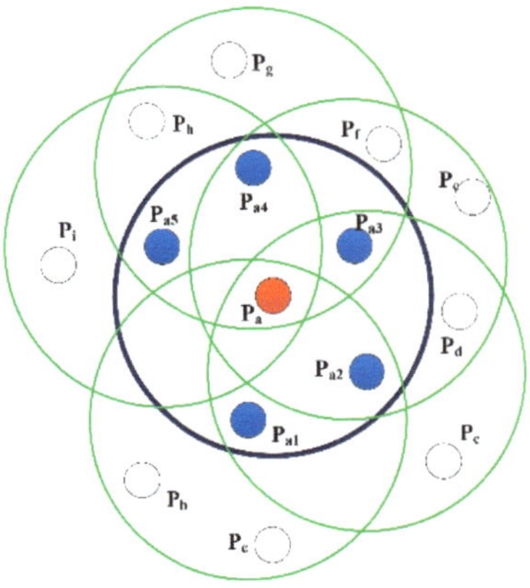

Figure 3. Schematic diagram of FPFH neighborhood.

The weight coefficient of the traditional FPFH algorithm is determined only by the Euclidean distance between the query point and the neighborhoods; often, there are huge differences among various weight coefficients. As is shown in Figure 4, the feature descriptor of the red query point q_r is mostly influenced by the closest black points q_a and q_b, and other points only have little effect, which may cause the feature descriptor to be inaccurate.

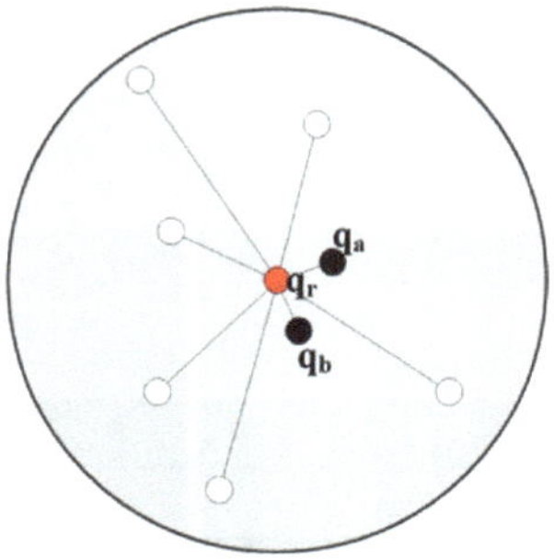

Figure 4. A neighborhood point diagram of a key point.

Therefore, an improved FPFH is proposed; each key point q_k in the key point set Q_t is taken as the query point, and based on the adaptive optimal neighborhood radius r_{opt}, all the neighborhood points q_{ki} within its neighborhood radius are found. The weight coefficient is redefined with the Gaussian distribution, as follows:

$$w_{Gi} = \frac{1}{\sqrt{2\pi\sigma}} \exp(-\frac{(w_i - \mu)^2}{2\sigma^2}) \quad (5)$$

where μ is the average distance of q_k and q_{ki}, and σ represents the standard deviation of the distance between q_k and q_{ki}.

The weight coefficient variation trend of the original approach and the proposed approach is shown in Figure 5. It can be seen that the weight coefficient of the original approach varies greatly, and the weight tends to infinity with the decrease of the distance. The proposed AGWF can avoid the problem that the weight coefficient difference between points is apparently large, and the weight coefficient reaches the maximum when the distance is close to the average value, which reasonably solves the unstable problem of the FPFH descriptor calculation.

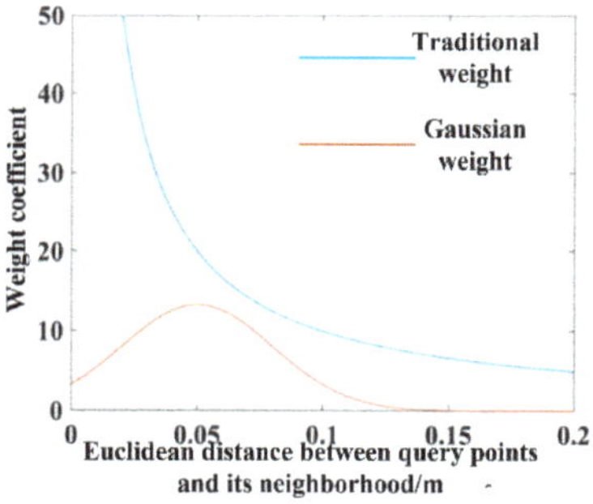

Figure 5. Weight variation trend of neighborhood points of a key point.

Combined with the adaptive optimal neighborhood radius selection criterion, an Adaptive Gaussian Weight-based Fast Point Feature Histogram (AGWF) is proposed. It can not only improve the efficiency of feature extraction but also improve the accuracy and robustness; the specific calculation steps are as follows:

1. For each key point q_k in the key point set Q_t, search all the neighborhood points q_{ki} within its optimal neighborhood radius r_{opt}.
2. Compute the normal vectors n_s and n_t corresponding to q_k and q_{ki}, calculate the relative position deviation (α, ϕ, θ) between n_s and n_t, and generate the Simple Point Feature Histograms (SPFH) of q_k(SPFH(q_k)). Local coordinate system (u,v,w) is defined to calculate this deviation, which is shown in Figure 6:

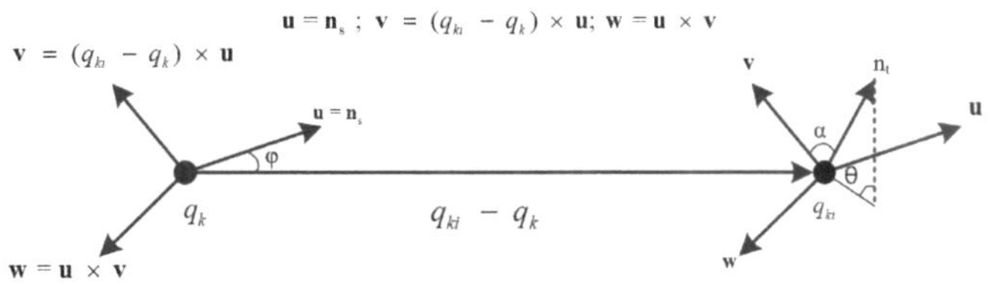

Figure 6. Diagram of local coordinate system (u,v,w).

The calculation formula of relative deviation is as follows:

$$\begin{aligned} \alpha &= v \cdot n_t \\ \varphi &= u \cdot \frac{(q_k - q_{ki})}{\|q_k - q_{ki}\|_2} \\ \theta &= \arctan(w \cdot n_t, u \cdot n_t) \end{aligned} \quad (6)$$

3. Then, search the neighborhood points of q_{ki} based on the adaptive optimal neighborhood radius r_{opt}, and generate the SPFH of q_{ki}(SPFH(q_{ki})). Based on the Gaussian weight w_{Gi}, the SPFH(q_{ki}) is weighted to obtain the Adaptive Gaussian Weight-based Fast Point Feature Histogram of key points q_k(AGWF(q_k)):

$$\mathrm{AGWFPFH}(q_k) = \mathrm{SPFH}(q_k) + \frac{1}{k}\sum_{i=1}^{k} w_{Gi}\mathrm{SPFH}(q_{ki}) \quad (7)$$

2.4. Point Cloud Registration

2.4.1. Coarse Registration

The purpose of coarse registration is to obtain the initial pose relationship between the source point cloud P_t and the target point cloud, so as to overcome the shortcomings of the ICP algorithm, which requires high initial pose and is easy to fall into local optimum. The SAC-IA algorithm can effectively adjust the initial pose relationship between the source point cloud P and the target point cloud Q and improve the accuracy of pose estimation. The specific steps are as follows:

1. Compute AGWF feature descriptors for all key points in the source point cloud P and the target point cloud Q.
2. N sample points P_u (u = 1, 2, . . . , N) are randomly selected from the source point cloud P, and the distance between two sample points is greater than the preset distance threshold d_{min}.
3. According to the AGWF, search the closest points Q_u (u = 1, 2, . . . , N) in the target point cloud Q to the sample points P_u, and obtain the initial match point pairs.
4. Obtain the rigid transformation matrix $\mathbf{M}_1$ between initial match point pairs by Singular Value Decomposition (SVD). Set a registration threshold e_l, and calculate the distance function $H(l_i)$ to evaluate the point cloud registration performance. The expression of the distance function $H(l_i)$ is as follows:

$$H(l_i) = \begin{cases} \frac{1}{2} l_i^2 & \|l\|_i < e_l \\ \frac{1}{2} e_l (2\|l\|_i - e_l) & \|l\|_i \geq e_l \end{cases} \quad (8)$$

where l is the average Euclidean distance between the source point cloud, and I is the number of iterations.

5. Repeat the above four steps; when $H(l_i)$ reaches the minimum, the corresponding transformation matrix is the coarse registration rigid transformation matrix $\mathbf{M}_c$. The rigid transformation of the source point cloud P is carried out based on $\mathbf{M}_c$ to obtain the point cloud P_r, and coarse registration is completed.

2.4.2. Accurate Registration

After coarse registration, the source point cloud Pr and the target point cloud Q can only roughly coincide, so it is necessary to improve the pose estimation accuracy by further accurate registration. The ICP algorithm is used for accurate registration. The algorithm obtains the nearest Euclidean point through exhaustive search and obtains accurate registration parameters based on the results of the optimal objective function. According to the registration parameters, 6 Degrees of Freedom (6 DOF) pose estimation result can be obtained [33].

1. Set a distance threshold e_f and the maximum number of iterations I_0. For each point p_{ri} in the source point set P_r, search for its corresponding closest point q_i in the target point set Q, and form the corresponding points pairs, set C_1.
2. Solve the rigid transformation matrix by SVD and obtain the rotation matrix $\mathbf{R}_n$ and the translation matrix $\mathbf{T}_n$, where n is the number of iterations. Convert the source point P_r by the translation matrix $(\mathbf{R}_n, \mathbf{T}_n)$ into P_{rn}, and form the corresponding point pairs C_n. Calculate the average Euclidean distance e_n between every corresponding point pair.
$$e_n = \frac{1}{k}\sum_{i=1}^{k}\|q_i - (p_{ri} \times \mathbf{R}_n + \mathbf{T}_n)\|^2 \quad (9)$$
where k is the number of the corresponding point pairs.
3. Repeat the above steps until e_n is smaller than e_f or the maximum number of iterations I_0 is reached, and finally obtain the optimal transformation matrix **R** and **T**.
4. Let $\mathbf{R}_x$, $\mathbf{R}_y$ and $\mathbf{R}_z$ be the rotation angles of the three coordinate axes, and $\mathbf{t}_x$, $\mathbf{t}_y$ and $\mathbf{t}_z$ be the translation vectors of the coordinate axes; the 6 DOF pose estimation can be represented as $(\mathbf{R}_x, \mathbf{R}_y, \mathbf{R}_z, \mathbf{t}_x, \mathbf{t}_y, \mathbf{t}_z)$. The optimal transformation matrix $\mathbf{M}_a$ can be expressed as

$$\mathbf{M}_a = \mathbf{T}(\mathbf{t}_x, \mathbf{t}_y, \mathbf{t}_z) \times \mathbf{R}(\mathbf{R}_x, \mathbf{R}_y, \mathbf{R}_z) \quad (10)$$

where $\mathbf{T}(\mathbf{t}_x, \mathbf{t}_y, \mathbf{t}_z)$ and $\mathbf{R}(\mathbf{R}_x, \mathbf{R}_y, \mathbf{R}_z)$ can be expressed as

$$\mathbf{T}(tx, ty, tz) = \begin{pmatrix} 1 & 0 & 0 & tx \\ 0 & 1 & 0 & ty \\ 0 & 0 & 1 & tz \\ 0 & 0 & 0 & 1 \end{pmatrix} \quad (11)$$

$$\mathbf{R}(\mathbf{R}_x, \mathbf{R}_y, \mathbf{R}_z) = \begin{pmatrix} r_{11} & r_{12} & r_{13} & 0 \\ r_{21} & r_{22} & r_{23} & 0 \\ r_{31} & r_{32} & r_{33} & 0 \\ 0 & 0 & 0 & 1 \end{pmatrix} \quad (12)$$

It can be concluded that the 6DOF pose estimation parameters of the target can be expressed as

$$(R_x, R_y, R_z, t_x, t_y, t_z) = (\arctan(r_{32}/r_{33}), \arcsin(-r_{13}), \arctan(r_{21}/r_{11}), t_x, t_y, t_z) \quad (13)$$

According to the actual situation of pallet fork taking in the storage environment, only the horizontal deviation t_x and t_z and the rotation angle R_y perpendicular to the ground need to be obtained. The driverless industrial trucks adjust the pallet fork taking path according to the deviation parameters (t_x, t_y, R_y).

3. Pallet Pose Estimation Experiment

3.1. Data Collection

An industrial vision sensor called the Percipio FM851-E2 3D vision sensor is adopted to acquire point source point cloud and scene point cloud, whose ranging principal is active binocular, and the operative range is 0.7–6.0 m. The structure of the Percipio FM851-E2 vision sensor is shown in Figure 7.

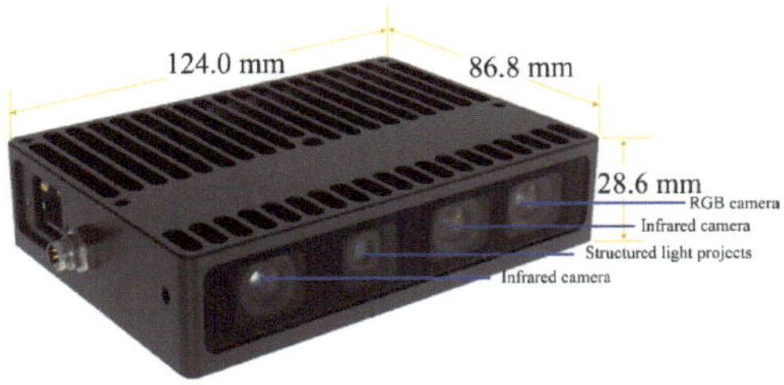

Figure 7. Structure of the Percipio FM851-E2 vision sensor.

To obtain the source point cloud of the pallet, a normal blue pallet is placed in a fixed position in a laboratory with normal brightness and no other obstructions, and the Percipio FM851-E2 vision sensor is fixed on the top of the fork frame to take pictures as shown in Figure 8. To meet the operational requirements of the production floor, the front of the fork is placed 500 mm away from the pallet, ensuring that the fork is perpendicular to the front of the pallet and the center of the sensor is aligned with the center of the pallet. The collected point cloud is processed by the point cloud filter and plain segmentation algorithm, and the remaining points are the source point cloud, which contains the position information of the pallet point cloud in the sensor coordinate system under the standard state after visualization, as shown in Figure 9. The key points of the source point cloud are extracted, and the adaptive optimal neighborhood radius is obtained; then, the AGWF features are calculated (the results are shown in Figure 10), and the results are saved in the database.

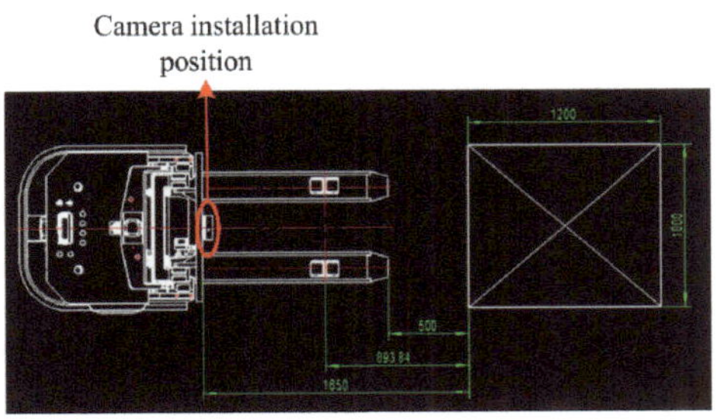

Figure 8. Relative position of unmanned industrial vehicle and pallet in standard state.

(a) (b)

Figure 9. (**a**) The source point cloud. (**b**) The key points of source point cloud.

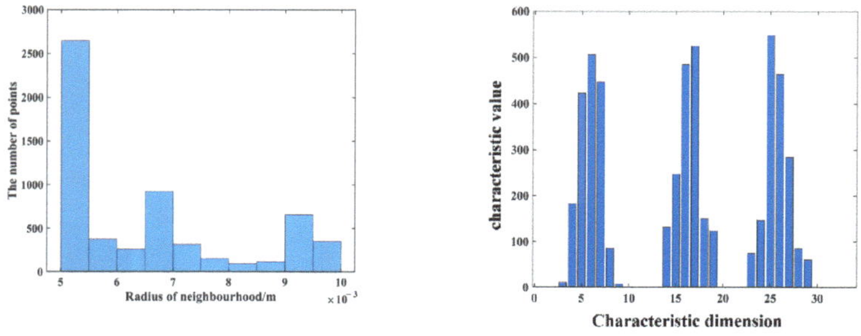

(**a**) (**b**)

Figure 10. Processing results of source point cloud. (**a**) Adaptive optimal neighborhood distribution. (**b**) The AGWF feature descriptor.

3.2. *Multiple Scenario Experiments*

In order to verify the effectiveness of the pose estimation algorithm, the pallet pose estimation algorithm is experimentally verified in several scenarios, mainly the common ground scenarios and shelf scenarios.

3.2.1. The Ground Scene

Considering the practical requirements of the pallet attitude estimation scenario, the relative attitude relationship between the storage pallet and the sensor is considered only for the horizontal lateral translation Δx, the longitudinal translation Δy, and the deflection angle φ. The deflection angles of 5°, 10°, 15° and 20° and the deviations of 0.05 m, 0.10 m, 0.15 m and 0.2 m in the horizontal direction, respectively, are selected for the experiments. The correction of the deflection angle needs to be completed through the rotation of the driverless industrial truck, so the rotation center of the deflection angle is actually the origin of the camera coordinates in this experiment. The scene point clouds are taken and preprocessed, the target pallet point clouds in the scene point clouds are extracted, and the adaptive optimal neighborhoods of key points and AGWF feature descriptors are calculated to match with the pallet point clouds. Taking the experiment at a deflection angle of 5° as an example, the visualization results of the scene point cloud, the relative position relationship between the target point cloud and the source point cloud, the key points of the target point cloud, and the rough and accurate registration are listed in Figure 11.

The number of point clouds and the time consumed for each step of the experiments are shown in Table 1, and the scene point clouds captured and the registration results are shown in Figure 12. The experiments' pose estimation results and errors are shown in Table 2. The experiments (d) and (h) are the experimental control groups with the out-of-limit deviations (bolded in Table 2), which are not taken into account for calculating the total average deviation and accuracy.

Table 1. Point cloud processing process.

Serial No.	Number of Scene Point Clouds	Number of Target Point Clouds	Number of Key Points	Time Consumed/s
a	55,827	6181	1154	1.6080
b	58,788	6500	1327	1.8263
c	64,656	7115	1442	1.9777
d	53,433	7155	1548	1.3595
e	50,085	5533	1014	0.7670
f	47,142	5217	1272	1.2597
g	44,118	4895	916	1.2393
h	36,054	4001	1173	1.0132

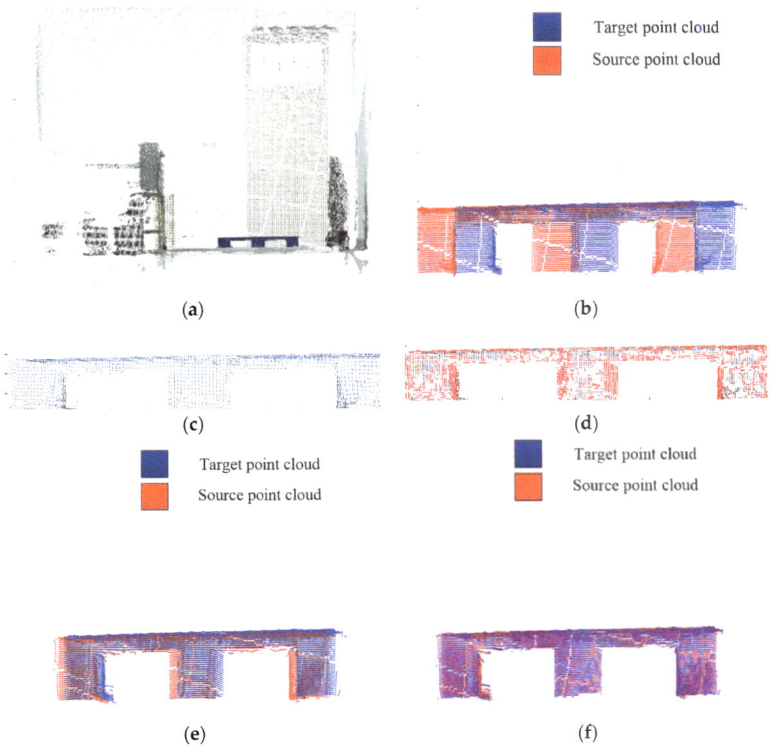

Figure 11. Point cloud processing results of 5° deflection angle. (**a**) Scene point cloud. (**b**) Initial relative position. (**c**) Target point cloud. (**d**) Key points of the target point cloud. (**e**) Coarse registration. (**f**) Accurate registration.

Table 2. Experimental results of the ground scene.

Serial No.	Actual Deviation			Estimated Deviation			Relative Deviation			Accuracy/%
	Δx	Δy	φ	Δx	Δy	φ	Δx	Δy	φ	
a	0	0	5	0.0102	0.0267	4.8876	0.0102	0.0267	0.1124	97.76
b	0	0	10	0.0109	0.0340	9.8656	0.0109	0.0340	0.1344	98.66
c	0	0	15	0.0212	0.0534	15.09	0.0212	0.0534	0.1900	98.73
d	**0**	**0**	**20**	**0.0536**	**0.0687**	**26.53**	**0.0536**	**0.0687**	**6.5300**	**67.30**
e	0.050	0.050	0	0.0535	0.0505	0.4060	0.0035	0.0005	0.4060	96.00
f	0.100	0.100	0	0.1096	0.0988	1.0620	0.0096	0.0012	1.0620	94.60
g	0.150	0.150	0	0.1536	0.1477	1.2014	0.0036	0.0023	1.2014	98.03
h	**0.200**	**0.200**	**0**	**0.1928**	**0.1579**	**3.2503**	**0.0072**	**0.0421**	**3.2503**	**87.68**
			Total Average				0.0098	0.0194	0.5010	97.30

As can be seen from the experimental results derived from the above table, the average estimation error of the horizontal direction is about 0.0098 m, the average estimation error is about 0.0194 m, and the average estimation error of the deflection angle is about 0.5°, with a total average accuracy of 97.3%. It can be seen that the algorithm has high accuracy and strong robustness when the horizontal deviation is within 0.15 m and the deflection angle is within 15°. However, as is shown the experiments (d) and (h) in Figure 12, when the horizontal deviation or deflection angle is too large, due to the field of view limitation of the depth camera and excessive initial pose deviation, the target point cloud and the source point cloud may fail to be aligned, which will affect the accuracy of the pallet pose estimation.

(a)

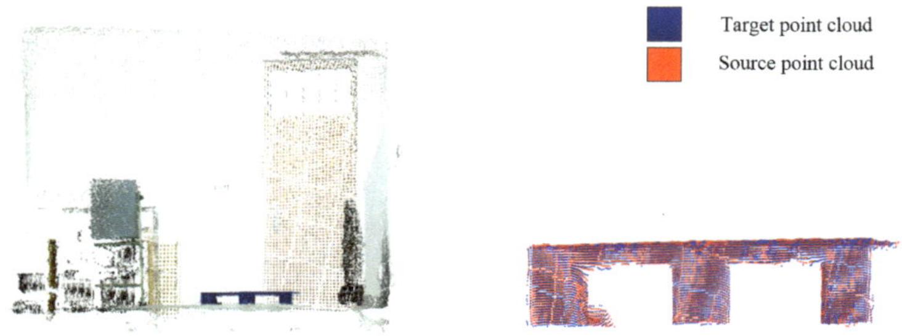

(b)

(c)

(d)

Figure 12. *Cont.*

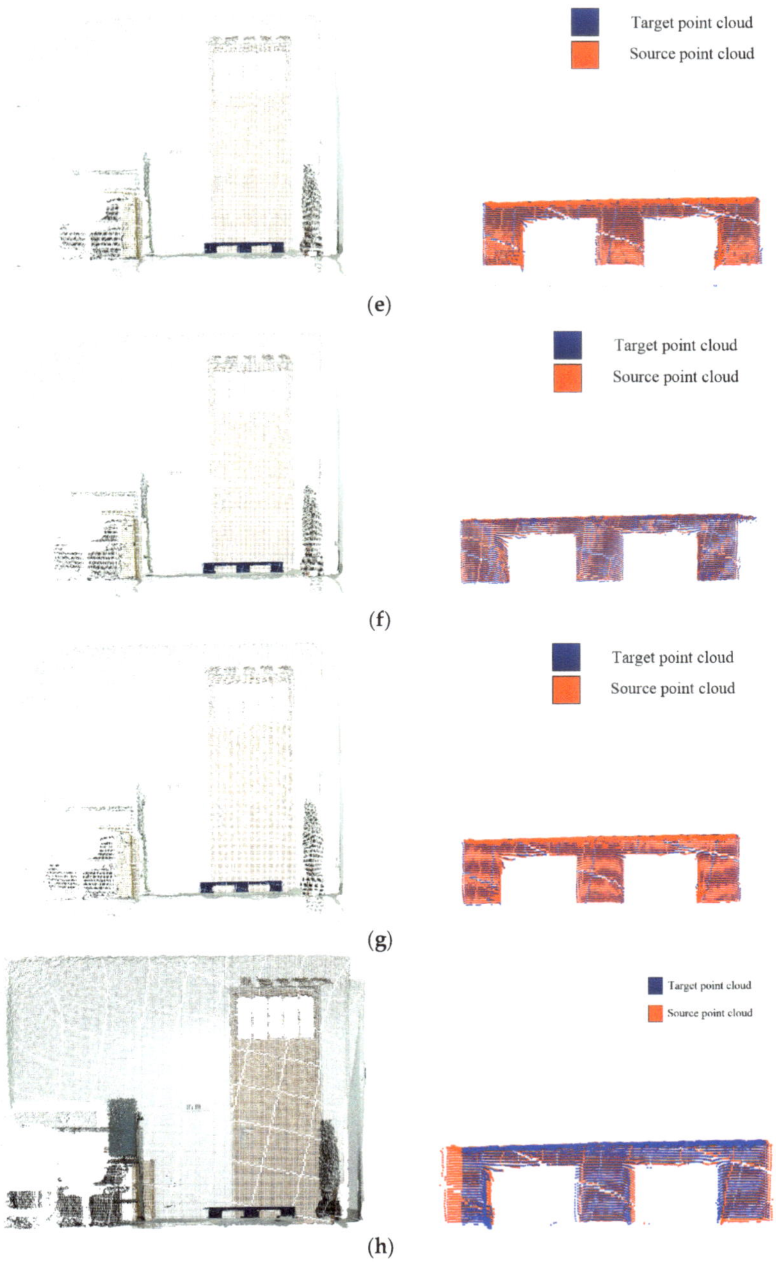

Figure 12. Scene point cloud and registration result. (**a**) 5° deflection angle. (**b**) 10° deflection angle. (**c**) 15° deflection angle. (**d**) 20° deflection angle. (**e**) Deviations of 0.05 m. (**f**) Deviations of 0.10 m. (**g**) Deviations of 0.15 m. (**h**) Deviations of 0.2 m.

3.2.2. The Shelf Scene

In a production workshop that has several shelves with three layers, a blue pallet is placed in the second layer, which has a certain position deviation with the standard state. The driverless industrial truck follows a preset path to a designated location, and the horizontal deflection and the size of the deflection angle must be calculated by the pose estimation algorithm to realize the accurate forklift of pallets for driverless industrial trucks.

The color image of the scene captured by the Percipio active binocular vision sensor is shown in Figure 13, and the scene point cloud is shown in Figure 14, among which the number of the scene points is 2,073,600. Redundant points were eliminated by pass-through filtering. According to the position relationship between the vision sensor and the pallet, the pass-through filtering parameters x, y and z are set as $x \in (-1.0, 1.0)$ m, $y \in (-0.5, 0.5)$ m, $z \in (1.5, 3.0)$ m to control the filtering interval of point cloud, and the number of remaining point clouds is 259,443; the results are shown in Figure 15. After plane segmentation and voxel grid downsampling, the target pallet point cloud in this shelf scene is obtained, and the remaining point cloud number is 7438, which is shown in Figure 16a. The search radius of key point extraction is set as 0.05 m, and the two thresholds are set as $r_1 = 0.4$ and $r_2 = 0.2$. A total of 1476 key points are extracted from the target point cloud, as shown in Figure 16b, the red points are the key points.

Figure 13. Color image of the shelf.

Figure 14. Shelf scene point cloud.

Figure 15. Pass-through filtering.

(**a**) (**b**)

Figure 16. Key point extraction for target point cloud. (**a**) Target pallet point cloud. (**b**) Target pallet key points.

Before computing the AGWF feature descriptors for the target point cloud, the adaptive neighborhood radius of each point needs to be determined and set to $r_{min} = 0.006$ m,

r_{max} = 0.012 m, and r_d = 0.001 m, and the average distance d_p = 0.006 m of the target point cloud is calculated. The adaptive optimal neighborhood radius of each point is obtained according to the minimum criterion of the neighborhood characteristic entropy function, and the adaptive optimal neighborhood radius distribution is shown in Figure 17. The horizontal coordinate indicates the different neighborhood radii, and the vertical coordinate indicates the number of points corresponding to each neighborhood radius. It can be seen that the optimal neighborhood radius of key points is concentrated on the given minimum neighborhood radius, which can significantly improve the efficiency of the feature descriptor.

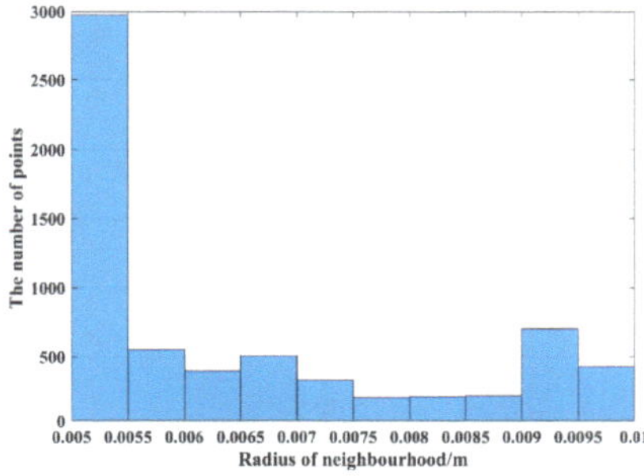

Figure 17. Shelf scene target pallet point cloud adaptive optimal neighborhood radius.

The AGWF feature descriptors of the source point cloud and the target pallet point cloud are calculated based on the minimum neighborhood radius, and the AGWF feature of a point is shown in Figure 18.

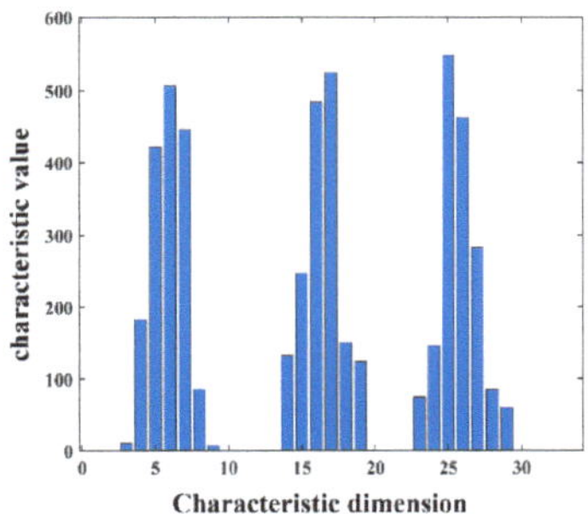

Figure 18. The AGWF feature descriptor for a point.

The SAC-IA algorithm is used for the coarse registration, and the ICP is used to complete the accurate registration; the accurate optimal rigid rotation matrix **R** and the translation matrix T are obtained as follows:

$$\mathbf{R} = \begin{bmatrix} 0.99221826 & -0.0056165964 & -0.12438513 \\ 0.0041089170 & 0.99991482 & -0.012374031 \\ 0.12444409 & 0.011766853 & 0.99215692 \end{bmatrix}$$

$$\mathbf{T} = \begin{bmatrix} 0.35200304 \\ -0.083008856 \\ 0.018321276 \end{bmatrix}$$

The result of the accurate registration is shown in Figure 19. The left figure shows the registration diagram of the source point cloud and the target pallet point cloud, where the red point cloud is the source point cloud and the blue point cloud is the pallet source point cloud. The figure on the right shows the pose of the source point cloud after transformation according to the accurate optimal rigid transformation matrix, where the red point cloud is the source point cloud, and the rest of the point clouds are the scene point clouds captured by the Percipio vision sensor while the driverless industrial truck is working. As can be seen in Figure 19, after the rigid transformation, the source point cloud and the target pallet point cloud in the field point cloud can achieve basic overlap, so it can be considered that the rotation matrix R and translation matrix T obtained from the accurate registration is reliable, and the source point cloud and the target point cloud have satisfactory registration results, which further verifies the accuracy of the proposed pose estimation algorithm. According to the rotation matrix R and the translation matrix T, it can be calculated that the horizontal deviations Δx and Δy are 0.35 m and 0.018, respectively, the deflection angel is 12.43°, and the pose estimation is completed.

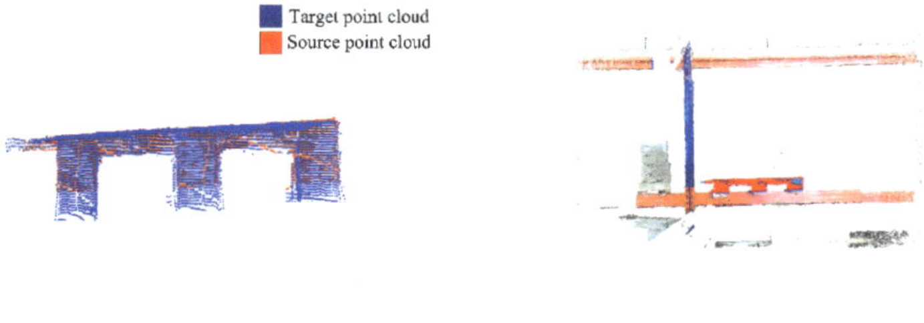

(a) (b)

Figure 19. Visualization of pallet pose estimation results. (**a**) Registration result (**b**) Pallet pose estimation.

3.3. Results and Discussion

To verify the rationality of Adaptive Gaussian Weight-based Fast Point Feature Histogram, a comparison is made with the traditional FPFH feature descriptor. A diagram of the comparison between AGWF and FPFH for a point is shown in Figure 20, with the horizontal coordinates indicating the feature dimensions and the vertical coordinates indicating the feature values on each dimension.

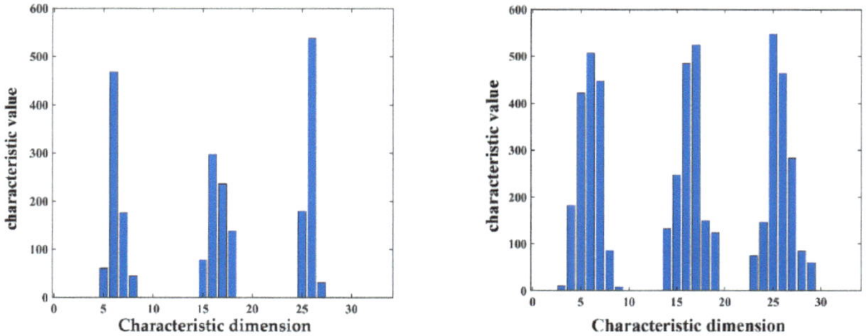

(a) (b)

Figure 20. Comparison of FPFH and AGWF at a point. (**a**) The FPFH feature descriptor for a point. (**b**) The AGWF feature descriptor for a point.

As can be seen in Figure 20, the proposed AGWF descriptor has more balanced values among dimensions and can describe more feature information as compared with the FPFH feature descriptor. The problem that the feature descriptor is overly influenced by a certain neighborhood point is avoided, and the robustness of the feature descriptor is improved.

The comparison of coarse registration results between FPFH and AGWF is shown in Figure 21.

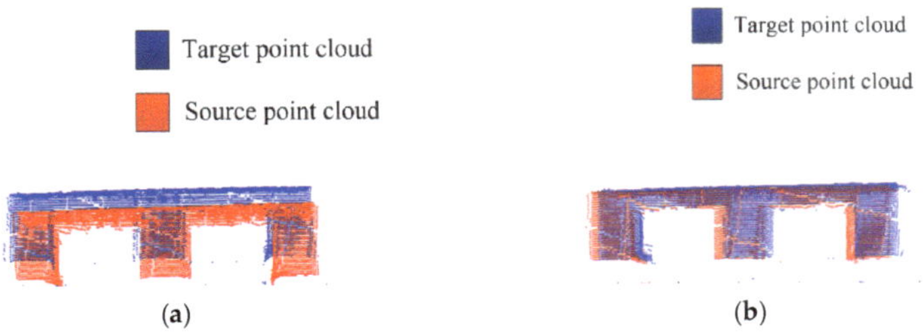

Figure 21. Comparison of coarse registration results between FPFH and AGWF. (**a**) Coarse registration based on FPFH. (**b**) Coarse registration based on AGWF.

The proposed approach is compared with the traditional PFH and FPFH feature descriptors as well as the advanced SHOT feature descriptors in terms of the feature extraction time consumption and the RMSE. The experimental results of different descriptors are shown in Table 3, where R is the neighborhood radius used for feature extraction, T is the time consumed for feature extraction, Dt is the percentage of time consumption reduction of the proposed method compared with the traditional methods, DRMSE is the percentage of error reduction of the proposed method compared with the traditional methods, and GWF is Gaussian Weight-based Fast Point Feature Histogram, which only improves the traditional FPFH by Gaussian weight without adaptive neighborhood. In Table 4, the neighborhood radius of the traditional descriptors is set as 0.014 m, and the neighborhood radius of the AGWF is set adaptively. The GWF and AGWF are the proposed feature descriptors in this paper, and they are bolded in the Table 3 and 4.

Table 3. Comparison of experimental results of different descriptors (R = 0.012 m).

Names	R/m	T/s	RMSE	Dt/%	DRMSE/%
SHOT	0.012	1.564	0.018234	−7.9	10.99
PFH	0.012	3.872	0.035647	56.6	54.47
FPFH	0.012	1.430	0.017559	−17.34	7.57
GWF	**0.012**	**1.678**	**0.016239**	/	/

The Dt and DRMSE are calculated as Dt = (FT − GWF)/FT, DRMSE = (FR − GWF)/FR, where FT is the feature extraction time of the traditional descriptors, and FR is the RMSE of traditional descriptors.

Table 4. Comparison of experimental results of different descriptors (R = 0.014 m).

Names	R/m	T/s	RMSE	Dt/%	DRMSE/%
SHOT	0.014	1.564	0.018234	46.61	36.35
PFH	0.014	3.872	0.035647	79.16	68.75
FPFH	0.014	1.430	0.017559	48.15	36.23
AGWF	**Adaptive radius**	**0.952**	**0.011715**	/	/

The Dt and DRMSE are calculated as Dt = (FT − AGWF)/FT, DRMSE = (FR − AGWF)/FR, where FT is the feature extraction time of the traditional descriptors, and FR is the RMSE of traditional descriptors.

As is shown in Table 3, the time consumption of the GWF is 7.9% and 17.34% longer than the SHOT and FPFH, respectively, and it is 56.6% faster than the PFH. Compared with the PFH, FPFH and SHOT, the RMSE of the GWF is reduced by 10.99%, 54.47% and 7.57%, respectively. The GWF does not have an absolute advantage in terms of time, because it needs to consider more information during feature extraction, but it has higher registration accuracy. It can be seen in Table 4, when the neighborhood radius is set to 0.014 m, the time consumption of the proposed method is reduced by 46.61%, 79.16% and 48.15%, and the

RMSE is reduced by 36.35%, 68.75% and 36.23%, respectively. Comparing Tables 3 and 4, the choice of the neighborhood radius can influence the time consumption of the feature extraction and the accuracy of the point cloud registration, and the proposed adaptive radius can significantly reduce the time consumption of feature extraction and the accuracy of the point cloud registration. Compared with the fixed radius (GWF), the proposed AGWF feature descriptor reduces the time consumption by 43.27%, and the RMSE by 27.86%. Above all, the proposed method can reduce the time consumption by more than 30% and the error by more than 35%.

Compared with the traditional FPFH feature descriptor, the proposed AGWF feature descriptor improves the weight coefficients in its calculation process, which can describe the local features of the point cloud better and fully consider the influence of each neighboring point on the key point features, thereby improving the accuracy and efficiency of the bit pose estimation. In addition, the methods based on traditional descriptors usually set the search radius of feature descriptors through experience when setting the radius of the key point neighborhood, which has the problem of inefficiency. The proposed approach adaptively sets the neighborhood radius according to the minimum criterion of the neighborhood information entropy function, which greatly reduces the computation time of feature descriptors and achieves efficient and high-precision pose estimation of the pallet.

4. Conclusions

A point cloud data-driven pallet pose estimation method using an active binocular vision sensor is proposed, which consists of point cloud preprocessing, adaptive Gaussian weight-based fast point feature histogram definition and point cloud registration, and improves the pose estimation accuracy by over 35% and reduces the feature extraction time by over 30%. The main contributions of the proposed method can be summarized as follows:

1. A point cloud-driven method for the driverless industrial trucks to estimate the pose of the pallet in the production shop is proposed, which solves the problem that the pallet cannot be accurately forked due to the position deviation, and improves the security and stability of the logistics system.
2. An adaptive optimal neighborhood radius selection criterion based on the minimum rule of the local neighborhood characteristic entropy function is proposed to determine the neighborhood radius of each key point adaptively, instead of selecting parameters based on experience manually, which significantly shortened the time of feature extraction and improved the accuracy.
3. Traditional descriptors only consider the Euclidean distance between the query key point and the neighborhood point as the traditional methods, and the weight of the proposed descriptor is optimized by the Gaussian distribution function. The change of the weights of each neighborhood point is smoother and can describe the key point features more accurately and completely, thereby effectively improving the robustness of the feature descriptors.

Nevertheless, since the proposed approach still requires a large amount of computation on the point cloud data, the real-time performance of the approach needs to be further improved. At the same time, due to the complex storage environment, the proposed approach also needs further discussion on problems such as vision occlusion and multi-target overlap caused by dynamic and static obstacles. The subsequent work will focus on improving the real-time performance of the algorithm, as well as research on scenes with different lighting and different degrees of occlusion.

Author Contributions: Conceptualization, J.L.; methodology, Y.S.; software, Z.F.; validation, Z.F.; formal analysis, Y.S.; investigation, B.Z.; resources, J.L. and Y.L.; data curation, Z.F. and Y.L.; writing—original draft preparation, Y.S.; writing—review and editing, Y.S. and Z.F.; visualization, Z.F.; supervision, J.L. and Y.L.; project administration, B.Z. and Y.L.; funding acquisition, Y.S. and J.L. All authors have read and agreed to the published version of the manuscript.

Funding: This work is supported by the Zhejiang Provincial Natural Science Foundation, grant number LQ22E050017, the Zhejiang Science and Technology Plan Project, grant number 2018C01003, the China Postdoctoral Science Foundation, grant number 2021M702894, and the Zhejiang Provincial Postdoctoral Science Foundation, grant number ZJ2021119. All experiments were performed at Noblelift Intelligent Equipment Co., Ltd. We are grateful to Noblelift engineers for their experimental support.

Institutional Review Board Statement: Not applicable.

Informed Consent Statement: Not applicable.

Data Availability Statement: Not applicable.

Conflicts of Interest: The authors declare no conflict of interest.

References

1. Baglivo, L.; Biasi, N.; Biral, F.; Bellomo, N.; Bertolazzi, E.; Da Lio, M.; De Cecco, M. Autonomous pallet localization and picking for industrial forklifts: A robust range and look method. *Meas. Sci. Technol.* **2011**, *22*, 085502. [CrossRef]
2. Hu, H.S.; Wang, L.; Luh, P. Intelligent manufacturing: New advances and challenges. *J. Intell. Manuf.* **2015**, *26*, 841–843. [CrossRef]
3. Shuai, L.; Mingkang, X.; Weilin, Z.; Huilin, X. Towards Industrial Scenario Lane Detection: Vision-Based AGV Navigation Methods. In Proceedings of the 2020 IEEE International Conference on Mechatronics and Automation, Beijing, China, 13–16 October 2020; pp. 1101–1106.
4. Baglivo, L.; Bellomo, N.; Marcuzzi, E.; Pertile, M.; Cecco, M. Pallet Pose Estimation with LIDAR and Vision for Autonomous Forklifts. In Proceedings of the IEEE 13th IFAC Symposium on Information Control Problems in Manufacturing IFAC-INCOM '09, Moscow, Russia, 3–5 June 2009; pp. 616–621.
5. Mohamed, I.S.; Capitanelli, A.; Mastrogiovanni, F.; Rovetta, S.; Zaccaria, R. Detection, localisation and tracking of pallets using machine learning techniques and 2D range data. *Neural Comput. Appl.* **2020**, *32*, 8811–8828. [CrossRef]
6. Zhang, Z.H.; Liang, Z.D.; Zhang, M.; Zhao, X.; Li, H.; Yang, M.; Tan, W.M.; Pu, S.L. RangeLVDet: Boosting 3D Object Detection in LIDAR with Range Image and RGB Image. *IEEE Sens. J.* **2022**, *22*, 1391–1403. [CrossRef]
7. Seelinger, M.; Yoder, J.D. Automatic visual guidance of a forklift engaging a pallet. *Robot. Auton. Syst.* **2006**, *54*, 1026–1038. [CrossRef]
8. Syu, J.L.; Li, H.T.; Chiang, J.S.; Hsia, C.H.; Wu, P.H.; Hsieh, C.F.; Li, S.A. A computer vision assisted system for autonomous forklift vehicles in real factory environment. *Multimed. Tools Appl.* **2017**, *76*, 18387–18407. [CrossRef]
9. Fan, R.Z.; Xu, T.B.; Wei, Z.Z. Estimating 6D Aircraft Pose from Keypoints and Structures. *Remote Sens.* **2021**, *13*, 663. [CrossRef]
10. Guo, K.; Ye, H.; Gao, X.; Chen, H.L. An Accurate and Robust Method for Absolute Pose Estimation with UAV Using RANSAC. *Sensors* **2022**, *22*, 5925. [CrossRef]
11. Varga, R.; Nedevschi, S. Robust Pallet Detection for Automated Logistics Operations. In Proceedings of the 11th International Conference on Computer Vision Theory and Applications, Rome, Italy, 27–29 February 2016; pp. 470–477.
12. Casado, F.; Lapido, Y.L.; Losada, D.P.; Santana-Alonso, A. Pose estimation and object tracking using 2D images. *Procedia Manuf.* **2017**, *11*, 63–71. [CrossRef]
13. Shao, Y.P.; Wang, K.; Du, S.C.; Xi, L.F. High definition metrology enabled three dimensional discontinuous surface filtering by extended tetrolet transform. *J. Manuf. Syst.* **2018**, *49*, 75–92. [CrossRef]
14. Zhao, C.; Du, S.C.; Lv, J.; Deng, Y.F.; Li, G.L. A novel parallel classification network for classifying three-dimensional surface with point cloud data. *J. Intell. Manuf.* 2021, *in press*. [CrossRef]
15. Zhao, C.; Lv, J.; Du, S.C. Geometrical deviation modeling and monitoring of 3D surface based on multi-output Gaussian process. *Measurement* **2022**, *199*, 111569. [CrossRef]
16. Shao, Y.P.; Du, S.C.; Tang, H.T. An extended bi-dimensional empirical wavelet transform based filtering approach for engineering surface separation using high definition metrology. *Measurement* **2021**, *178*, 109259. [CrossRef]
17. Zhao, C.; Lui, F.C.; Du, S.C.; Wang, D.; Shao, Y.P. An Earth Mover's Distance based Multivariate Generalized Likelihood Ratio Control Chart for Effective Monitoring of 3D Point Cloud Surface. *Comput. Ind. Eng.* **2023**, *175*, 108911. [CrossRef]
18. Wang, X.D.; Liu, B.; Mei, X.S.; Wang, X.T.; Lian, R.H. A Novel Method for Measuring, Collimating, and Maintaining the Spatial Pose of Terminal Beam in Laser Processing System Based on 3D and 2D Hybrid Vision. *IEEE Trans. Ind. Electron.* **2022**, *69*, 10634. [CrossRef]

19. Lee, H.; Park, J.M.; Kim, K.H.; Lee, D.H.; Sohn, M.J. Accuracy evaluation of surface registration algorithm using normal distribution transform in stereotactic body radiotherapy/radiosurgery: A phantom study. *J. Appl. Clin. Med. Phys.* **2022**, *23*, e13521. [CrossRef]
20. Xie, X.X.; Wang, X.C.; Wu, Z.K. 3D face dense reconstruction based on sparse points using probabilistic principal component analysis. *Multimed. Tools Appl.* **2022**, *81*, 2937–2957. [CrossRef]
21. Liu, W.L. LiDAR-IMU Time Delay Calibration Based on Iterative Closest Point and Iterated Sigma Point Kalman Filter. *Sensors* **2017**, *17*, 539. [CrossRef]
22. Yang, J.L.; Li, H.D.; Campbell, D.; Jia, Y.D. Go-ICP: A Globally Optimal Solution to 3D ICP Point-Set Registration. *IEEE Trans. Pattern Anal.* **2016**, *38*, 2241–2254. [CrossRef]
23. Wu, P.; Li, W.; Yan, M. 3D scene reconstruction based on improved ICP algorithm. *Microprocess. Microsyst.* **2020**, *75*, 103064. [CrossRef]
24. Fotsing, C.; Menadjou, N.; Bobda, C. Iterative closest point for accurate plane detection in unorganized point clouds. *Automat. Constr.* **2021**, *125*, 103610. [CrossRef]
25. Guo, N.; Zhang, B.H.; Zhou, J.; Zhan, K.T.; Lai, S. Pose estimation and adaptable grasp configuration with point cloud registration and geometry understanding for fruit grasp planning. *Comput. Electron. Agric.* **2020**, *179*, 105818. [CrossRef]
26. Rusu, R.B.; Blodow, N.; Beetz, M. Fast Point Feature Histograms (FPFH) for 3D Registration. In Proceedings of the 2009 IEEE International Conference on Robotics and Automation, Kobe, Japan, 12–17 May 2009; pp. 1848–1853.
27. Tombari, F.; Salti, S.; Di Stefano, L. Unique Signatures of Histograms for Local Surface Description. *Lect. Notes Comput. Sci.* **2010**, *6313*, 356–369.
28. Li, D.p.; Liu, N.; Guo, Y.L.; Wang, X.M.; Xu, J. 3D object recognition and pose estimation for random bin-picking using Partition Viewpoint Feature Histograms. *Pattern Recogn. Lett.* **2019**, *128*, 148–154. [CrossRef]
29. Toumieh, C.; Lambert, A. Voxel-Grid Based Convex Decomposition of 3D Space for Safe Corridor Generation. *J. Intell. Robot. Syst.* **2022**, *105*, 87. [CrossRef]
30. Khanna, N.; Delp, E.J. Intrinsic Signatures for Scanned Documents Forensics: Effect of Font Shape and Size. In Proceedings of the Proceedings of 2010 IEEE International Symposium on Circuits and Systems, Paris, France, 30 May–2 June 2010; pp. 3060–3063.
31. Xu, G.X.; Pang, Y.J.; Bai, Z.X.; Wang, Y.L.; Lu, Z.W. A Fast Point Clouds Registration Algorithm for Laser Scanners. *Appl. Sci.* **2021**, *11*, 3426. [CrossRef]
32. Shao, Y.P.; Fan, Z.S.; Zhu, B.C.; Zhou, M.L.; Chen, Z.H.; Lu, J.S. A Novel Pallet Detection Method for Automated Guided Vehicles Based on Point Cloud Data. *Sensors* **2022**, *22*, 8019. [CrossRef] [PubMed]
33. Hu, H.P.; Gu, W.K.; Yang, X.S.; Zhang, N.; Lou, Y.J. Fast 6D object pose estimation of shell parts for robotic assembly. *Int. J. Adv. Manuf. Technol.* **2022**, *118*, 1383–1396. [CrossRef]

sensors

MDPI

Article

High-Precision Tribometer for Studies of Adhesive Contacts

Iakov A. Lyashenko [1,2,*], Valentin L. Popov [1,*], Roman Pohrt [1] and Vadym Borysiuk [1,3]

1 Department of System Dynamics and Friction Physics, Institute of Mechanics, Technische Universität Berlin, 10623 Berlin, Germany
2 Department of Applied Mathematics and Complex Systems Modeling, Faculty of Electronics and Information Technology, Sumy State University, 40007 Sumy, Ukraine
3 Department of Nanoelectronics and Surface Modification, Faculty of Electronics and Information Technology, Sumy State University, 40007 Sumy, Ukraine
* Correspondence: i.liashenko@tu-berlin.de (I.A.L.); v.popov@tu-berlin.de (V.L.P.); Tel.: +49-(0)30-314-75917 (I.A.L.)

Abstract: Herein, we describe the design of a laboratory setup operating as a high-precision tribometer. The whole design procedure is presented, starting with a concept, followed by the creation of an exact 3D model and final assembly of all functional parts. The functional idea of the setup is based on a previously designed device that was used to perform more simple tasks. A series of experiments revealed certain disadvantages of the initial setup, for which pertinent solutions were found and implemented. Processing and correction of the data obtained from the device are demonstrated with an example involving backlash and signal drift errors. Correction of both linear and non-linear signal drift errors is considered. We also show that, depending on the research interests, the developed equipment can be further modified by alternating its peripheral parts without changing the main frame of the device.

Keywords: force sensor; tribometer; drift; hysteresis; contact mechanics; contact area; adhesion

1. Introduction

High-quality scientific experiments require precise laboratory equipment. Experimental devices designed for studies of mechanical interactions typically contain force sensors and positioning motors. An example of a high-precision experimental setup that operates with mechanical contact is an atomic force microscope (AFM), which has lateral resolution up to a few nanometers and a normal resolution up to 0.1 nm. AFM technology also uses force sensors, which allow for measurement of van der Waals forces in the range of 10^{-10} to 10^{-12} N [1–3]. Experimental facilities that deal with nanotechnology often use a variety of high-sensitivity electrical, mechanical, optical and other sensors to measure physical quantities of low magnitude [4,5].

High-accuracy force measurements are very important in macroscopic experiments with mechanical contacts, especially in cases in which the smallest changes in macro parameters need to be detected. For investigation of mechanical contacts of adhesive surfaces, high-precision force and positioning sensors are used as main components [6–8]. A number of tribometers and force sensors with a variety of measuring features have been designed recently for this purpose [9–17]. A central part of these experimental setups are force sensors manufactured by various electronics companies (see, for example, [18]). Many of them are compatible with personal computers (PCs) and other laboratory equipment through standard interfaces and can be easily adopted to measure desirable quantities. However, to create a fully operational device that performs complete cycles of specific experiments, sensors, motors and other parts must be assembled into one apparatus. There are various potential sources of errors in complete systems that have to be compensated for, e.g., signal drift, temperature fluctuations and hysteresis in the force sensor [14].

In this paper, we provide an overview of laboratory setups equipped with KD24S and K3D40 force sensors manufactured by ME systems, which were designed to perform experiments on the indentation of a rigid indenter into a soft substrate [19–21]. The designed facility operates as a high-precision tribometer. It performs indentation and retraction in the normal direction, as well as displacement in the tangential direction, with precise positioning of the sample in all three dimensions, in addition to measuring loading forces, enabling observation of dynamical processes occurring in the contact area in real time with photo and video recording.

During a series of experiments, the designed device was upgraded several times according to progress in research plans. Thus, the earliest version of the apparatus was able to perform experiments only with normal contact, whereas the current version operates with three-dimensional positioning. Based on our experience with the design and use of the developed equipment, herein, we outline its pros and cons and discuss the well-known peculiarities of tribometers, such as backlash in motor and signal drift in the force sensor. We also suggest some practical solutions that helped us to overcome certain issues in the performed experiments.

Our work will be interesting to scientists and engineers who work in the field of contact mechanics and deal with similar instruments.

2. Materials and Methods

In our experiments, we studied physical phenomena within mechanical contacts that involve highly adhesive surfaces. Adhesion often plays a crucial role in the natural environment and numerous technologies. As an illustration of adhesive processes in a biological system, we can refer to the ability of gecko lizards to climb vertical surfaces and move on ceilings [22–24]. Adhesion is also important for microorganisms, viruses and living cells [25–27]. The study of specific adhesion processes is an important part of research on novel SARS-CoV-2 (COVID-19) coronavirus infection [28]. Technologies based on adhesive phenomena have found numerous applications in industry, where various surfaces come into contact, such as coatings, welding, granulation and other applications [29,30].

In order to study contact between a rigid body and an elastomer under laboratory conditions and obtain insights into the physical processes in the contact area, an appropriate laboratory setup was designed and fabricated. The concept of the designed laboratory device was inspired by an instrument developed in [19]. Figure 1 shows the progress in designing and assembling the facility, from the idea of a device pictured as a sketch (Figure 1a) to an exactly matching 3D model of all its parts (Figure 1b) and, eventually, a fully operable piece of equipment (Figure 1c). All functional parts of the device that are denoted in the figure are the same for all three panels, namely (1) and (2) high-precision M-403.2DG motorized linear stages (manufactured by PI), which are handled by PI C-863 one-axis servo controllers; (3) an ME K3D40 three-axis force sensor; (4) an indenter that is mounted on the force sensor; (5) the sample being indented; (6) a tilt mechanism; and (7) a digital camera. A GSV-1A4 SubD37/2 four-channel analog amplifier was used to amplify the electrical signal from the force sensor. The amplifier is coupled with a computer via an NI USB-6211 16-bit analog-to-digital converter (ADC). A photo of the contact region with a clear view of the indenter (4), force sensor (3) and sample (5) is shown in Figure 2 (note the different configuration of the experiment and indenter compared to Figure 1c). In our experiments on adhesive contacts we often use optically transparent rubber sheets produced by Japanese company TARNAC (see [31]) as a substrate, because this rubber is optically transparent and has high adhesive properties. As shown in the Figure 1c, the device has a cylindrical steel indenter (wheel). This particular device was used to perform experiments investigating adhesion in rolling and sliding friction in [32]. Moreover, equipment with different indenters has been used to perform a series of experiments [20, 21,32–39].

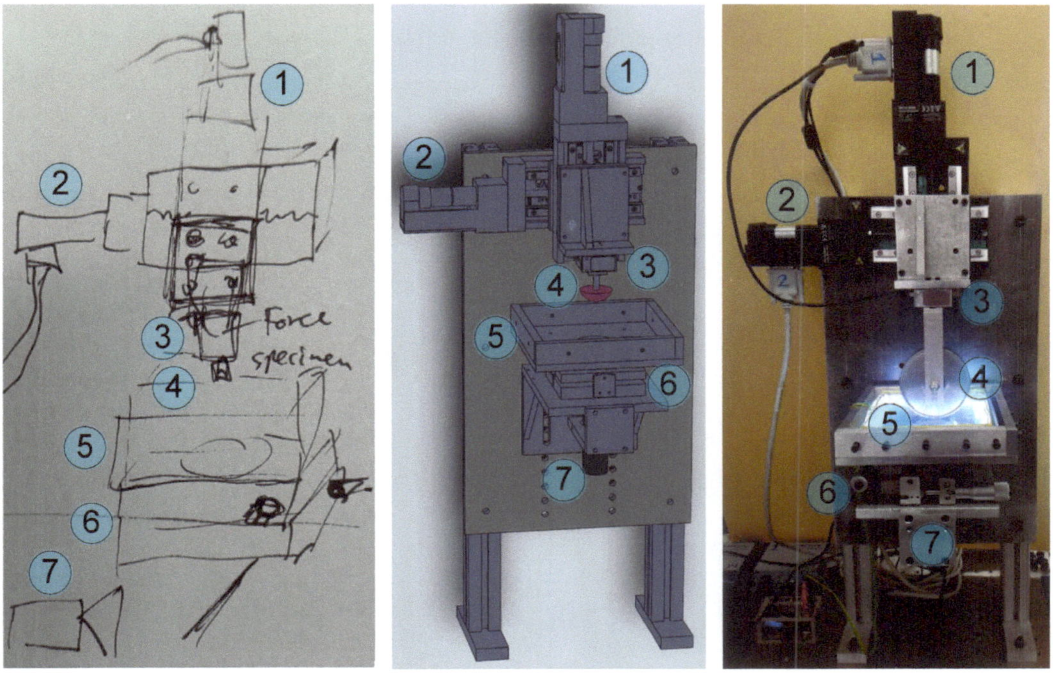

(**a**) (**b**) (**c**)

Figure 1. The process of development of the laboratory equipment: design starts with the concept, shown as a sketch panel (**a**), which later is transformed into a detailed blueprint panel (**b**) and then assembled as a laboratory tribometer panel (**c**). The labeling of the equipment is the same in all panels: (1) and (2) M-403.2DG high-precision motorized linear stages (manufactured by PI), (3) three-axis force sensor ME K3D40 with (4) a mounted indenter, (5) the sample being indented, (6) a tilt mechanism and (7) a digital camera. (**b**,**c**) Device with different indenters.

The developed device is shown in Figure 1c. The main disadvantage of this device is the backlash in the M-403.2DG motorized linear stages. These drives are capable of high-precision positioning (up to 1 μm) of the indenter in unidirectional repeatability mode; however, after changing the direction of motion, their backlash can reach 10 μm. In our experiments, we observed backlash of about 6 μm for both drivers (1) and (2) (see Figure 1). Technically, this means that after changing the direction of motion, the device shows that the indenter is still moving (the screw attached to the motion platform is rotating), even though for the first six micrometers, the indenter remains physically motionless. This effect must be considered during data analysis; otherwise, it will lead to inadequate experimental results for both indentation and pull-off. To illustrate the impact of a backlash on experimental data, let us consider the following example.

Figure 3 shows the results of an experiment in which a steel sphere with a radius of $R = 33$ mm was indented into a layer of TARNAC CRG N3005 rubber with a thickness of $h = 5$ mm. After the indentation depth reached its maximum value of $d_{\text{max}} = 0.2$ mm, the indenter was moved in the opposite direction until it lost contact with the substrate. The curve plotted in red represents the indentation phase. It starts from zero and is located above the pull-off curve, which is indicated in blue. Owing to the strong adhesive interaction between the steel indenter and rubber substrate, negative forces of $F_N < 0$ N are present in the system during detachment. This means that additional force must be applied to the indenter to detach it from the sample compared to contact force at the same position of an indenter during the indentation phase. The presence of backlash in the M-403.2DG drive leads to a situation in which the indenter physically remains in a stationary position, whereas the equipment shows its motion for six micrometers after the change in direction. Moreover, the magnitude of normal force (F_N) remains constant, as there is no actual movement in the contact. Therefore, the pull-off curve becomes shifted to the left along the abscissa axis and crosses the loading curve related to indentation at point *A* (Figure 3a).

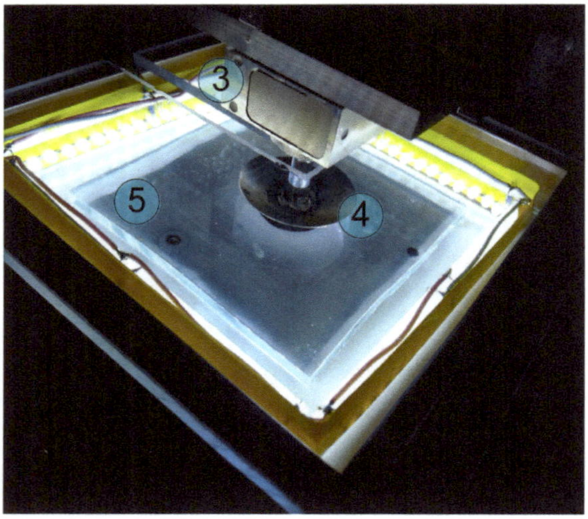

Figure 2. Detailed photo of the contact region: indenter (4), sensor (3) and elastomer (5). The all-sides lighting system consisting of 80 LEDs is also clearly visible.

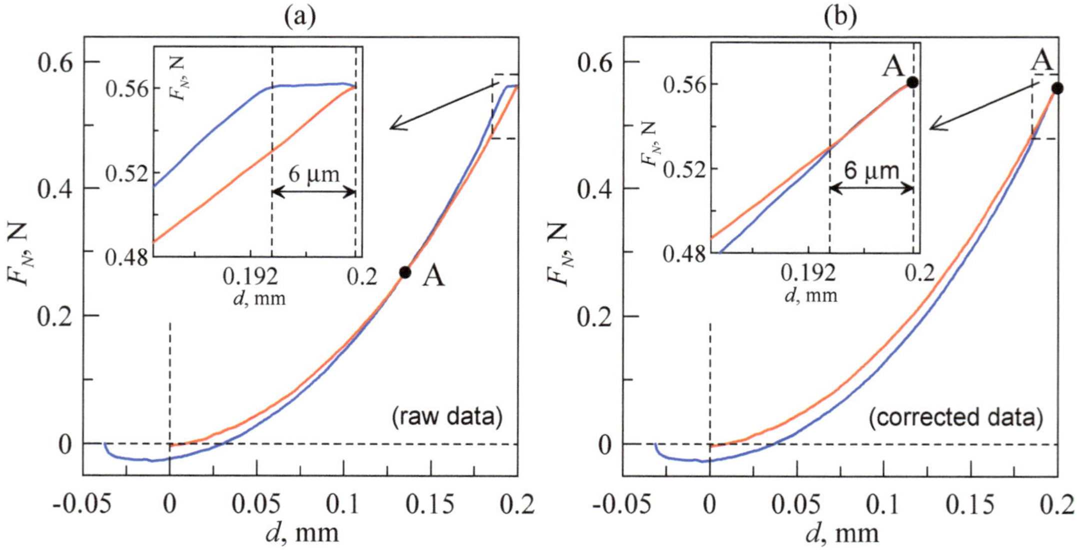

Figure 3. Dependencies of the normal force (F_N) on indentation depth (d) measured during the indentation of the spherical indenter with a radius of R = 33 mm into a rubber (*TARNAC CRG N3005*) substrate with a thickness of h = 5 mm [33]. Experiments were performed with the equipment shown in Figure 1. (**a**) Raw measured data; (**b**) dependencies with compensation for a backlash of the drive.

To obtain an accurate diagram of a contact load, backlash correction must be introduced in data analysis. In our case, such a correction is a simple addition of the compensational term of 6 μm (backlash magnitude) to the pull-off curve, which shifts it along the x axis in the positive direction, as shown in Figure 3b. In general, the need for data correction can be determined according to the goal of the experiments. If the goal of an experiment is to determine the adhesive strength of a contact (minimal external force needed to completely destroy the contact), as in [20], then adjustment is unnecessary, as the shift along the abscissa axis does not change the magnitudes of the acting force. On the other hand, if the dissipation of the mechanical energy in the full cycle of loading during indentation and pull-off (see also [20]) has to be determined, then correction plays a crucial role, as the dissipation is defined by the area of a region bounded by both indentation and pull-off curves ($F_N(d)$). However, such data corrections are not always possible. For instance, the presence of backlash makes the interpretation of the results of experiments even more difficult when the indenter performs a cycling motion (see [34]).

Figure 3 illustrates a simple adjustment of the data by shifting to the right the pull-off curve, and it is completely sufficient to obtaining an adequate result from the experiment. However, assuming that our experiment requires changing the direction of motion more than one time during the pull-off phase prior to complete disruption of the contact, we would observe a backlash that starts at the turning point and lasts for 6 μm as the indentation depth (d) increases according to the equipment measurements. Thus, the obtained curve must be shifted left instead of right. As a result, during cycling motion, dependence related to a change in direction from indentation to pull-off must be shifted to the right, whereas dependence that corresponds to the change from pull-off to indentation must be shifted to the left. As a result of such data manipulations, the total load curve ($F_N(d)$) becomes non-closed. Technically, this occurs because, owing to the backlash, the indenter is not physically present in certain segments of its expected trajectory. Therefore, analysis of the $F_N(d)$ curves related to the cycling load requires extrapolation of the obtained dependencies, which may significantly affect results. Studying oscillations with small amplitudes becomes virtually impossible, as within the backlash magnitudes, the indenter does not move at all.

It is worth mentioning that backlash magnitude depends on the temperature and load applied to the driver, so the automatic correction of this particular feature is a very complicated challenge, and every new load scenario requires additional analysis and corrections. One possible solution to the described problem is a direct measurement of the indenter coordinate by a laser vibrometer; however, this dramatically increases the cost of the equipment, as well as its size. Another solution is to use different types of drives that have improved positioning mechanisms without backlash. Although such drives are more expensive than the M-403.2DG described above, they are cheaper than using a laser vibrometer or similar high-quality equipment. In the upgraded version of the developed equipment, we adopted the second solution (see description below).

Experience reveals another peculiarity of the equipment shown in Figure 1 related to the Conrad USB digital camera. This camera has a resolution of 1600 × 1200 pixels, which is enough for precise monitoring of the contact area. However, a disadvantage of this camera as a part of the designed tribometer is the brightness correction, which is performed automatically if the intensity of external illumination changes. In our experiments, the indenter moved in the normal and tangential directions, as well as in both directions simultaneously; as a result, the distribution of luminous flux was affected. Under stationary conditions, illumination of the sample is homogeneous, as it is provided by a static comprehensive lighting system (see Figures 1c and 2). Light redistribution causes the camera to switch into a new mode of photographing, and as a result, pictures taken of the contact area have different contrast. This situation does not change the general view of the contact zone; however, it makes automatic processing of the photos and determination of the size of the contact area much more complicated [20,33,34]. The software developed to analyze the digital photos of the indentation zone and to calculate the contact area uses an algorithm that calculates the size of a contact by analyzing the difference in the intensity of pixel color, which determines whether a given pixel is included in the contact area or not (for a detailed description of the software performance, see [20]). If the intensity of pixels constantly changes due to the different settings of the camera mode, such analysis becomes extremely difficult.

Assembling the equipment shown in Figure 1 was a significant advancement from the initial facility [19], as the options for tangential motion of the indenter and more precise positioning of the indented sample, as well as the possibility of considerably increasing lateral sizes of the substrate, were added to the device. However, as mentioned above, the device shown in Figure 1 has its downsides as well. With the goal of rectifying these disadvantages and increasing the quality of the obtained results, the equipment was upgraded. An exact 3D model of the upgraded equipment and a photo of the assembled device are shown in Figure 4a,b, respectively. As shown in the figure, the device has several differences from the previous version.

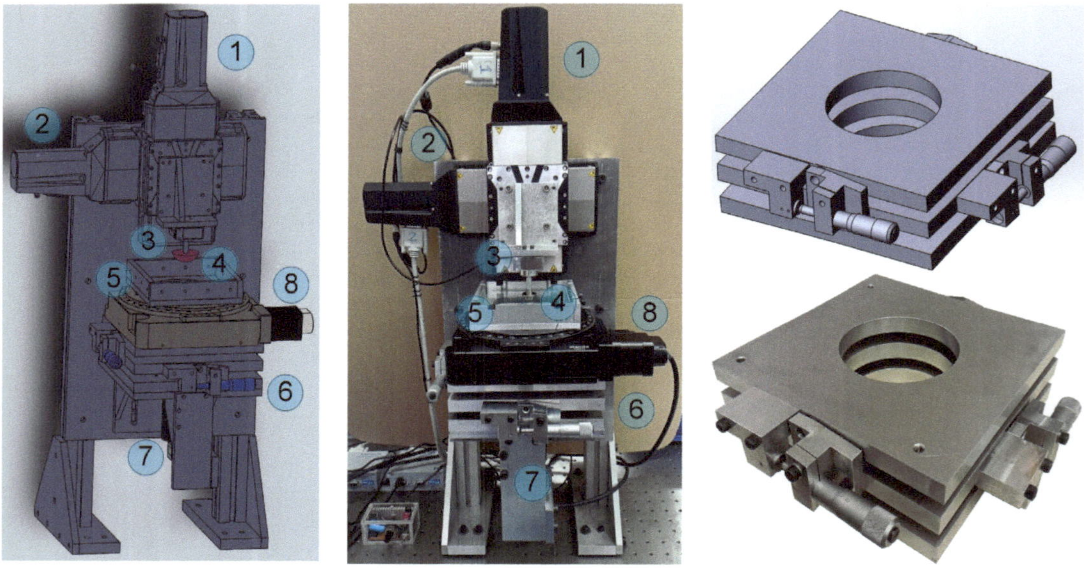

(**a**) (**b**) (**c**)

Figure 4. Upgraded version of the device shown in Figure 1. (**a**) Exact 3D model of the facility with an individual blueprint for every part. (**b**) Photo of the assembled and fully operational equipment. (**c**) A 3D model (top) and photo (bottom) of the tilt mechanism fabricated from aluminum plates, which is one of the most significant improvements added to the device. Position (8) denotes the 8MR190-90-4247-MEn1 motorized rotation stage in (**a**,**b**); all other labels are the same as in Figure 1.

The main novelties of the proposed equipment are:

(a) More precise L-511.24AD00 drives with bidirectional repeatability up to ± 0.2 μm. The main attribute of these drives is the absence of backlash after changing the direction of motion. This feature is achieved using a ball-screw transmission and an incremental linear encoder inserted for feedback. The encoder performs direct measurements of the moving platform coordinate during motion and corrects it. Thus, these drives do not create artefacts after changing the direction of motion and have high-precision positioning that is more than enough for our experiments. However, the presence of feedback makes the positioning system vulnerable to equipment vibrations. These vibrations could prevent the correct positioning of the motion platform, as the drive repeatably searches for correct coordinates and, as a result, works as a generator of mechanical oscillations, making measurements impossible. Therefore, an oscillation damper for the whole device must be introduced to the facility. To dampen oscillations of the equipment, we used rubber (TARNAC CRG N0505 [31]) sheets placed between the stands of the device and the laboratory table with pneumatic support (see Figure 4b), serving as a foundation on which the device was assembled. That particular rubber is optically transparent and has highly adhesive properties, so it was typically used as a substrate for indentation in our experiments, as it allows for direct observation of the processes that occur in the contact area through the layer of rubber via a camera. However, this material was initially designed as an oscillation damper [31], so in this case, it was used as an intended. It is also worth mentioning that maximal velocity of the table movement in L-511.24AD00 drives is 90 mm/s, which is significantly higher than the 2.5 mm/s in M-403.2DG.

(b) A high-quality XimeaMQ022CG-CM color camera with a FUJINON HF16SA-1, 2/3″ lens. This camera is USB 3.1-compatible and has a resolution of 2048 × 1088 pixels at 170 frames per second, making observation of fast processes in the contact area possible. In our future experiments, we plan to use it to study the propagation of elastic waves in the contact area during its reconfiguration. The presence of these waves was detected in experiments with the bare eye; therefore, more detailed study with special equipment is needed. Another advantage of this camera as a part of the designed tribometer relative to previously used cameras is its manual brightness setup instead of an automatic mode. The brightness is regulated by a built-in mechanical diaphragm; as a result, all taken photos have the same brightness level, which makes it easier to automatically calculate the contact area using image analysis software.

(c) A large STANDA 8MR190-90-4247-MEn1 motorized rotation stage (position (8) in Figure 4a,b) operated by a single-channel 8SMC5-USB-B8-1 controller was added to the facility, which allowed the horizontal orientation of the substrate to be changed. The maximal angular speed of the stage is 36°/s (6 full circles in a minute). This stage has a clear aperture of 89 mm, which is almost twice as large as the 45 mm of the previous version of the equipment, and allows for observation of contact processes in a wider range of tangential displacement of the indenter. According to its specification, the stage is characterized by a backlash of 40 arcsec and bidirectional repeatability of 72 arcsec, which are related to the displacement at the edge of an aperture 8.67 μm and 15.54 μm, respectively. These features must be taken into account during experiments in which the direction of circular motion changes. However, this was not an issue in our experiments. The rotation stage was added to the facility with the intention of studying a stationary sliding mode without changing the rotation direction.

(d) A completely new tilt mechanism (position (6) in Figure 4a,b). The biggest difference between the upgraded version and the previous mechanism shown in Figure 1 is a much wider aperture (equal to the 8MR190-90-4247-MEn1 stage aperture), which was enlarged to observe larger portion of the contact area. We could not find the required in the available markets, so it was completely designed and manufactured in-house at the TU Berlin workshop from aluminum plates. The 3D blueprint and a photo of the fabricated mechanism are shown in Figure 4c.

The device shown in Figure 4 was designed in a such way that drives (1) and (2), together with the rotation stage (8), are capable of positioning the indenter over the substrate in three dimensions (as shown in Video S1 in Supplementary Material). Although described equipment was designed for experiments on adhesive contact with complex indentation trajectories, it can also be used for other purposes depending on what is installed as an "indenter". For instance, if a laser emitter is mounted instead of an indenter, the device will work as a basic laser engraver. Similarly, a solid ink printhead transforms the device into a simple 3D printer, and with any writing tool, the equipment operates as a plotter. All mentioned (and similar) modifications can be introduced to the facility without any significant changes to the main part of the equipment. To demonstrate the possibility of modification, we created a simple analog of a plotter using an ordinary pen as a writing tool. An example of its performance is presented in Video S1 in Supplementary Material.

It is worth noting that a further important aspect relates to measurements. Experiments described in the original work [19] were performed with an ME KD24S force sensor operating within the range of ±2 N. The signal from this sensor was amplified by a single-channel strain gauge amplifier/DM1 DMS amplifier (manufacturer LEG–Industrie-Elektronik GmbH). Furthermore, the upgraded version of the facility (Figures 1 and 4) was equipped with an ME K3D40 three-axillar force sensor (version with a force range of ±10 N) and a GSV-1A4 SubD37/2 4-channel amplifier. In all versions of the device, the output signal from the amplifier was transmitted to a personal computer through the NI USB-6211 16-bit ADC. Many force sensors (including that used in our equipment and described above) have signal drift, due to which even unloaded sensors display increasing (or decreasing) force over time. Figure 5 shows the time dependencies of force measured by unloaded (i.e., the magnitude of actual force acting on the sensor during the whole measurement equals zero) sensors.

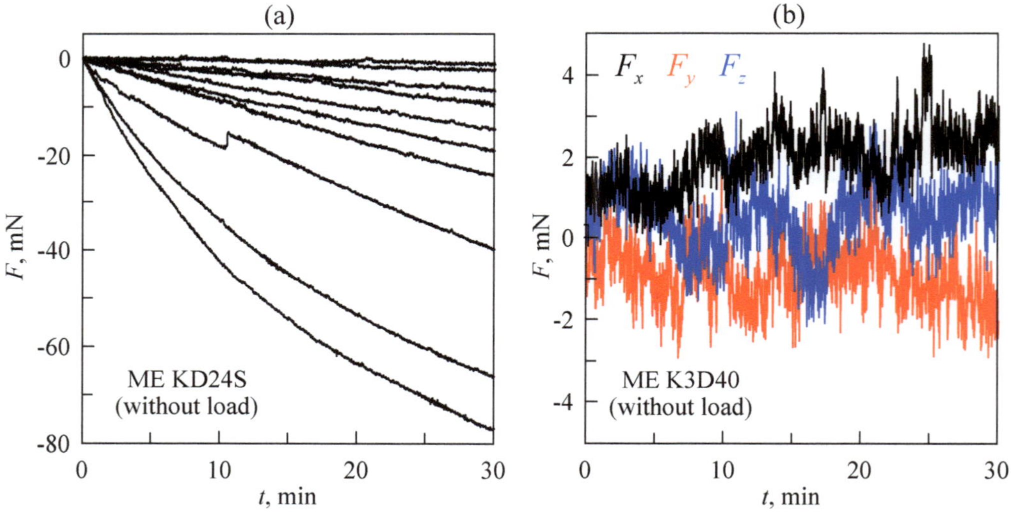

Figure 5. Time dependencies of "force" measured by an unloaded sensor (signal drift). (**a**) Ten dependencies corresponding to different measurements obtained by the ME KD24S sensor. (**b**) Time dependencies of three components of a force measured by the ME K3D40 three-axis sensor in one experiment.

Dependencies in panel (a) were measured by the ME KD24S sensor. As shown in the figure, apart from three bottom lines, the signal drift is almost linear, with different slopes. Linear drift can be easily subtracted from obtained data; however, the presence of non-linear progressive changes in the signal make data analysis more complicated. Figure 5b shows similar data but measured by another ME K3D40 sensor. Here, all three components of a force were measured in one experiment. As the figure shows, the signal drift is also present in the measurements; however it is significantly lower than in most data measured by the ME KD24S sensor (see Figure 5a).

Thus, sensors with small or no drift are always preferable as a part of laboratory equipment (for instance, ME K3D40 instead ME KD24S). Furthermore, the impact of signal drift on a measured force magnitude can be reduced by various methods. For instance, it can be subtracted from the measured data if the shape of the signal drift is known. Another way to reduce drift influence is to perform the indentation cycle as fast as possible or measure the data within the range of forces that significantly exceed the drift magnitude so it will not affect the results. However, these methods often contradict the required conditions of the experiments. It is worth noting that dependencies shown in both panels of Figure 5 were obtained using a GSV-1A4 amplifier; however, in experiments with a DMS DM1 amplifier, similar problems with drift signal were observed. Additionally, in our experiments, two different ADCs (NI USB-6211 and NI USB-6000) were used for the PC interface, and signal drift was observed with both of them.

Let us consider an illustrative example of how to handle signal drift from the sensor and to correct the results. Among the two available sensors, ME KD24S has the larger signal drift; thus, we will demonstrate the data processing procedure with this sensor. Figure 6 shows the measured dependencies of the normal force (F_N) on the indentation depth (d).

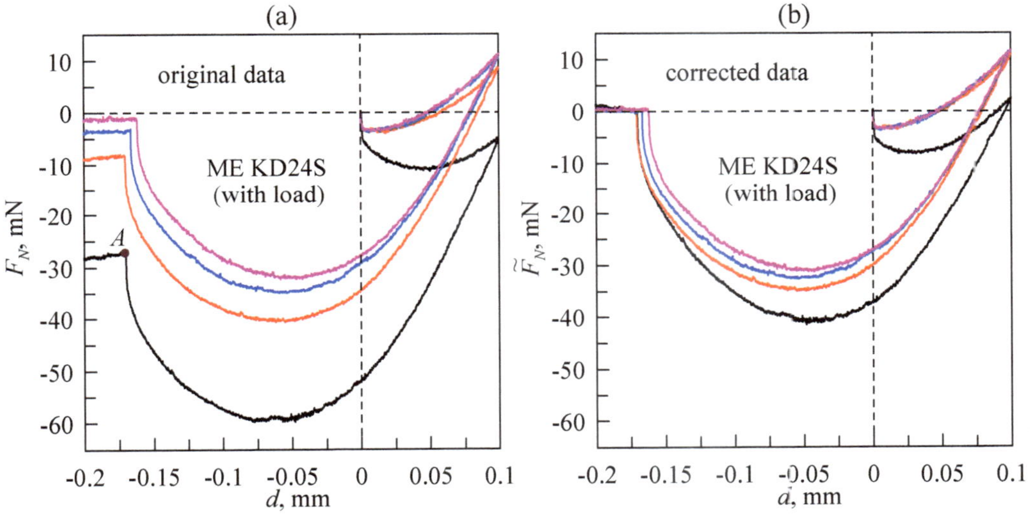

Figure 6. Dependencies of the normal force (F_N) on the indentation depth (d) obtained during the indentation of a steel spherical indenter with a radius of R = 30 mm into a TARNAC CRG N0505 rubber substrate with a thickness of h = 5 mm. (**a**) The original data measured in four experiments performed under the same conditions. (**b**) The corrected dependencies after subtraction of the supposed signal drift. Curves, corresponding to the same experiments, are denoted by the same color in both panels.

The presented dependencies were measured in an experiment on indentation of a steel sphere with a radius of R = 30 mm into a highly adhesive TARNAC CRG N0505 rubber substrate. In the experiments, the indenter reached an indentation depth of d_{max} = 0.1 mm, then moved in the opposite direction up to d_{min} = −0.2 mm. Complete detachment of the indenter was observed in all experiments before it was pulled off to its highest position above the substrate (d_{min}). It is worth noting that none of the dependencies shown in Figure 6 exhibit hysteresis after a change in the direction of motion (compared to the data presented in Figure 3) because of the L-511.24AD00 backlash-free drives used in the upgraded equipment. The points of complete detachment during the pull-off phase can be detected for all dependencies shown in Figure 6a. On the bottom curve, it is denoted as point A. Although there was no contact at point A and the indenter was completely detached from the substrate, a non-zero normal force was detected by a force sensor due to its signal drift. The magnitude of this force and the related indentation depth are denoted as F_A and d_A, respectively. If during the experiment, drift is considered to be linear (as in top the 7 dependencies shown in Figure 5), then the corrected time dependence of the normal force ($F_N(t)$) can be easily defined as:

$$\widetilde{F}_N(t) = F_N(t) - \frac{F_A}{t_A}t, \tag{1}$$

where t_A is the time interval from the beginning of the experiment until complete detachment (point A). In the experiments shown in Figure 6, the velocity of the indenter was chosen in such a way that the full cycle of indentation for every curve lasted for 30 min, similarly to the signals shown in Figure 5. Thus, the path of the indenter during the measurements was $s = 2d_{max} + |d_{min}| = 0.4$ mm, whereas its velocity was $v = s/t = 0.4/1800 \approx 2.22 \cdot 10^{-4}$ mm/s or 0.222 μm/s. At this velocity magnitude, the contact can be considered quasi-static [35]. Dependencies corrected according to procedure (1) are shown in Figure 6b. As shown in the figure, all curves except the bottom line almost overlap in the indentation phase. Therefore, we can conclude that the bottom curve cannot be corrected through expression (1), as the signal drift during its measurements was non-linear (see also the three bottom lines in Figure 5a). The difference between three upper curves in Figure 6b during the pull-off phase can be explained by adhesion phenomena. Thus, adhesion slightly affects contact propagation during the indentation phase, whereas during detachment, it plays a crucial role [19,20,33–35]. During the indentation phase, dependence ($F(d)$) is determined only by elastic parameters (elastic modulus (E) and Poisson's ratio (ν)) of the rubber layer. These parameters determine the rigidity of a contact and are constant over time. Moreover, the specific work of adhesion significantly affects the pull-off phase. The specific work of adhesion, as a function of surface energies of contacting bodies, decreases for each new indentation cycle because of dirt and oxidation on the surfaces [21]. Therefore, the differences between the three upper $F_N(d)$ curves shown in Figure 6 in the pull-off phase are the result of natural causes that are not related to signal drift.

Figure 6 shows the results of an experiment on indentation of a smooth spherical indenter with monotonic $F_N(d)$ dependencies. However, if the indenter has noticeable roughness or an irregular geometric shape, the dependence of a normal force on indentation depth may be non-monotonic during the pull-off phase [20,21]. Therefore, it is important to understand whether such drastic changes in force are caused by physical processes in the contact area or whether these changes are measurement artefacts (an example of such an artefact is the third curve from the bottom in Figure 5a). For this purpose, it is necessary to analyze additional characteristics such as the configuration of the contact area. Any sharp changes in force (F_N) are, naturally, caused by reconfiguration of the contact area; thus, such reconfiguration must be observable. In our experiments, used a transparent substrate, which allows for direct observation of the contact area. Supplementary Video S2 shows an example of such an experiment (corresponding measurements are shown in Figure 6). In the video, both the original and corrected dependencies ($F_N(d)$) are shown, together with images of the contact area. It is worth mentioning that this example shows an experiment with a mirror-polished indenter; thus, the camera, lights and other equipment parts are reflected on its surface. This makes automatic calculation of the size of the contact area based on numerical analysis of digital photos impossible. Therefore, if the information on the size of a contact area is critically important (for example, for calculation of shear stresses), an indenter with low-amplitude roughness is preferred for experiments. In this case, the contact area can be identified more easily due to diffuse light scattering, and its size can be calculated automatically by software through analysis of the taken photos (for example, see [21] and related Supplementary Material).

Figure 7 shows the dependencies measured in the experiment under the same conditions as the data presented in Figure 6, only with a K3D40 force sensor. As the K3D40 force sensor has significantly smaller signal drift compared to the KD24S sensor (see also Figure 5), the dependencies ($F_N(d)$) that are shown in Figure 7a already depict the contact processes precisely enough. Removing the drift does not significantly affect the quality of the obtained results (see Figure 7b). It is worth noting that installed K3D40 force sensor operates within a force range of ±10 N; thus it is not as precise as the KD24S sensor (version with a range of ±2 N). Therefore the $F(d)$ dependencies measured by a K3D40 sensor exhibit a significant amount of noise. Nevertheless, in our experiments, the K3D40 force sensor was the preferred choice over the KD24S, as it does not require additional analysis of the obtained data or compensation for signal drift.

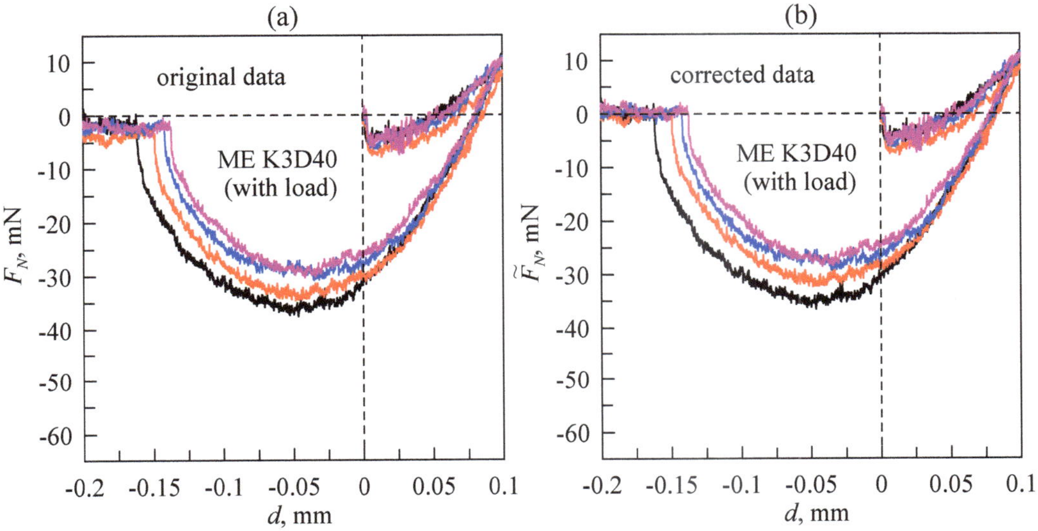

Figure 7. Dependencies of the normal force (F_N) on the indentation depth (d), similar to curves shown in Figure 6 but measured by the K3D40 force sensor. Original (**a**) and corrected (**b**) data. Curves, corresponding to the same experiments, are denoted by the same color in both panels.

3. Data Analysis and Interpretation

The Figures 3, 6 and 7 presented above contain the results of experiments involving normal indentation of steel spherical indenters into a soft elastomer. A general feature of all performed experiments is the difference between $F_N(d)$ dependencies related to the indentation phase and pull-off. Such behavior is typical for adhesive systems and assumed to be caused by several factors: roughness of the surfaces, viscoelasticity and the humidity effect, among others [40]. The exact cause of hysteresis after changing the direction of indenter movement has not been clearly established yet. JKR theory [41] is frequently used to describe the adhesive contact during the indentation of rigid spheres into soft elastomers. JKR determines the relations between the normal force (F_N), indentation depth (d) and contact radius (a) as [41]:

$$d(a) = \frac{a^2}{R} - \sqrt{\frac{2\pi a \Delta\gamma}{E^*}}, \tag{2}$$

$$F_N(a) = \frac{4E^* a^3}{3R} - \sqrt{8\pi a^3 E^* \Delta\gamma}, \tag{3}$$

where $\Delta\gamma$ is a specific work of adhesion, and $E^* = E/(1-\nu^2)$ is a reduced elastic modulus; parameters E and ν is the elastic modulus and Poisson's ratio of the elastomer. According to Equation (2), after the first contact at $d = 0$ mm, the contact area immediately propagates in size up to equilibrium radius

$$a(d=0) = \left(\frac{2\pi R^2 \Delta\gamma}{E^*}\right)^{1/3}, \tag{4}$$

where the corresponding adhesive force, according to Equation (3), is negative:

$$F_N(d=0) = -\frac{4}{3}\pi R\Delta\gamma. \tag{5}$$

The experiments described above consider the indentation of a steel indenter into a soft substrate, so the indenter can be considered as absolutely rigid, i.e., no strains occur in the indenter. However in the case in which the indenter undergoes deformation, relations (2) and (3) can still be used to describe the contact mechanics if the reduced elastic modulus is defined as $1/E^{*} = (1-\nu_1^2)/E_1 + (1-\nu_2^2)/E_2$, where indices 1 and 2 denote the materials of the indenter and half-space, respectively. It is worth noting that expressions (2) and (3) are valid within the half-space approximation when the contact radius (a) is significantly smaller than the thickness of the elastomer (h); otherwise, the elastomer thickness (h) must be taken into account. Several methods are available for this procedure, including, for instance, analytical [42] or numerical [43] modeling.

Figure 8 shows the $F_N(d)$ dependencies obtained in our experiments and discussed above. Each panel of the figure shows one cycle of indentation; Figure 8a relates to Figure 3b, Figure 8b relates to Figure 6b, and Figure 8c relates to Figure 7b. Besides the experimental dependencies (solid lines), Figure 8 also shows the results of a numerical simulation (dashed lines) performed according to the boundary elements method (BEM), where the thickness of the elastomer (h) is also taken into account [43]. In the BEM model, main parameters such as indenter radius (R), elastomer thickness (h) and indentation depth (d) are exactly the same as in the experiment.

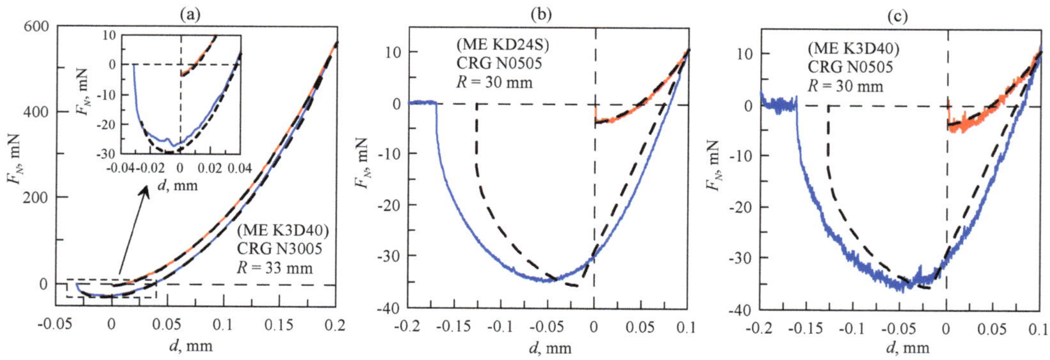

Figure 8. (**a–c**) Dependencies ($F_N(d)$) obtained in experiments with the designed facility (solid lines) compared to numerical simulation (dashed lines) performed within the BEM model. Each panel shows a different experiment. Dependencies shown in panels (**b**,**c**) were measured under the same conditions with different force sensors.

In the simulation, we assumed that the specific work of adhesion ($\Delta\gamma_1$) during the indentation phase was significantly smaller than $\Delta\gamma_2$ during the pull-off. The choice of different magnitudes of $\Delta\gamma$ for indentation and pull-off is a rather phenomenological approach; nevertheless it enables description of the behavior of the system that is observed in experiments, although the exact reasons of this behavior remain unknown for now. As follows from Figure 8, the results of simulations and experiments almost overlap in the indentation phase, whereas during the pull-off, there are clearly noticeable differences. The main reasons for these differences are the simulation conditions; the indenter has an ideal spherical shape, and the elastomer is absolutely flat. Moreover, the adhesion-specific work ($\Delta\gamma$) is evenly distributed over the surface (the same value at every point), and factors such as surface roughness, viscoelastic effects and others are not taken into account.

In numerical simulations for CRG N3005 material (Figure 8a), the following values of elastic parameters were used: E = 0.31 MPa and ν = 0.47, $\Delta\gamma_1$ = 0.027 J/m^2 during the indentation phase and $\Delta\gamma_2$ = 0.19 J/m^2 during pull-off. For CRG N0505 material (Figure 8b,c) the corresponding values were: E = 0.031 MPa, ν = 0.47, $\Delta\gamma_1$ = 0.027 J/m^2 and $\Delta\gamma_2$ = 0.24 J/m^2. In both cases, the magnitude of $\Delta\gamma_1$ related to the indentation phase is significantly smaller than $\Delta\gamma_2$ related to pull-off. However, even a low value of $\Delta\gamma_1$ ensures the propagation of the contact due to adhesion. After the moment of the first contact, the contact area expands (see Video S2), and the normal force is negative ($F_N(d = 0) < 0$ N). Negative force at zero indentation depth (d) during the indentation phase is clearly visible in Figure 8b,c. Dependencies in Figure 8a are shown in a wider range of a normal force so that the mentioned behavior is not visible; however it is clearly visible in the inset. Note that the adhesive strength of the contact (the maximum absolute value of the adhesive force during pull-off at $d < 0$ mm) is approximately the same for all cases and equals $F_A \approx 30$ mN. In the case of half-space when the JKR theory (2), (3) is valid, the maximum of the adhesive force (corresponding to the minimum of the $F_N(d)$ dependence during pull-off) is defined as:

$$|F_{\min}| = \frac{3}{2}\pi R\Delta\gamma. \qquad (6)$$

If a rigid sphere is indented into a relatively thin layer of elastomer (the contact radius is significantly larger than the elastomer thickness), the adhesive strength increases up to a magnitude of $|F_{\min}| = 2\pi R\Delta\gamma$ in the case in which the elastic film is frictionless on the interface between the film and the substrate. When the elastic film is perfectly bonded on the interface between the film and the substrate, the adhesive strength became even larger, i.e., $|F_{\min}| = 3\pi R\Delta\gamma$ [44,45]. Therefore, in the general case, adhesive force can be defined as $|F_{\min}| = \alpha\pi R\Delta\gamma$, where parameter α depends on the condition of the experiment. For the dependencies shown in Figure 8, $\Delta\gamma \approx 0.2$ J/m^2, $R \approx 30$ mm and $|F_{\min}| \approx 30$ mN, so $\alpha \approx 1.6$, which is very close to the JKR estimation (6).

4. Investigation of the Tangential Contact

In the proposed work, we describe the design and performance of the facility, the main purpose of which is to study tangential contacts. In the previous sections, we considered typical problems of the equipment (backlash of the drives, drift of the signal from the force sensors and the peculiarity of the lighting system) and suggested possible solutions. However, all the examples of device performance presented above were related to normal indentation. Hence, to complete our description, in this section, we consider an experiment involving tangential motion. In the mentioned experiment, an indenter with a radius of R = 40 mm was indented into a layer of rubber to a depth of d = 0.4 mm, followed by a shift in the tangential direction for x = 5 mm before returning to the previous position of x = 0 mm. Then, the indenter was lifted up in the normal direction to complete detachment.

The results of the experiment are shown in Figure 9. The figure shows time dependencies of all three components of the contact force, along with images of the contact area taken at certain moments of time.

In general, the presented dependencies are similar to the data described in our previous work [37]. Thus, after some time, the system exhibits a sliding mode with constant friction force (F_x). As the indenter moves only along the x direction, the lateral component (F_y) is expected equal zero in an ideal case; however, we observed non-zero values of this component. This situation arises for several reasons, such as non-precise positioning of the indenter, inhomogeneities in the contact area, etc. In real experiments, F_y should be small enough compared to F_x. As shown by the presented dependencies, this condition is fulfilled in our experiments. The main difference between the experiment shown in Figure 9 from that described in [37] is the chemical inhomogeneities on the surface of the indenter.

Inhomogeneous distribution of the surface energy was created by a short (approximately 1 min) treatment of the surface of the steel indenter with a water solution of $FeCl_3$; then, the indenter was washed with water, then washed with alcohol and dried. Treated areas of the indenter can be seen as dark spots in the photos shown in Figure 9. These spots are characterized by a stronger adhesive strength compared to the untreated area [21]. As it follows from the Figure 9, inhomogeneous distribution of the surface energy leads to more complex contact behavior; elastic waves and pores appear in the contact area, and contact is no longer simply connected. Figure 9 shows only part of images of the contact area, which are not enough to track the dynamics of the tangential shift. A complete illustration of this process is presented in the Supplementary Material (Video S3). The description and analysis of this processes is a non-trivial task, which is not the aim of this work and will be provided elsewhere.

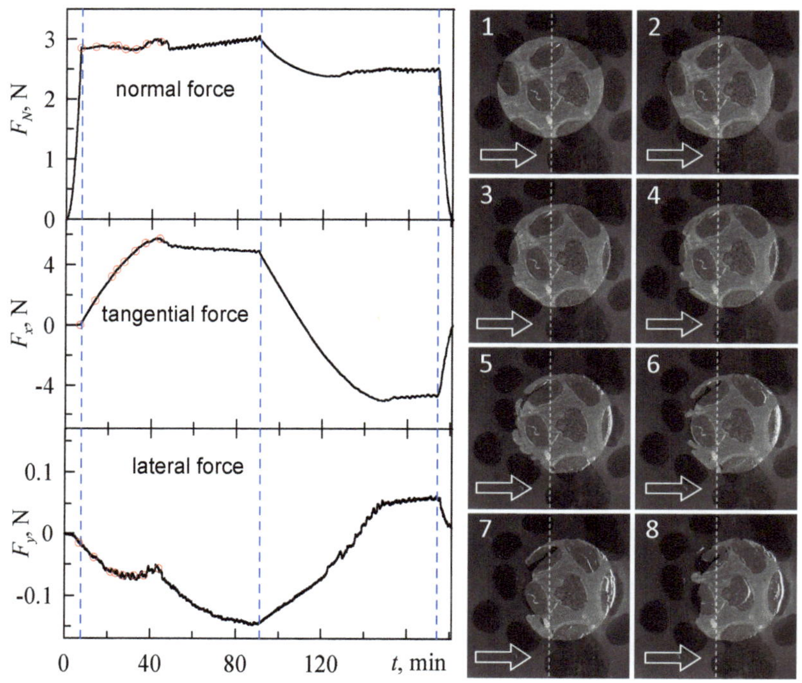

Figure 9. Time dependencies of all three components of the contact force ((**left**) panels) and related images of the contact area ((**right**) panels). Chemical inhomogeneities on the surface of the indenter in the (**right**) panel are pictured as dark spots. 8 red dots on the curves (**left** panel) show the time moments when the snapshots (**right** panel) were taken. Increasing numbers of the snapshots relate to growth of both tangential shift of the indenter and time from the start of the experiment.

It is worth noting that behavior similar to that described above was observed in experiments more than 50 years ago; waves that propagate within the contact area during tangential shift are known as Schallamach waves [46]. Although several theoretical and experimental studies of Schallamach waves have been conducted (see for example [47–49]), no complete explanation of the process exists yet. This situation is caused by the fact that well-known and popular theories such as JKR [41] are built for quasistatic contact and cannot be applied for description of fast dynamical processes. Moreover, as experiments show, the most complex behavior during wave propagation is observed at high values of the adhesion-specific work, as in this case, the rubber undergoes severe non-linear deformation prior to propagation of the first wave. Additional factors, such as liquid or particles in the contact area, destruction of the material, chemical inhomogeneities (see Figure 9 and Video S3), etc., also affect wave propagation. In our future works, we plan to study the behavior shown in Figure 9 more precisely.

5. Summary and Discussion

Various modifications of the described device were used in several experiments, the results of which have already been described in the literature [19–21,32–39]. In particular, in [19], experiments on contacts of rigid punches that have a flat but oddly shaped face with a soft adhesive layer were performed; in [20], the influence of the roughness of contacting surfaces on adhesive strength was studied; surfaces with artificially created chemical inhomogeneity were studied in [21]. Additionally, in [21], the presence of a third body in the contact area was considered. This particular case simulates wear debris in contacting parts of various mechanisms. Peculiarities of the sliding and rolling friction modes were studied in [32]; in [33], the dependence of adhesive strength on the duration of the contact was investigated; dissipation of the mechanical energy within oscillating adhesive contact was studied in [34]. Therein, it was shown that dissipation is not observed if the oscillation amplitude does not exceed a certain critical magnitude. In [35], an experimental study of the secondary adhesive hysteresis that occurs after changing the direction of motion of the indenter was conducted; contact of ellipsoidal flat forms with an elastomer substrate was studied in [36]. In particular, the influence of eccentricity of the ellipsoidal form on contact processes was investigated. How adhesion affects sliding friction (tangential contact) was studied in [37]. The transportation of particles driven by surfaces with periodic relief in tangential contact was studied in [38]. All experiments mentioned in this paragraph involved adhesive contacts.

It is worth mentioning that the sensitivity of the designed equipment is limited by the abilities of the installed force sensor, which can be increased by using more capable sensors. Furthermore, although functional modification of the device for certain types of experiment is possible, more advanced studies or experiments under unique conditions require a special type of tribometer. As notable examples of such tribometers, we can refer to a millitribometer designed to study small friction units and small loads [50], a six-axis force/torque sensor for slow friction measurements under ultra-high vacuum and other conditions [51] and a setup for the study of friction-induced vibrations [52] and haptics [53].

Application of the designed facility is not limited to studying adhesion. Thus, experiments performed in [39] considered capillary forces and their effect on the strength and shape of a contact. For this purpose, a drop of liquid was placed between contacting surfaces (a steel sphere and a glass sheet).

In the articles cited above, the designed equipment was used to study both normal and tangential contacts. However, as we described it in the previous section, a motorized rotation stage was recently on the facility. This particular upgrade adds additional degrees of freedom by making a rotation of the indented sample possible. Therefore, the upgraded setup will be used in our future experiments on the transition between friction and adhesion modes. Our previous experiments on tangential motion revealed the presence of a sliding mode with constant tangential mechanical stresses in the contact area. In this mode, the friction force depends on the size of the contact area, whereas the friction coefficient depends on the indentation depth; in this case, the friction coefficient cannot be considered a parameter characterizing friction. This is a well-known and typical behavior of friction contact of surfaces with high adhesion interaction [37,54]. However, in [54], it was experimentally shown that in such contacts, the friction mode (which can be characterized by a friction coefficient) is observable under large loads. Thus, in upcoming experiments, we plan to study the transition between two the mentioned modes by gradually increasing the indentation depth (contact load). Such experimental an condition can be achieved with the installed rotation stage. The rotation stage can provide the possibility of performing formally infinite motion, as it is not limited by the sliding range of a linear stage.

The designed facility will also be used in experiments with more complex load scenarios, in which contact modes with non-trivial behavior may occur. For instance, in a theoretical study [55], it was shown that the friction coefficient in a dry contact might depend on load history if rotation is present in the system. We plan to verify these particular theoretical results experimentally.

6. Conclusions

Herein, we proposed an intuitive strategy for designing and assembling a laboratory tribometer that uses a force sensor as the main measuring tool. The proposed approach based on previous experience concerning design of laboratory equipment and conducting tribology and contact mechanics experiments. We also discussed particular problems and suggested possible solutions that emerged during the use of the equipment. In particular, the signal drift of force sensors and the backlash of the indenter drives was discussed with respect to a number of measurements.

Supplementary Materials: The following supporting information can be downloaded at: https://www.mdpi.com/article/10.3390/s23010456/s1, Video S1: Example of a modification of the designed tribometer; Video S1 demonstrates the ability of the designed tribometer to function as a simple plotting device. For this purpose, an ordinary ballpoint pen was mounted instead of an indenter to act as a plotting tool, and a piece of paper was used as the substrate. The plotter was operated through an original software interface that was specially designed to manipulate the plotter from a desktop PC. As the video shows, the writing tool, operated by software, repeats the trajectory of a computer mouse pointer dragged within the special field in the screen. The resulting plot is shown at the end of the video. Video S2: Demonstration of the data corrections during indentation and detachment with visualization of the contact area in real time. In Video S2, images of the contact area configuration are shown, together with normal force and indentation depth measured simultaneously; thus, the contact area configuration and load curve are synchronized in real time. The video also demonstrates the difference between the corrected data and data measured with error due to signal drift during one full cycle of indentation. Video S3: Demonstration of tangential contact and sliding with visualization of the dynamics of the contact area synchronized with measured normal force, shear force and the friction coefficient in real time. Propagation of waves is clearly visible. The surface of the steel indenter had inhomogeneous distribution of the surface energy created through a short treatment of the indenter surface with water solution of ferric chloride. Treated areas of the indenter can be seen in the video as dark spots; these spots are characterized by a stronger adhesive strength compared to the untreated area.

Author Contributions: Conceptualization, supervision, project administration and writing—review and editing, V.L.P.; methodology, hardware, software, validation, experimental data analysis, visualization and writing—original draft preparation, I.A.L.; conceptualization, hardware and software, R.P.; data analysis and writing—original draft preparation, V.B. All authors have read and agreed to the published version of the manuscript.

Funding: This research was funded by Deutsche Forschungsgemeinschaft (Project DFG PO 810-55-3). V.B. is grateful to Technische Universität Berlin for support.

Institutional Review Board Statement: Not applicable.

Informed Consent Statement: Not applicable.

Data Availability Statement: The datasets generated for this study are available upon request from the corresponding authors.

Acknowledgments: The authors are grateful to Dipl.-Ing. Bernhard Zientek and Daniel Heppner for manufacturing custom parts for the equipment according to provided blueprints.

Conflicts of Interest: The authors declare no conflict of interest.

References

1. Binnig, G.; Quate, C.F.; Gerber, C. Atomic Force Microscope. *Phys. Rev. Lett.* **1986**, *56*, 930. [CrossRef] [PubMed]
2. Ohnesorge, F.; Binnig, G. True Atomic Resolution by Atomic Force Microscopy Through Repulsive and Attractive Forces. *Science* **1993**, *260*, 1451–1456. [CrossRef] [PubMed]
3. Farré, M.; Barceló, D. Chapter 1—Introduction to the Analysis and Risk of Nanomaterials in Environmental and Food Samples. In *Comprehensive Analytical Chemistry*; Farré, M., Barceló, D., Eds.; Elsevier: Amsterdam, The Netherlands, 2012; Volume 59, pp. 1–35. [CrossRef]
4. Brugger, J. Nanotechnology impact on sensors. *Nanotechnology* **2009**, *20*, 430206. [CrossRef] [PubMed]

5. Giessibl, F.J. High-speed force sensor for force microscopy and profilometry utilizing a quartz tuning fork. *Appl. Phys. Lett.* **1998**, *73*, 3956–3958. [CrossRef]
6. Stachowiak, G.W.; Batchelor, A.W.; Stachowiak, G.B. 3—Tribometers. In *Tribology Series*; Stachowiak, G.W., Batchelor, A.W., Stachowiak, G.B., Eds.; Elsevier: Amsterdam, The Netherlands, 2004; Volume 44, pp. 25–78. [CrossRef]
7. Singhal, S.; Agarwal, R.; Kumar, R.; Dwivedi, R. Critical Review on Tribometers and Their Contact Mechanism. In *Optimization of Industrial Systems*; Panchal, D., Tyagi, M., Sachdeva, A., Pamucar, D., Eds.; Scrivener Publishing LLC: Beverly, MA, USA, 2022; pp. 217–225. [CrossRef]
8. Randall, N.X. Experimental Methods in Tribology. In *Tribology for Scientists and Engineers*; Menezes, P., Nosonovsky, M., Ingole, S., Kailas, S., Lovell, M., Eds.; Springer: New York, NY, USA, 2013; pp. 141–175. [CrossRef]
9. Waters, J.F.; Guduru, P.R. Mode-mixity-dependent adhesive contact of a sphere on a plane surface. *Proc. R. Soc. A* **2010**, *466*, 1303–1325. [CrossRef]
10. Liu, Z.; Lu, H.; Zheng, Y.; Tao, D.; Meng, Y.; Tian, Y. Transient adhesion in a non-fully detached contact. *Sci. Rep.* **2018**, *8*, 6147. [CrossRef]
11. Degrandi-Contraires, E.; Poulard, C.; Restagno, F.; Léger, L. Sliding friction at soft micropatterned elastomer interfaces. *Faraday Discuss.* **2012**, *156*, 255–265. [CrossRef]
12. Ku, I.S.Y.; Reddyhoff, T.; Choo, J.H.; Holmes, A.S.; Spikes, H.A. novel tribometer for the measurement of friction in MEMS. *Tribol. Int.* **2010**, *43*, 1087–1090. [CrossRef]
13. Gautschi, G. Piezoelectric Sensors. In *Piezoelectric Sensorics*; Springer: Berlin/Heidelberg, Germany, 2002; pp. 73–91. [CrossRef]
14. Koike, R.; Sakaino, S.; Tsuji, T. Hysteresis Compensation in Force/Torque Sensors Using Time Series Information. *Sensors* **2019**, *19*, 4259. [CrossRef]
15. Prost, J.; Boidi, G.; Lebersorger, T.; Varga, M.; Vorlaufer, G. Comprehensive review of tribometer dynamics-Cycle-based data analysis and visualization. *Friction* **2022**, *10*, 772–786. [CrossRef]
16. Meng, Y.; Xu, J.; Jin, Z.; Prakash, B.; Hu, Y. A review of recent advances in tribology. *Friction* **2020**, *8*, 221–300. [CrossRef]
17. Valdastri, P.; Roccella, S.; Beccai, L.; Cattin, E.; Menciassi, A.; Carrozza, M.C.; Dario, P. Characterization of a novel hybrid silicon three-axial force sensor. *Sens. Actuators A Phys.* **2005**, *123-124*, 249–257. [CrossRef]
18. Available online: https://www.me-systeme.de (accessed on 17 November 2022).
19. Popov, V.L.; Pohrt, R.; Li, Q. Strength of adhesive contacts: Influence of contact geometry and material gradients. *Friction* **2017**, *5*, 308–325. [CrossRef]
20. Lyashenko, I.A.; Pohrt, R. Adhesion between rigid indenter and soft rubber layer: Influence of roughness. *Front. Mech. Eng.* **2020**, *6*, 49. [CrossRef]
21. Lyashenko, I.A.; Li, Q.; Popov, V.L. Influence of chemical heterogeneity and third body on adhesive strength: Experiment and simulation. *Front. Mech. Eng.* **2021**, *7*, 658858. [CrossRef]
22. Maderson, P.F.A. Keratinized epidermal derivatives as an aid to climbing in gekkonid lizards. *Nature* **1964**, *203*, 780–781. [CrossRef]
23. Autumn, K.; Liang, Y.A.; Tonia Hsieh, S.; Zesch, W.; Chan, W.P.; Kenny, T.W.; Fearing, R.; Full, R.J. Adhesive force of a single gecko foot-hair. *Natue* **2000**, *405*, 681–685. [CrossRef]
24. Autumn, K.; Sitti, M.; Liang, Y.A.; Peattie, A.M.; Hansen, W.R.; Sponberg, S.; Kenny, T.W.; Fearing, R.; Israelachvili, J.N.; Full, R.J. Evidence for van der Waals adhesion in gecko setae. *PNAS* **2002**, *99*, 12252–12256. [CrossRef]
25. Springer, T.A. Adhesion receptors of the immune system. *Natue* **1990**, *346*, 425–434. [CrossRef]
26. Hermansson, M. The DLVO theory in microbial adhesion. *Colloids Surf. B Biointerfaces* **1999**, *14*, 105–119. [CrossRef]
27. Berne, C.; Ellison, C.K.; Ducret, A.; Brun, Y.V. Bacterial adhesion at the single-cell level. *Nat. Rev. Microbiol.* **2018**, *16*, 616–627. [CrossRef] [PubMed]
28. Brake, S.J.; Barnsley, K.; Lu, W.; McAlinden, K.D.; Eapen, M.S.; Sohal, S.S. Smoking Upregulates Angiotensin-Converting Enzyme-2 Receptor: A Potential Adhesion Site for Novel Coronavirus SARS-CoV-2 (Covid-19). *J. Clin. Med.* **2020**, *9*, 841. [CrossRef] [PubMed]
29. Lin, H.; Ho, H.; Chen, C.; Yeh, T.; Sheu, M. Process and formulation characterizations of the thermal adhesion granulation (TAG) process for improving granular properties. *Int. J. Pharm.* **2008**, *357*, 206–212. [CrossRef] [PubMed]
30. da Silva, L.F.M.; Öchsner, A.; Adams, R.D. *Handbook of Adhesion Technology*, 2nd ed.; Springer: Berlin, Germany, 2018.
31. TARNAC. Available online: https://www.k-tanac.co.jp/crystalnone (accessed on 17 November 2022).
32. Lyashenko, I.A.; Popov, V.L. The Influence of Adhesion on Rolling and Sliding Friction: An Experiment. *Tech. Phys.* **2022**, *67*, 203–214. [CrossRef]
33. Lyashenko, I.A.; Popov, V.L. The effect of contact duration and indentation depth on adhesion strength: Experiment and numerical simulation. *Tech. Phys.* **2020**, *65*, 1695–1707. [CrossRef]
34. Lyashenko, I.A.; Popov, V.L. Dissipation of mechanical energy in an oscillating adhesive contact between a hard indenter and an elastomer. *Tech. Phys. Lett.* **2020**, *46*, 1092–1095. [CrossRef]
35. Lyashenko, I.A.; Popov, V.L. Hysteresis in an Adhesive Contact upon a Change in the Indenter Direction of Motion: An Experiment and Phenomenological Model. *Tech. Phys.* **2021**, *66*, 672–690. [CrossRef]
36. Li, Q.; Wilhayn, J.; Lyashenko, I.A.; Popov, V.L. Adhesive contacts of rough elliptical punches. *Mech. Res. Commun.* **2022**, *122*, 103880. [CrossRef]

37. Popov, V.L.; Li, Q.; Lyashenko, I.A.; Pohrt, R. Adhesion and friction in hard and soft contacts: Theory and experiment. *Friction* **2021**, *9*, 1688–1706. [CrossRef]
38. Li, Q.; Lyashenko, I.A.; Starcevic, J. An experimental study on third-body particle transport in sliding contact. *Facta Univ. Ser. Mech. Eng.* **2021**, *19*, 1–5. [CrossRef]
39. Popov, V.L.; Lyashenko, I.A.; Starcevic, J. Shape of a sliding capillary contact due to the hysteresis of contact angle: Theory and experiment. *Facta Univ. Ser. Mech. Eng.* **2021**, *19*, 175–185. [CrossRef]
40. Greenwood, J.A. Refections on and Extensions of the Fuller and Tabor Theory of Rough Surface Adhesion. *Tribol. Lett.* **2017**, *65*, 159. [CrossRef]
41. Johnson, K.L.; Kendall, K.; Roberts, A.D. Surface energy and the contact of elastic solids. *Proc. Royal Soc. Lond. A* **1971**, *324*, 301–313. [CrossRef]
42. Argatov, I.I.; Sabina, F.J. Asymptotic analysis of the substrate effect for an arbitrary indenter. *Q. J. Mech. Appl. Math.* **2013**, *66*, 75–95. [CrossRef]
43. Li, Q.; Pohrt, R.; Lyashenko, I.A.; Popov, V.L. Boundary element method for nonadhesive and adhesive contacts of a coated elastic half-space. *Proc. IMechE Part J J. Eng. Tribol.* **2020**, *234*, 73–83. [CrossRef]
44. Yang, F. Adhesive contact between a rigid axisymmetric indenter and an incompressible elastic thin film. *J. Phys. D: Appl. Phys.* **2002**, *35*, 2614–2620. [CrossRef]
45. Papangelo, A. Adhesion between a power-law indenter and a thin layer coated on a rigid substrate. *Facta Univ. Ser. Mech. Eng.* **2018**, *16*, 19–28. [CrossRef]
46. Schallamach, A. How does rubber slide? *Wear* **1971**, *17*, 301–312. [CrossRef]
47. Barquins, M. Sliding Friction of Rubber and Schallamach Waves—A Review. *Mater. Sci. Eng.* **1985**, *73*, 45–63. [CrossRef]
48. Rand, C.J.; Crosby, A.J. Insight into the periodicity of Schallamach waves in soft material friction. *Appl. Phys. Lett.* **2006**, *89*, 261907. [CrossRef]
49. Koushik, V.; Srinivasan, C. Fifty years of Schallamach waves: From rubber friction to nanoscale fracture. *Phil. Trans. R. Soc. A* **2022**, *380*, 20210339. [CrossRef]
50. Myshkin, N.K.; Goryacheva, I.G.; Grigoriev, A.Y.; Kavaliova, I.N.; Makhovskaya, Y.Y. Contact Interaction in Precision Tribosystems. *J. Frict. Wear* **2020**, *41*, 191–197. [CrossRef]
51. Guibert, M.; Oliver, C.; Durand, T.; Le Mogne, T.; Le Bot, A.; Dalmas, D.; Scheibert, J.; Fontaine, J. A versatile flexure-based six-axis force/torque sensor and its application to tribology. *Rev. Sci. Instrum.* **2021**, *92*, 085002. [CrossRef]
52. Tonazzi, D.; Passafiume, M.; Papangelo, A.; Hoffmann, N.; Massi, F. Numerical and experimental analysis of the bi-stable state for frictional continuous system. *Nonlinear Dyn.* **2020**, *102*, 1361–1374. [CrossRef]
53. Vardar, Y.; Güçlü, B.; Basdogan, C. Effect of Effect of Waveform in Haptic Perception of Electrovibration on Touchscreens. In *Haptics: Perception, Devices, Control, and Applications*; Bello, F., Kajimoto, H., Visell, Y., Eds.; Springer: Cham, Switzerland, 2016; Volume 9774, pp. 190–203. [CrossRef]
54. Homola, A.M.; Israelachvili, J.N.; McGuiggan, P.M.; Gee, M.L. Fundamental experimental studies in tribology: The transition from "interfacial" friction of undamaged molecularly smooth surfaces to "normal" friction with wear. *Wear* **1990**, *136*, 65–83. [CrossRef]
55. Pohrt, R. Friction Influenced by Vibrations: A Refined Contact-Mechanics View on Lateral and Rotational Oscillations. *Front. Mech. Eng.* **2020**, *6*, 566440. [CrossRef]

sensors

Article

3D Metrology Using One Camera with Rotating Anamorphic Lenses

Xiaobo Chen [1,*], Jinkai Zhang [2,*] and Juntong Xi [1]

[1] School of Mechanical Engineering, Shanghai Jiao Tong University, Shanghai 200240, China
[2] School of Mechanical Engineering, Jinan University, Jinan 250022, China
* Correspondence: xiaoboc@sjtu.edu.cn (X.C.); me_zhangjk@ujn.edu.cn (J.Z.)

Abstract: In this paper, a novel 3D metrology method using one camera with rotating anamorphic lenses is presented based on the characteristics of double optical centers for anamorphic imaging. When the anamorphic lens rotates $-90°$ around its optical axis, the 3D data of the measured object can be reconstructed from the two anamorphic images captured before and after the anamorphic rotation. The anamorphic lens imaging model and a polynomial anamorphic distortion model are firstly proposed. Then, a 3D reconstruction model using one camera with rotating anamorphic lenses is presented. Experiments were carried out to validate the proposed method and evaluate its measurement accuracy. Compared with stereo vision, the main advantage of the proposed 3D metrology approach is the simplicity of point matching, which makes it suitable for developing compact sensors for fast 3D measurements, such as car navigation applications.

Keywords: 3D reconstruction; anamorphic lens; anamorphic stereo vision

1. Introduction

In modern industry, 3D metrology is an important technology, based on various methods. Optical metrology and non-optical metrology are the two categories of 3D metrology [1]. Non-optical 3D metrology methods may use a coordinate measuring machine (CMM), a scanning electron microscope, and a scanning probe microscope [2,3], whereas optical 3D metrology methods may employ chromatic confocal microscopy, point autofocus instruments, focus variation instruments, phase-shifting interferometry, coherence scanning interferometry, imaging confocal microscopy, and stereo vision. An overview of such methods was presented in detail [4–6]. Each 3D metrology method has its own properties and differs in application areas, measurement accuracy, scale, efficiency, cost, etc. Optical metrology procedures are often fast, precise, and non-destructive of the measured objects. Stereo vision, which uses two cameras with spherical lenses, is a frequently used optical metrology approach [7–9]. The measuring precision of stereo vision can range from several microns to several millimeters, depending on the measured area and the spherical lenses used. The most challenging aspect of stereo vision involves the extraction of the corresponding points from stereo images. Projectors are widely used to generate structured light to achieve high-quality corresponding points [10]. Stereo vision has been widely used in many areas, such as industry metrology, agriculture, daily living, etc. [11,12]

Anamorphic lenses are increasingly employed in the film industry to capture wide images for broad screens [13–15]. Anamorphic lenses are distinguished by two characteristics: double focal lengths and double optical centers [16–18]. Double focal lengths means the focal lengths differ greatly in the tangential and sagittal planes. This characteristic of anamorphic lenses is often used in the film industry, as it allows images to be compressed horizontally. Metrology, physics, and some optical imaging applications also use anamorphic lenses [19–22]. Double optical centers means there are two optical centers, one in the tangential plane and the other in the sagittal plane. The distance between the two optical centers is the anamorphic distance, which is an intrinsic quantity for anamorphic lenses. The anamorphic distance will provide anamorphic images with more 3D information, making anamorphic lenses considerably more suitable for 3D metrology. H. Durko used anamorphic lenses for measuring parts with large aspect ratios [23]. F. Blais performed an exploratory calibration of anamorphic lenses based on a pinhole imaging model [24]. In our previous work, a high-precision anamorphic lens calibration method was proposed with 3D and 2D calibration targets [16,25].

In this study, a novel 3D metrology method using one camera with rotating anamorphic lenses is proposed based on its characteristic of double optical centers. When the anamorphic lens rotates -90° around its optical axis, the 3D data of the measured object can be reconstructed from the two anamorphic images captured before and after the anamorphic rotation. The rest of this paper is organized as follows. Section 2 presents the anamorphic lens imaging model as well as the anamorphic distortion model. Section 3.1 describes the 3D reconstruction process employing rotating anamorphic lenses. Section 3.2 describes the point matching approach, and Section 3.3 depicts the rotating anamorphic stereo vision. Section 4 describes the experiments and evaluates the proposed method's measurement accuracy. Finally, Section 5 concludes the paper.

2. Anamorphic Imaging Model

This section presents the imaging model and the distortion model of anamorphic lenses, which differ significantly from those based on spherical lenses. Anamorphic lenses are typically composed of rear spherical lenses on the back and an anamorphic attachment on the front. The anamorphic attachment consists of cylindrical lenses that have no optical power in planes parallel to the cylindrical axis but have optical power in planes perpendicular to it.

2.1. Anamorphic Imaging Model

The imaging model of anamorphic lenses differs from the traditional pinhole imaging model of spherical lenses. The anamorphic distance and the anamorphic angle are introduced as new extra intrinsic parameters [16]. Figure 1 depicts the imaging model of anamorphic lenses, where *ad* denotes the anamorphic distance, and *aa* denotes the anamorphic angle. If the anamorphic lens is precisely assembled, the anamorphic angle *aa*, as shown in Figure 1, can be as small as zero. The cylindrical lenses or the anamorphic attachment determine the anamorphic distance *ad*, which must be precisely calibrated. O_W-$X_WY_WZ_W$ are the world coordinates, O_{ci}-$X_{ci}Y_{ci}Z_c$ are the camera coordinates centered in the CCD plane, and O_p-$X_pY_pZ_c$ are the pixel coordinates centered in the CCD plane. O_{cx} and O_{cy} are the optical centers in the horizontal plane and the vertical plane, and f_x and f_y are the focal lengths in the two planes. The imaging model for anamorphic lenses can be expressed as follows:

$$\begin{gathered}\begin{bmatrix} X_C \\ Y_C \\ Z_C \end{bmatrix} = \begin{bmatrix} R & T \end{bmatrix} \begin{bmatrix} X_W \\ Y_W \\ Z_W \\ 1 \end{bmatrix} \\ \frac{Y_I}{f_y} = \frac{Y_C}{Z_C+ad} \\ \frac{X_I}{f_x} = \frac{X_C}{Z_C} \\ \begin{bmatrix} X_P \\ Y_P \end{bmatrix} = \begin{bmatrix} \cos(aa) & \sin(aa) \\ -\sin(aa) & \cos(aa) \end{bmatrix} \begin{bmatrix} X_I \\ Y_I \end{bmatrix}\end{gathered} \tag{1}$$

where $[X_w; Y_w; Z_w]$ is a point in the world coordinates, $[X_c; Y_c; Z_c]$ is the point expressed in the anamorphic coordinates. R and T denote the rotational and translating matrix from the world coordinates to the anamorphic coordinates. (X_I, Y_I) are the image coordinates in the camera coordinates in the image plane, and (X_P, Y_P) are the image coordinates in the pixel coordinates in the image plane.

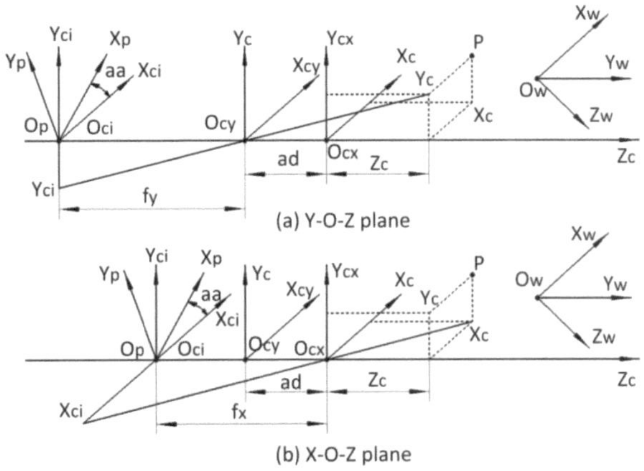

Figure 1. Anamorphic imaging model. (**a**) Imaging rays in the Y_c-O_{cy}-Z_c plane; (**b**) imaging rays in the X_c-O_{cx}-Z_c plane.

2.2. Anamorphic Distortion Model

Anamorphic lenses have a distinct distortion model with respect to spherical lenses. The front anamorphic attachment introduces many more aberrations [26]. Based on the aberration theory and numerical experiments for anamorphic lenses, we proposed a polynomial distortion model for anamorphic lenses, which allows for high-precision camera calibration [25]. As shown in Equation (2), anamorphic distortions can be divided into radial distortions (X_{rad}, Y_{rad}), third-order distortions (X_3, Y_3), and second-order distortions (X_2, Y_2). (x_d, y_d) denotes the distorted image coordinates, and (x_c, y_c) denotes the undistorted image coordinates. $[k_1, k_2, n_1, n_2, m_1, m_2, x_{21}, x_{12}, x_{03}, x_{20}, x_{11}, x_{02}, y_{30}, y_{21}, y_{12}, y_{03}, y_{20}, y_{11}, y_{02}]$ are the 19 distortion coefficients for anamorphic lenses that must be calibrated.

$$
\begin{aligned}
x_c = {} & \underbrace{x_d\left(1+k_1\left(x_d^2+y_d^2\right)+k_2\left(x_d^2+y_d^2\right)^2+m_1\left(a^2x_d^2+y_d^2\right)+m_2\left(a^2x_d^2+y_d^2\right)^2\right)}_{Xrad} \\
& \underbrace{+x_{21}x_d^2y_d+x_{12}x_dy_d^2+x_{03}y_d^3}_{X_3} \\
& \underbrace{+x_{20}x_d^2+x_{11}x_dy_d+x_{02}y_d^2}_{X_2} \\
y_c = {} & \underbrace{y_d\left(1+n_1\left(x_d^2+y_d^2\right)+n_2\left(x_d^2+y_d^2\right)^2+m_1\left(a^2x_d^2+y_d^2\right)+m_2\left(a^2x_d^2+y_d^2\right)^2\right)}_{Yrad} \\
& \underbrace{+y_{30}x_d^3+y_{21}x_d^2y_d+y_{12}x_dy_d^2+y_{03}y_d^3}_{Y_3} \\
& \underbrace{+y_{20}x_d^2+y_{11}x_dy_d+y_{02}y_d^2}_{Y_2}
\end{aligned}
\tag{2}
$$

3. 3D Metrology Using Rotating Anamorphic Lenses

3.1. Description of 3D Metrology Using Rotating Anamorphic Lenses

As shown in Figure 2a, the vertical position is the first anamorphic position, and the cylindrical axis of the anamorphic attachment is vertical. After a $-90°$ rotation, the cylindrical axis of the anamorphic lens is horizontal, and this anamorphic position is known as the horizontal position. The anamorphic lens can switch between the two positions by rotating. The imaging model of anamorphic lenses shown in Section 2 can be reduced to an ideal anamorphic lens for ease of analysis, which implies that the anamorphic angle is zero, there are no distortions, and the rotation angle is precisely $-90°$ without decentering. As shown in Figure 1 and Equation (1), the following equations describe the vertical position of an ideal anamorphic lens:

$$\begin{aligned} x_v &= \frac{X_C}{Z_C} f_x \\ y_v &= \frac{Y_C}{(Z_C + ad)} f_y \end{aligned} \tag{3}$$

where (x_v, y_v) refer to the image position of the object point $P_V = (X_C, Y_C, Z_C)$, and P_V is in the anamorphic coordinates in the vertical position. After the anamorphic rotation, the coordinates of the object point in the anamorphic coordinates change as follows:

$$P_H = RP_V \tag{4}$$

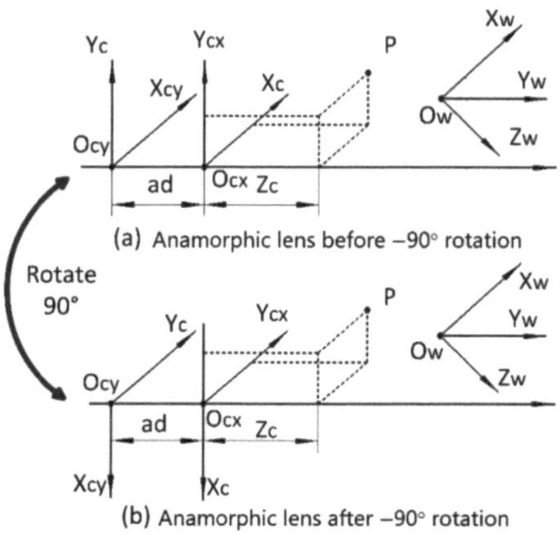

Figure 2. The two anamorphic positions. (**a**) Vertical position and (**b**) horizontal position which is achieved by a rotation of the anamorphic lens in the vertical position by $-90°$.

If the anamorphic lens rotates alone its optical axis by $-90°$, R can be expressed as:

$$R = \begin{bmatrix} 0 & 1 & 0 \\ -1 & 0 & 0 \\ 0 & 0 & 1 \end{bmatrix} \tag{5}$$

Thus, when the anamorphic lens is rotated to its horizontal position, the object coordinates can be calculated as $P_H = [Y_C; -X_C; Z_C]$, and we have the following equations:

$$\begin{aligned} x_h &= \frac{Y_C}{Z_C} f_x \\ y_h &= -\frac{X_C}{(Z_C + ad)} f_y \end{aligned} \tag{6}$$

Given Equations (3) and (6), the coordinates of the object point can be calculated as:

$$\begin{aligned} X_C &= ad \frac{x_v y_v}{f_y x_h - f_x y_v} \\ Y_C &= ad \frac{x_h y_v}{f_y x_h - f_x y_v} \\ Z_C &= ad \frac{f_x y_v}{f_y x_h - f_x y_v} \end{aligned} \tag{7}$$

Thus, given the internal parameters f_x, f_y, and ad, as well as the image coordinates (x_v, y_v) and (x_h, y_h) of an object point in an ideal anamorphic lens, the 3D coordinates of this point can be easily reconstructed using Equation (7).

3.2. Point Matching

Point matching is crucial for stereo vision but is also challenging because the two images are captured by two different cameras. Points in different positions will have different image points in stereo images. Epipolar constraints and projectors are commonly adopted to achieve robust point matching, which requires a complex calculation. Point matching limits the applications of stereo vision in areas requiring fast or real-time 3D reconstruction, such as car navigation. Compared with stereo vision, point matching is much simpler in our proposed 3D metrology method. A simulation is provided to demonstrate its image point matching, as shown in Figures 3–6. The simulated anamorphic lens has the following internal parameters: ad = 30 mm, $aa = 0°$, f_x = 12 mm, f_y = 16 mm, and no distortions. As shown in Figure 3, the object to be built is a collection of points on a spherical surface, with their coordinates given in the vertical position of the anamorphic lens. The simulated image points on the image planes are shown in Figure 4, where the dot points and the circle points express the image points when the anamorphic lens is in the vertical and horizontal positions, respectively. It is not easy to match the image points from these two positions in the current stage. In our method, the point matching can be greatly simplified by using a parameter of an anamorphic lens known as the anamorphic ratio AR, which is:

$$AR \approx \frac{f_y}{f_x} \tag{8}$$

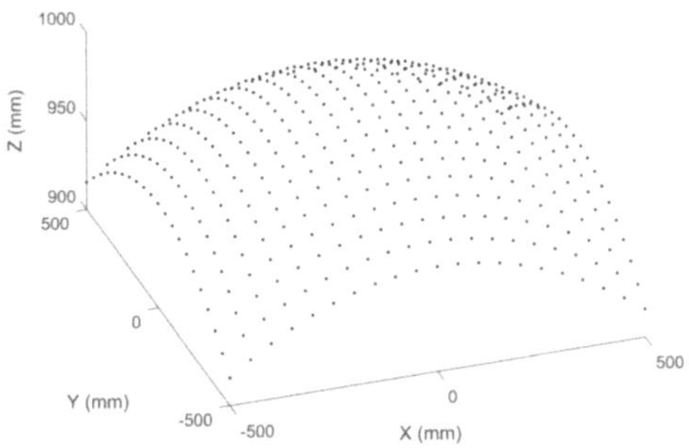

Figure 3. Simulated object points on a spherical surface in anamorphic coordinates of the vertical position.

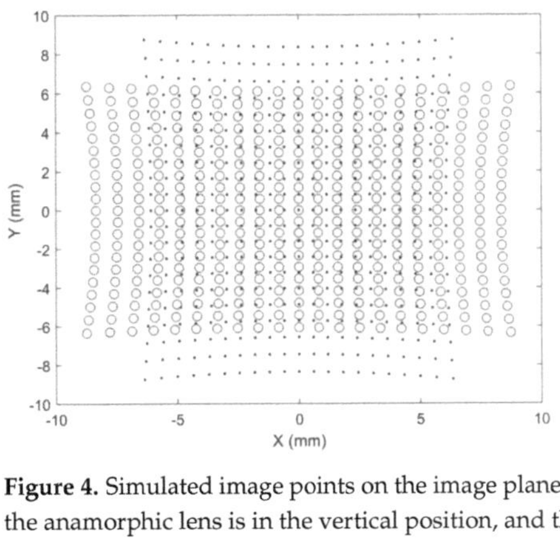

Figure 4. Simulated image points on the image plane. The dot points refer to the image points when the anamorphic lens is in the vertical position, and the circle points refer to the image points when the anamorphic lens is in the horizontal position.

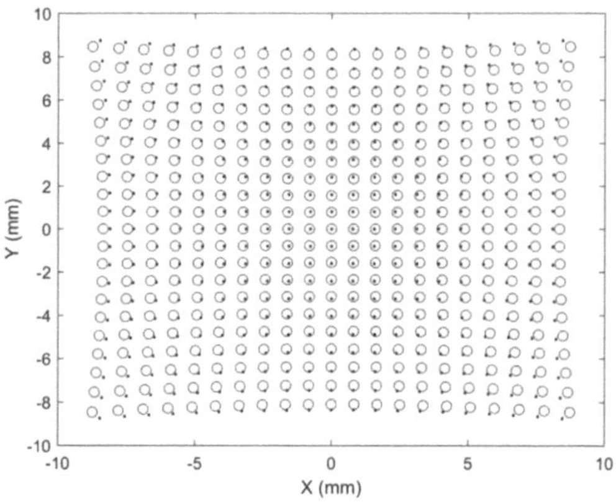

Figure 5. Simulated image points after anamorphic ration (AR) expansion. The dot points are rectified horizontally by AR, and the circle points are rectified vertically by AR.

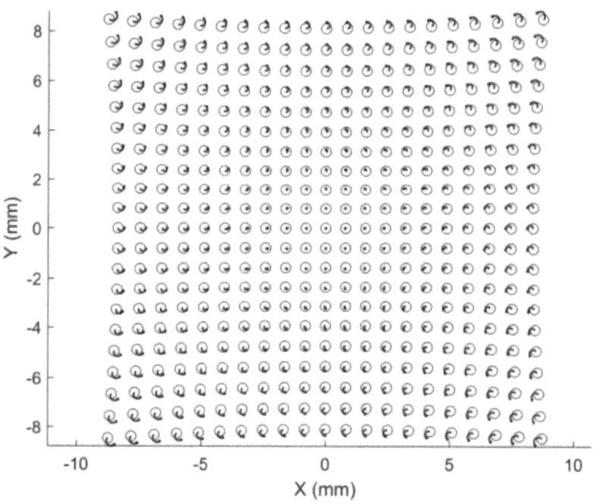

Figure 6. Simulated image points tracing from the vertical position to the horizontal position.

As a result, if the dot points and circle points in Figure 4 are expanded horizontally and vertically by AR, respectively, we obtain the corresponding image points shown in Figure 5. Because the corresponding points are nearby dot points and circle points, point matching in rotating anamorphic lenses becomes a simple and straightforward task. In addition, as illustrated in Figure 6, we can trace the image positions during the anamorphic rotation from the dot points to the circle points for point matching. Point matching in anamorphic rotating 3D metrology is much easier than stereo vision with spherical lenses, as shown in Figures 5 and 6. The deviations between the dot points and the circle points are due to the anamorphic distance ad. In this example, the anamorphic distance of 30 mm was much smaller than the object distance of 1500 mm. The deviation between the dot points and the circles in the image planes will increase when the object points are imaged using an anamorphic lens with a large anamorphic distance.

3.3. Stereo Vision with Anamorphic Lenses

Inevitably, there will be a misalignment between the optical axis of the anamorphic lens and the rotation axis of the rotary table, which will be magnified if the object is far away from the lens. As a result, large errors will occur if Equation (7) is applied directly for 3D reconstruction. A rational approach for high-precision 3D metrology is to treat the rotating anamorphic lenses as two distinct anamorphic lenses, which constitute a stereo vision with anamorphic lenses. The main differences between stereo vision and stereo vision with anamorphic lenses are that spherical lenses are replaced by anamorphic lenses and the baseline distance between the two anamorphic lenses is extremely small.

For 3D metrology using rotating anamorphic lenses, we should first calibrate the internal parameters of the anamorphic lenses according to the anamorphic imaging model and the anamorphic distortion model. A 3D calibration target can be used to calibrate the anamorphic lens [16]. As can be seen in Equation (2), the distortion model of the anamorphic lens has 19 unknown distortion coefficients. If only one image of a 3D calibration target is used for anamorphic lens calibration, local convergence is likely to occur for the distortion coefficients. Commonly, the 3D calibration target cannot cover the entire imaging area, which will lead to a poor calibration result for other imaging areas. Therefore, to achieve a stable calibration result for the entire imaging area, mixed calibration targets are adopted, which means 3D calibration targets and 2D calibration targets are both required [25]. The initial values of the internal parameters are calibrated using the 3D calibration target, and those parameters are refined using 2D calibration targets. After anamorphic calibration using mixed calibration targets, the internal anamorphic lens parameters [f_x, f_y, u_0, v_0, ad, aa] and the 19 distortion coefficients [k_1, k_2, n_1, n_2, m_1, m_2, x_{21}, x_{12}, x_{03}, x_{20}, x_{11}, x_{02}, y_{30}, y_{21}, y_{12}, y_{03}, y_{20}, y_{11}, y_{02}] can be determined.

After anamorphic lens calibration, we should calibrate the relative position of the anamorphic lens in the vertical position and the horizontal position. Two images of the 3D calibration target are captured, one for the vertical position and one for the horizontal position. Based on the calibrated internal parameters of the anamorphic lens and the image of the 3D calibration target, the relative position [t_x, t_y, t_z, a, b, r] between the 3D calibration target and the anamorphic lens can be easily calibrated:

$$T = \begin{bmatrix} t_x \\ t_y \\ t_z \end{bmatrix} \tag{9}$$

$$R = \begin{bmatrix} \cos(b)\cos(r) & \sin(a)\sin(b)\cos(r) - \cos(a)\sin(r) & \sin(a)\sin(r) + \cos(a)\sin(b)\cos(r) \\ \cos(b)\sin(r) & \cos(a)\cos(r) + \sin(a)\sin(b)\sin(r) & \cos(a)\sin(b)\sin(r) - \sin(a)\cos(r) \\ -\sin(b) & \cos(b)\sin(a) & \cos(a)\cos(b) \end{bmatrix} \tag{10}$$

where T and R refer to the translation vector and the rotation matrix. For the two anamorphic positions, there are the following equations between the coordinates of the 3D calibration target and the two anamorphic coordinates in the two positions:

$$\begin{aligned} P_{CV} &= R_{CV}^{W} P_W + T_{CV}^{W} \\ P_{CH} &= R_{CH}^{W} P_W + T_{CH}^{W} \end{aligned} \tag{11}$$

where the superscript W refers to the coordinates of the 3D calibration target, and the subscripts CV and CH refer to the coordinates of the anamorphic lens in the vertical and horizontal positions. Given Equation (11), the relationship between the vertical and the horizontal anamorphic coordinates can be determined as follows:

$$P_{CH} = R_{CH}^{CV} P_{CV} + T_{CH}^{CV} \tag{12}$$

where R_{CH}^{CV} and T_{CH}^{CV} are determined as follows:

$$\begin{aligned} R_{CH}^{CV} &= R_{CH}^{W} \left(R_{CV}^{W}\right)^{-1} \\ T_{CH}^{CV} &= T_{CH}^{W} - R_{CH}^{CV} T_{CV}^{W} \end{aligned} \tag{13}$$

The horizontal position is rotated from the vertical position by $-90°$ along the optical axis; thus, the relative position between the two anamorphic positions is similar to [t_x, t_y, t_z, a, b, r] = [0 mm, 0 mm, 0 mm, 0°, 0°, 90°]. After the anamorphic lens calibration and the stereo calibration for the rotating anamorphic lenses, the 3D reconstruction using rotating anamorphic stereo vision is similar to that in stereo vision with spherical lenses.

4. Experiments

4.1. Experiments

An experiment was performed to validate the proposed 3D metrology method using one camera with rotating anamorphic lenses. The design of the experiment is shown in Figure 7. The camera was MER-504-10Gx-P from Daheng China, and its resolution was 2448 × 2448 with a pixel size of 3.45 μm × 3.45 μm. The anamorphic lens was mounted on the rotary table HK130-10. The servo motor was SGM7J-02A7C6E, and the servo controller was SGD7S-1R6A00A002, both from YASKAWA, Japan. As shown in Figure 7, the 3D calibration target was composed of two 2D planar calibration targets, and the relative position of the two planar calibration targets was determined by the 3D calibration target camera, which was previously calibrated. As shown in Figure 8, the anamorphic lens was composed of a front anamorphic attachment and a 16 mm Computar spherical lens. As shown in Figure 9, the anamorphic attachment comprised three cylindrical lens elements, and the effective focal length in the YOZ plane was −2489 mm. A paraxial lens design for anamorphic lenses can be seen in [27,28]. The lens parameters for the anamorphic attachment are shown in Table 1.

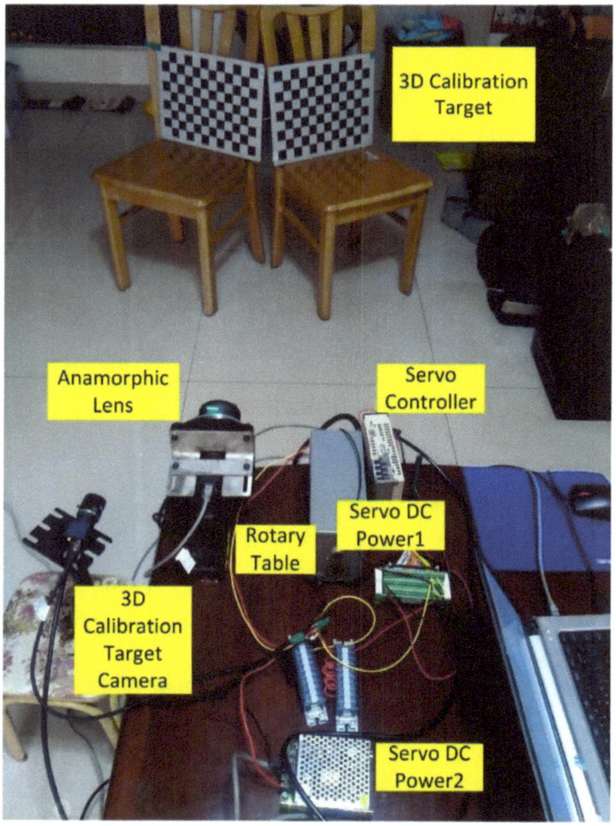

Figure 7. Experiment of 3D metrology using rotation anamorphic lenses.

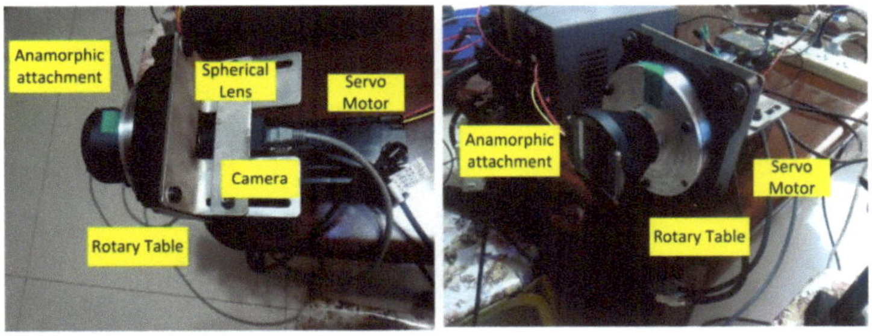

Figure 8. Anamorphic lens composed of a front anamorphic attachment and a rear spherical lens. The anamorphic lens was mounted on a rotary table which could rotate the anamorphic lens by −90° along the optical axis. (**a**) Side view; (**b**) Front view.

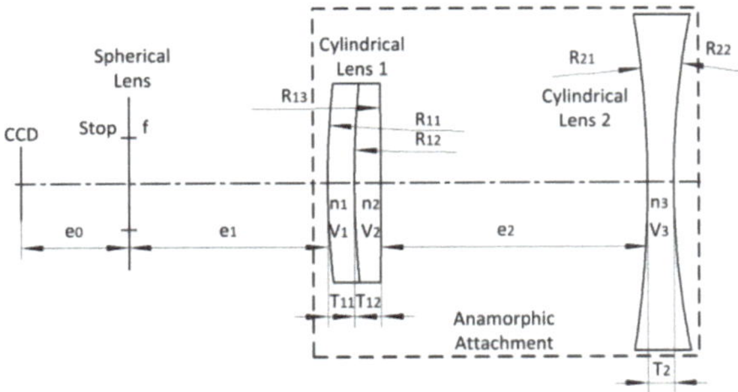

Figure 9. Anamorphic attachment for a Computar 16 mm spherical lens in the YOZ plane.

Table 1. Parameters of the anamorphic attachment used for the anamorphic lens.

e_0 (mm)	e_1 (mm)	e_2 (mm)	T_{11} (mm)	T_{12} (mm)	T_2 (mm)
16.163062	30	40	4	4	4
f (mm)	R_{11} (mm)	R_{12} (mm)	R_{13} (mm)	R_{21} (mm)	R_{22} (mm)
16	155	156.4	−857.6	−174.8	124.5
n_1	V_1	n_2	V_2	n_3	V_3
1.516797	64.212351	1.672702	32.17888	1.516797	64.212351

The anamorphic lens was calibrated using 3D and 2D calibration targets, and the internal parameters are shown in Table 2. And as shown in Figure 7, the relative position of the vertical position with respect to the horizontal position was calibrated using the 3D calibration target, and the results were $[t_x, t_y, t_z, a, b, r]$ = [−0.8943 mm, 0.3591 mm, −0.1712 mm, −0.5586°, −0.8074°, 90.0059°]. After the anamorphic lens calibration and the rotating anamorphic stereo calibration, we conducted the 3D metrology using the rotating anamorphic lens. As shown in Figure 10, to achieve dense 3D points, a checkerboard was adopted with a square's length of 5 mm. The original anamorphic images are shown in Figure 10. Figure 10a shows the image when the anamorphic lens was in the vertical position, and Figure 10b shows the image when the anamorphic lens was in the horizontal position. Figure 10b was achieved by rotating the anamorphic lens by −90° anticlockwise along the optical axis. Once the corresponding corners in Figure 10a,b were determined, the 3D coordinates of the corners could be reconstructed from the rotating anamorphic stereo vision.

Table 2. Internal parameters of the calibrated anamorphic lens.

f_x (mm)	f_y (mm)	u_0 (pixel)	v_0 (pixel)	ad (mm)
12.0520	16.1026	1.2790×10^3	1.0023×10^3	26.5516
aa (°)	k_1	k_2	n_1	n_2
−0.5789	0.0304	-2.7151×10^{-5}	0.0183	-3.3372×10^{-5}
m_1	m_2	x_{21}	x_{12}	x_{03}
−0.0155	1.277e-5	9.1986×10^{-5}	−0.0139	1.8928×10^{-5}
y_{30}	y_{21}	y_{12}	y_{03}	x_{20}
1.6727×10^{-5}	0.011	1.0611×10^{-4}	−0.0023	1.3802×10^{-4}
x_{11}	x_{02}	y_{20}	y_{11}	y_{02}
-4.7071×10^{-5}	5.8709×10^{-5}	9.3657×10^{-5}	-3.9944×10^{-4}	-2.9851×10^{-4}

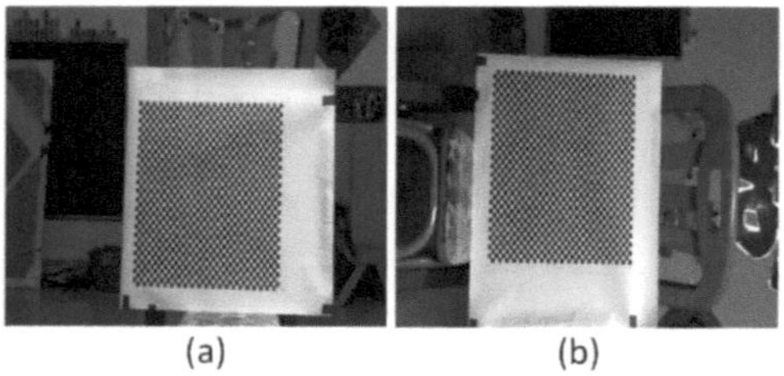

Figure 10. Anamorphic images. (**a**) Image when the anamorphic lens was in the vertical position, and (**b**) image when the anamorphic lens was in the horizontal position.

One possible method to determine the corresponding corners between Figure 10a,b is to trace the corners while rotating the anamorphic lens, similar to Figure 6. In order to trace the corners correctly, the rotating speed must be limited. Another method is to rectify the two anamorphic images by the anamorphic ratio *AR*, as shown in Equation (6), after which the corresponding points are very close, as in Figure 5. Figure 11a was obtained by expanding Figure 10a horizontally by the *AR*. Rotating Figure 10b by −90° anticlockwise and expanding the height of the rotated image by the *AR*, we obtained Figure 11b. As can be seen in Figure 11a,b, the two images are very similar after *AR* rectification, and the corresponding points are nearby, for the overlapping areas in Figure 11a,b.

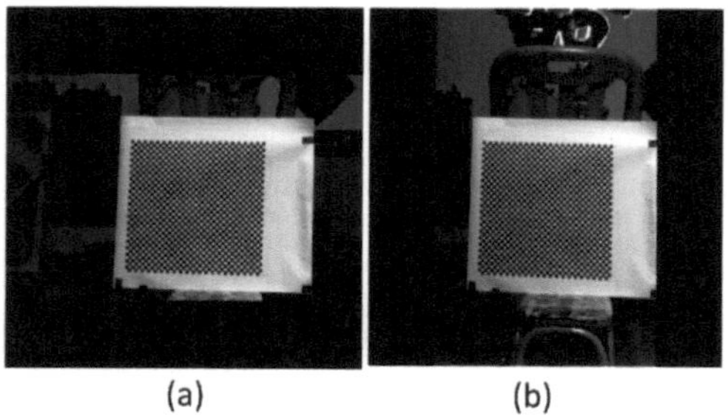

Figure 11. Anamorphic images after anamorphic ratio (*AR*) rectification. (**a**) Rectified image when the anamorphic lens was in the vertical position, and (**b**) rectified image when the anamorphic lens was in the horizontal position.

The chessboard corners in Figure 11a were extracted and represented as red points in Figure 12; the corners in Figure 11b were represented by green circles in Figure 12. The red points and green circles in Figure 12a are in their original positions. It can be seen that the corresponding corners were not adjacent and were separated from one another. This is because there was a misalignment between the optical axis and the rotation axis of the rotary table. This misalignment was further amplified by the large object distance. If

Equation (7) is directly applied for 3D reconstruction, there will be significant errors. Thus, the rotating anamorphic stereo vision was adopted for the 3D reconstruction. As can be seen in Figure 11, the two images are similar, the only difference between Figure 11a,b being a decentering. If we shift Figure 11b by (d_u, d_v) = (105.0792 pixels, −28.8297 pixels), the two images will overlap. In Figure 12b, the corresponding points are very close, as shown in Figure 5.

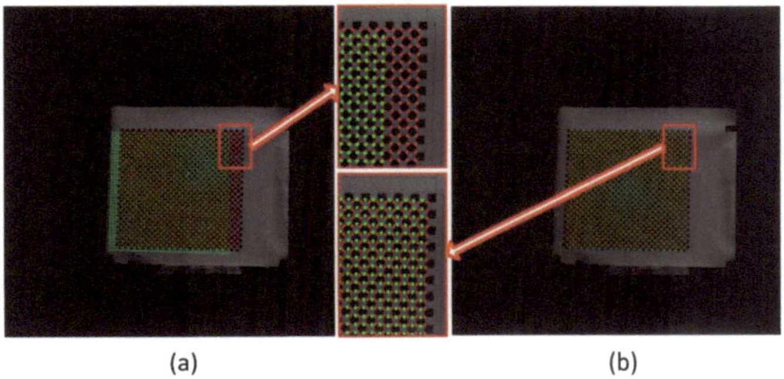

Figure 12. Corners after anamorphic ratio (*AR*) rectification. The red points indicate the corners in Figure 11a, and the green points indicate the corners in Figure 11b. (**a**) Image points in their original position, (**b**) green points shift their positions entirely, after which the two images almost overlap. This pixel decentering was due to the deviation between the optical axis and the axis of the rotary table.

Given the internal parameters of the anamorphic lens, the relative positions between the vertical anamorphic lens and the horizontal anamorphic lens, and the corresponding image points, the 3D coordinates of the corners could be reconstructed using the rotating anamorphic stereo vision. As shown in Figures 13 and 14, the 3D positions of the corners in checkerboards were constructed. Figure 13 shows the images of the checkerboards when the anamorphic lens was in the vertical position, and Figure 14 shows the images of the calibration targets when the anamorphic lens was in the horizontal position. When constructing the eight objects, the first two objects were 3D objects, and the other six objects were 2D objects. The constructed 3D points are shown in Figure 15, while not all corners in the checkerboard were reconstructed. Some points were no constructed for they lay in the anamorphic gas where the construction accuracy was very low. The anamorphic gas is discussed in detail in Section 4.2.2. As can be seen in Figure 15, the reconstruction accuracy was low compared with that of the stereo vision. The accuracy of this metrology method is also discussed in Section 4.2.

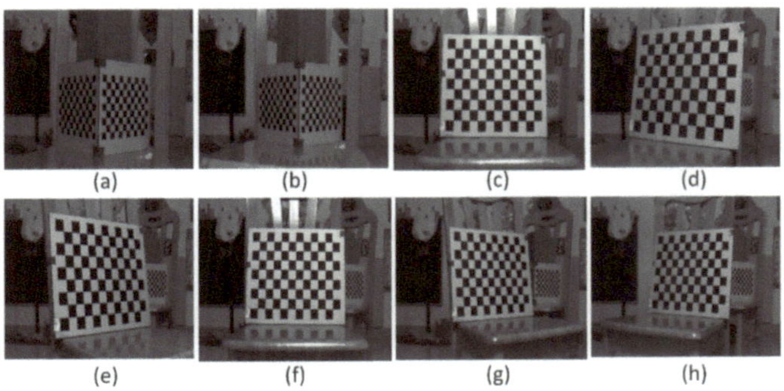

Figure 13. Images for 3D construction when the anamorphic lens was in the vertical position. (**a**,**b**) refer to images of 3D targets and (**c**–**h**) refer to the images of a 2D target.

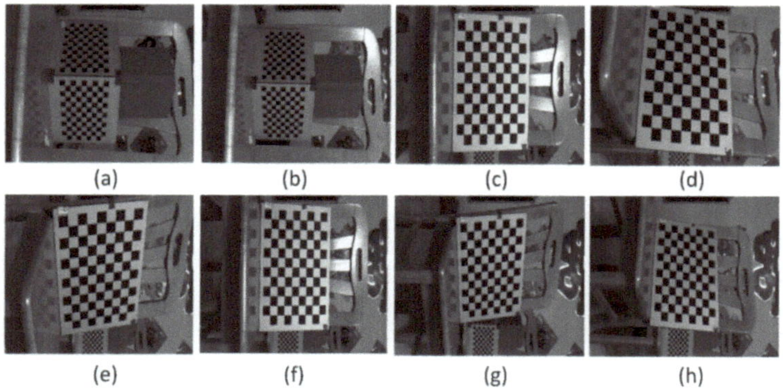

Figure 14. Images for 3D construction when the anamorphic lens was in the horizontal position. (**a**,**b**) refer to images of 3D targets and (**c–h**) refer to the images of a 2D target.

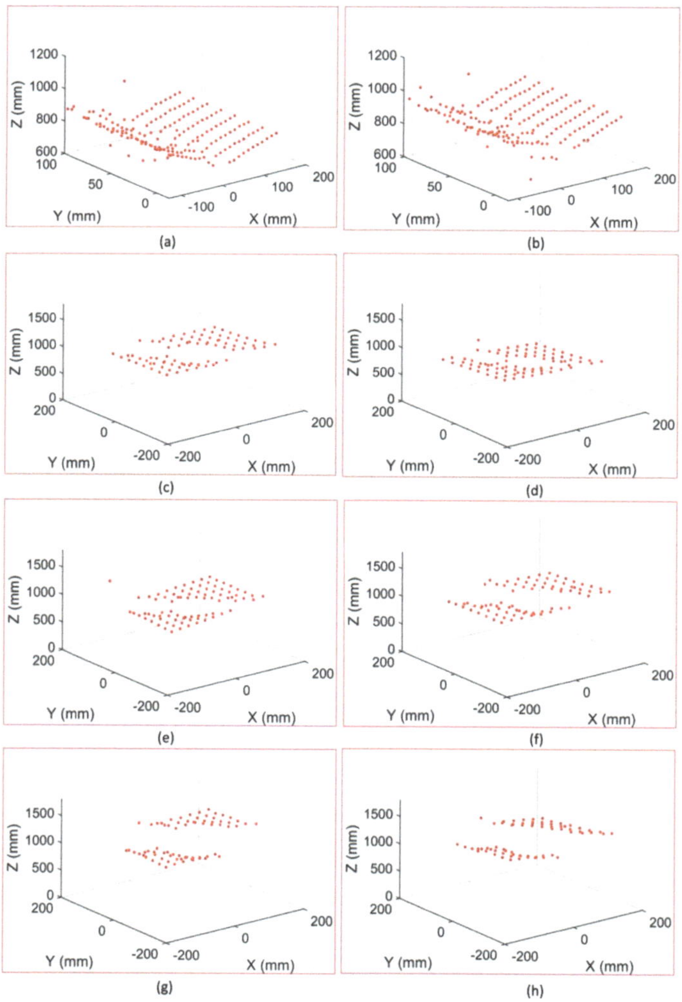

Figure 15. Constructed 3D points for images (**a–h**) in Figures 13 and 14.

4.2. Accuracy Analysis

As shown in Equation (14), the pixel error was the main source of error for this metrology, though other parameters such as ad, f_x, and f_y are also very important. This method's measuring principle is based primarily on the anamorphic distance, which changes the image position compared to spherical lenses. The anamorphic distance ad shown in Section 4.1 was small in comparison to the object distance; thus, any small pixel position error would have a significant impact on the measurement results. From Equation (7), we have:

$$\begin{aligned} dX_C &= \frac{\partial X_C}{\partial x_h}\delta x_h + \frac{\partial X_C}{\partial x_v}\delta x_v + \frac{\partial X_C}{\partial y_v}\delta y_v \\ dY_C &= \frac{\partial Y_C}{\partial x_h}\delta x_h + \frac{\partial Y_C}{\partial y_v}\delta y_v \\ dZ_C &= \frac{\partial Z_C}{\partial x_h}\delta x_h + \frac{\partial Z_C}{\partial y_v}\delta y_v \end{aligned} \quad (14)$$

where δx_h, δx_v, and δy_v are the pixel errors. If we substitute the pixel errors in Equation (14) with δ, the point error can be deduced from Equation (14) as follows:

$$\begin{aligned} \Delta xyz &= \sqrt{dX_C{}^2 + dY_C{}^2 + dZ_C{}^2} \\ &= ad \cdot \delta \cdot \frac{\sqrt{x_2y_2 \cdot f_x^2 \cdot f_y^2 + x_2 \cdot f_x^2 + x_1y_1 \cdot f_x \cdot f_y + y_2 \cdot f_y^2}}{\left(f_y \cdot x_h - f_x \cdot y_v\right)^2} \end{aligned} \quad (15)$$

where the coefficients x_2y_2, x_2, x_1y_1, and y_2 in Equation (15) are given by:

$$\begin{aligned} x_2y_2 &= (x_h - y_v)^2 \\ x_2 &= 2y_v^4 \\ x_1y_1 &= -2y_v^2\left(x_h^2 + (x_v + y_v)x_h - x_vy_v\right) \\ y_2 &= x_h^4 + (x_v + y_v)^2x_h^2 - 2x_vy_v(x_v + y_v)x_h + x_v^2y_v^2 \end{aligned} \quad (16)$$

To further evaluate the measurement accuracy, we conducted two simulation experiments. The simulation anamorphic lens was the same as the anamorphic lens shown in Section 3.2: f_x = 12 mm, f_y = 16 mm, aa = 0°, and ad = 30 mm.

4.2.1. Accuracy Analysis for a Point

First, we simulated an object point with the coordinates $[X_C, Y_C, Z_C]$ = [500 mm, 500 mm, 1500 mm] in anamorphic coordinates for the vertical position. The pixel coordinates of the point in the vertical and the horizontal anamorphic positions were calculated using Equations (3)–(6). Given the pixel coordinates and the calibrated parameters of the rotating anamorphic stereo vision, the 3D coordinates of the point could be calculated. The pixel errors in the image plane were assumed to have a Gaussian normal distribution with a standard deviation of 0.2 pixels, which was the calibration result of the anamorphic lens. Then, 5000 3D points were reconstructed with varying pixel errors, and the reconstructed 3D points are shown in Figure 16. The standard deviation of the reconstructed 3D point was 17.3378 mm.

4.2.2. Accuracy Analysis for a Surface

We also simulated a set of object points in a plane. The X_C and Y_C coordinates ranged from −1000 mm to 1000 mm, with an interval of 50 mm, and all the Z_C were 1500 mm. The pixel errors in the image plane satisfied a Gauss normal distribution, and the standard deviation of the pixel error was set to 0.2 pixels. The standard deviation for each reconstructed 3D point is shown in Figure 16. As can be seen in Figure 17, the measuring accuracy for the points away from the X_C axis was high, while it was very low if the object points were near the X_C axis. The points near the X_C axis were removed, which left a gap in Figure 17. Thus, there was a measuring gap if rotating anamorphic lenses were used for 3D construction. We named this gap the anamorphic gap.

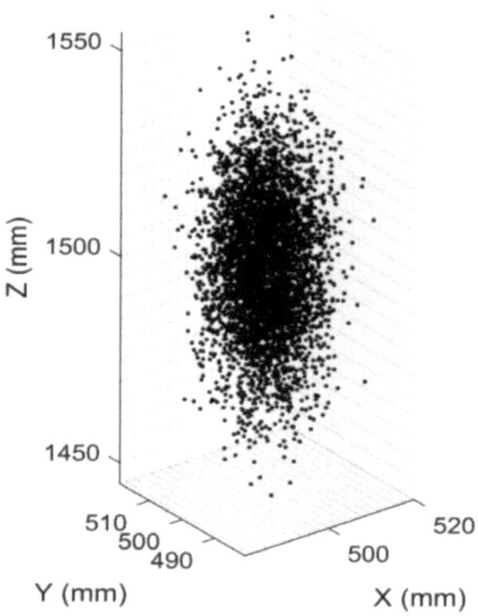

Figure 16. Image showing 5000 constructed 3D points for an object point in [500 mm, 500 mm, 1500 mm] with different pixel errors. The standard deviation of the constructed 3D points was 17.3378 mm.

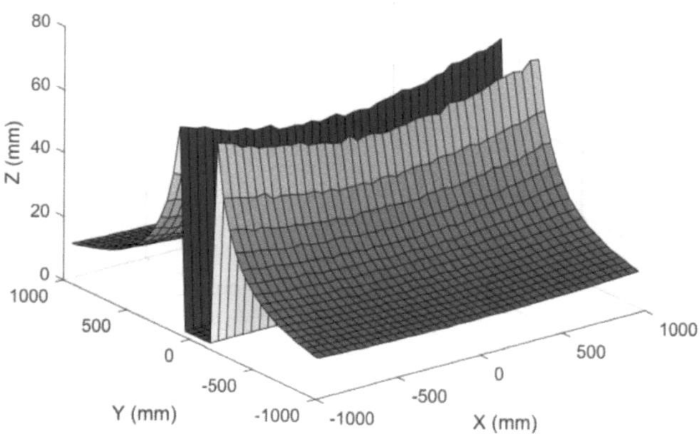

Figure 17. Standard deviation for object points in a plane. The standard deviations for object points with small Y_C coordinates were removed for large reconstruction errors. The standard deviation of the pixel position was 0.2 pixels, the ad was 30 mm, and the Z_C was 1500 mm.

If we individually increase the anamorphic distance from 30 mm to 100 mm, the measuring accuracy will increase correspondingly, as shown in Figure 18. The lens parameters ad, f_x, and f_y are closely related because they are determined by the anamorphic lens structure, and it is not possible to simply change one parameter independently [25]. We proposed paraxial lens designs for anamorphic lenses with zero anamorphic distance [26], but designing an anamorphic attachment with an extremely large anamorphic distance appeared to be difficult.

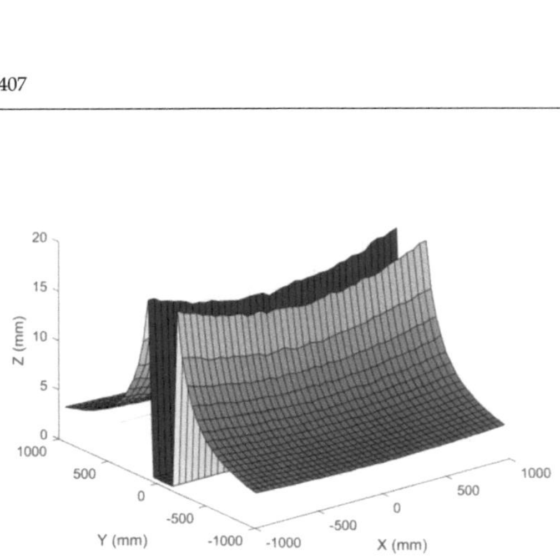

Figure 18. Standard deviation for object points in a plane. The standard deviations for object points with small Y_C coordinates were removed for large reconstruction errors. The standard deviation of the pixel position was 0.2 pixels, the ad was 100 mm, and the Z_C was 1500 mm.

The measuring precision of this 3D metrology method appeared to be lower than that of stereo vision with spherical lenses. The fundamental reason for this is that, in comparison to stereo vision, the baseline distance between the two anamorphic positions was quite small. The measuring precision will be as good as that of stereo vision with spherical lenses if the two anamorphic lenses are sufficiently apart from each other. Because the two images are captured with the same revolving anamorphic lens, the main advantage of this 3D metrology is its ease of point matching. The appropriate matching points are very close if the two images are rectified with the anamorphic ratio, as shown in Figures 5 and 12. Alternatively, as illustrated in Figure 6, it is possible to trace the matching points by rotating the anamorphic lens. When the anamorphic lens is rotated at a rapid speed, it can measure 3D data four times in one circle, which is ideal for a quick 3D reconstruction with low measurement accuracy. One potential application of this would be car navigation.

5. Conclusions

A 3D metrology method using one camera with rotating anamorphic lenses is proposed in this paper. The anamorphic lens can rotate by -90° along the optical axis, capturing two anamorphic images of an object before and after the rotation. This setup is similar to that of stereo vision; therefore, it can be considered a stereo vision with anamorphic lenses. The main differences between stereo vision and rotating anamorphic lenses are that anamorphic lenses replace spherical lenses, and the baseline distance between the two anamorphic lenses is very small. This feature will make rotating anamorphic lenses a compact sensor suitable for 3D metrology in a constrained space. The feasibility of the proposed method in 3D reconstruction is based on one internal parameter of anamorphic lenses, i.e., the anamorphic distance, which will slightly change the image position of the object. The corresponding points between the two anamorphic images are close if the two anamorphic images are rectified by the anamorphic ratio, which substantially simplifies the corresponding point matching and reduces the calculation amount. When compared to stereo vision, the measurement accuracy of the rotating anamorphic stereo vision is lower, especially for the image points near the X_C axis of the anamorphic lens in the vertical position. These characteristics make the rotating anamorphic stereo vision suitable for a fast 3D reconstruction without a high demand for measurement accuracy, such as car navigation applications. Further research might include high precision anamorphic lens calibration, error compensation, point matching for rotating anamorphic lenses, and anamorphic lens designs with a large anamorphic distance.

Author Contributions: Conceptualization, J.Z., X.C. and J.X.; methodology, X.C. and J.Z.; software, X.C and J.Z.; validation, J.Z.; data curation, J.Z.; writing—original draft preparation, J.Z.; writing—review and editing, X.C. and J.Z. All authors have read and agreed to the published version of the manuscript.

Funding: This research was funded by the National Natural Science Foundation of China (52175478), the Defense Industrial Technology Development Program (JCKY2020203B039), the Science and Technology Commission of Shanghai Municipality (21511102602), the Doctoral Foundation of the University of Jinan (XBS1641), and the Joint Special Project of Shandong Province (ZR2016EL14).

Data Availability Statement: The data underlying the results presented in this paper are not publicly available at this time but may be obtained from the authors upon reasonable request.

Conflicts of Interest: The authors declare no conflict of interest.

References

1. Leach, R. *Optical Measurement of Surface Topography*; China Science Publishing & Media Ltd.: Beijing, China, 2012; pp. 154–196.
2. Gaha, R.; Durupt, A.; Eynard, B. Towards the implementation of the Digital Twin in CMM inspection process: Opportunities, challenges and proposals. *Procedia Manuf.* **2021**, *54*, 216–221. [CrossRef]
3. Sutton, M.A.; Li, N.; Joy, D.C.; Reynolds, A.P.; Li, X. Scanning electron microscopy for quantitative small and large deformation measurements part I: SEM imaging at magnifications from 200 to 10,000. *Exp. Mech.* **2007**, *47*, 775–787. [CrossRef]
4. Harding, K. Challenges and opportunities for 3D optical metrology: What is needed today from an industry perspective. *Two-Three-Dimens. Methods Insp. Metrol. VI* **2008**, *7066*, 112–119.
5. Marrugo, A.G.; Gao, F.; Zhang, S. State-of-the-art active optical techniques for three-dimensional surface metrology: A review. *JOSA A* **2020**, *37*, B60–B77. [CrossRef] [PubMed]
6. Soid, S.N.; Zainal, Z.A. Spray and combustion characterization for internal combustion engines using optical measuring techniques—A review. *Energy* **2011**, *36*, 724–741. [CrossRef]
7. O'Riordan, A.; Newe, T.; Dooly, G.; Toal, D. December. Stereo vision sensing: Review of existing systems. In Proceedings of the 2018 12th International Conference on Sensing Technology (ICST), Limerick, Ireland, 4–6 December 2018.
8. Lazaros, N.; Sirakoulis, G.C.; Gasteratos, A. Review of stereo vision algorithms: From software to hardware. *Int. J. Optomech.* **2008**, *2*, 435–462. [CrossRef]
9. Hussmann, S.; Ringbeck, T.; Hagebeuker, B. A performance review of 3D TOF vision systems in comparison to stereo vision systems. In *Stereo Vision*; Intechopen: London, UK, 2008.
10. Zhang, S. High-speed 3D shape measurement with structured light methods: A review. *Opt. Lasers Eng.* **2018**, *106*, 119–131. [CrossRef]
11. Chen, M.; Tang, Y.; Zou, X.; Huang, Z.; Zhou, H.; Chen, S. 3D global mapping of large-scale un-structured orchard integrating eye-in-hand stereo vision and SLAM. *Comput. Electron. Agric.* **2021**, *187*, 106237. [CrossRef]
12. Tang, Y.; Zhou, H.; Wang, H.; Zhang, Y. Fruit detection and positioning technology for a Camellia oleifera C. Abel orchard based on improved YOLOv4-tiny model and binocular stereo vision. *Expert Syst. Appl.* **2023**, *211*, 118573. [CrossRef]
13. Navarro, A.V.; Navarro, A.V.; Garrido, C.A. Anamorphic Lens. US 9063321 B2, 23 June 2015.
14. Lippman, D.H.; Teverovsky, D.S.; Bentley, J.L. Monte Carlo first-order design method for anamorphic cinema zoom lenses. *Opt. Eng.* **2021**, *60*, 051203. [CrossRef]
15. Xu, C.; Song, W.; Wang, Y. Design of a miniature anamorphic lens with a freeform front group and an aspheric rear group. *Opt. Eng.* **2021**, *60*, 065104. [CrossRef]
16. Zhang, J.; Wang, X.; Ma, M.; Li, F.; Liu, H.; Cui, H. Camera calibration for anamorphic lenses with three-dimensional targets. *Appl. Opt.* **2020**, *59*, 324–332. [CrossRef] [PubMed]
17. Dodoc, A. Anamorphic prime and zoom lenses. In Proceedings of the Zoom Lenses VI, SPIE, San Diego, CA, USA, 9 September 2019; Volume 11106, pp. 21–40.
18. Zhang, J.; Chen, X.; Liu, H.; Li, F.; Sun, X. Thin lens aberrations for anamorphic lenses. *Appl. Opt.* **2019**, *58*, 182–188. [CrossRef] [PubMed]
19. Soskind, M.; Soskind, Y.G. Propagation invariant laser beams for optical metrology applications. In Proceedings of the Modeling Aspects in Optical Metrology V, SPIE, Munich, Germany, 21 June 2015; Volume 9526, pp. 392–398.
20. Ma, M.; Shao, H.; Zhang, J.; Wang, X.; Li, G. A Calibration Method of Anamorphic Lens Camera Based on Virtual 3D Target. In Proceedings of the 2019 IEEE/ASME International Conference on Advanced Intelligent Mechatronics (AIM), Hong Kong, China, 8 July 2019.
21. Towers, C.E.; Towers, D.P.; Campbell, H.I. Three-dimensional particle imaging by wavefront sensing. *Opt. Lett.* **2006**, *31*, 1220–1222. [CrossRef] [PubMed]
22. Cao, Z.; Wang, K.; Wu, Q. Aspherical anamorphic lens for shaping laser diode beam. *Opt. Commun.* **2013**, *305*, 53–56. [CrossRef]
23. Durko, H.L.; Barrett, H.H.; Furenlid, L.R. High-Resolution Anamorphic SPECT Imaging. *IEEE Trans. Nucl. Sci.* **2014**, *61*, 1126–1135. [CrossRef] [PubMed]

24. Blais, F.; Beraldin, J.-A. Calibration of an anamorphic laser based 3-D range sensor. In Proceedings of the SPIE Videometrics V., San Diego, CA, USA, 27 July–1 August 1997; Volume 3174, pp. 113–122.
25. Chen, X.; Zhang, J. High-precision anamorphic lens calibration with 3D and 2D calibration targets. *Appl. Opt.* **2022**, *61*, 6062–6075. [CrossRef] [PubMed]
26. Yuan, S.; Sasian, J. Aberrations of anamorphic optical systems. II. Primary aberration theory for cylindrical anamorphic systems. *Appl. Opt.* **2009**, *48*, 2836–2841. [CrossRef] [PubMed]
27. Chen, X.; Zhang, J. Lens design for parallel cylindrical anamorphic attachments with finite object distance. *Appl. Opt.* **2022**, *61*, 4610–4619. [CrossRef] [PubMed]
28. Jinkai, Z.; Chen, X. Paraxial lens design of anamorphic lenses with a fixed anamorphic ratio. *OSA Contin.* **2019**, *2*, 1430–1454.

MDPI AG
Grosspeteranlage 5
4052 Basel
Switzerland
Tel.: +41 61 683 77 34

Sensors Editorial Office
E-mail: sensors@mdpi.com
www.mdpi.com/journal/sensors

www.ingramcontent.com/pod-product-compliance
Lightning Source LLC
Chambersburg PA
CBHW041359210926
51850CB00013B/3132